AF885819

Lecture Notes in Energy

Volume 91

Lecture Notes in Energy (LNE) is a series that reports on new developments in the study of energy: from science and engineering to the analysis of energy policy. The series' scope includes but is not limited to, renewable and green energy, nuclear, fossil fuels and carbon capture, energy systems, energy storage and harvesting, batteries and fuel cells, power systems, energy efficiency, energy in buildings, energy policy, as well as energy-related topics in economics, management and transportation. Books published in LNE are original and timely and bridge between advanced textbooks and the forefront of research. Readers of LNE include postgraduate students and non-specialist researchers wishing to gain an accessible introduction to a field of research as well as professionals and researchers with a need for an up-to-date reference book on a well-defined topic. The series publishes single- and multi-authored volumes as well as advanced textbooks.

Indexed in Scopus and EI Compendex The Springer Energy board welcomes your book proposal. Please get in touch with the series via Anthony Doyle, Executive Editor, Springer (anthony.doyle@springer.com)

Ibrahim Dincer · Dogan Erdemir ·
Muhammed Iberia Aydin · Huseyin Karasu ·
Greg Vezina

Ammonia Energy Technologies

Springer

Ibrahim Dincer
Ontario Tech University
Oshawa, ON, Canada

Dogan Erdemir
Ontario Tech University
Oshawa, ON, Canada

Muhammed Iberia Aydin
Ontario Tech University
Oshawa, ON, Canada

Huseyin Karasu
Ontario Tech University
Oshawa, ON, Canada

Greg Vezina
Hydrofuel Canada Inc.
Mississauga, ON, Canada

ISSN 2195-1284 ISSN 2195-1292 (electronic)
Lecture Notes in Energy
ISBN 978-3-031-13531-6 ISBN 978-3-031-13532-3 (eBook)
https://doi.org/10.1007/978-3-031-13532-3

© The Editor(s) (if applicable) and The Author(s), under exclusive license to Springer Nature Switzerland AG 2022
This work is subject to copyright. All rights are solely and exclusively licensed by the Publisher, whether the whole or part of the material is concerned, specifically the rights of translation, reprinting, reuse of illustrations, recitation, broadcasting, reproduction on microfilms or in any other physical way, and transmission or information storage and retrieval, electronic adaptation, computer software, or by similar or dissimilar methodology now known or hereafter developed.
The use of general descriptive names, registered names, trademarks, service marks, etc. in this publication does not imply, even in the absence of a specific statement, that such names are exempt from the relevant protective laws and regulations and therefore free for general use.
The publisher, the authors, and the editors are safe to assume that the advice and information in this book are believed to be true and accurate at the date of publication. Neither the publisher nor the authors or the editors give a warranty, expressed or implied, with respect to the material contained herein or for any errors or omissions that may have been made. The publisher remains neutral with regard to jurisdictional claims in published maps and institutional affiliations.

This Springer imprint is published by the registered company Springer Nature Switzerland AG
The registered company address is: Gewerbestrasse 11, 6330 Cham, Switzerland

Preface

We are in a critical era where increased levels of carbon dioxide emissions, global warming issues and local and global climate change problems require urgent responses and hence immediate actions. The way that energy is produced, stored, converted and consumed directly affects these issues and problems. One thing is clear that energy consumption has been increasing exponentially in the world due to the increasing population and demand, technological advancements and increased life standards and comfort requirements. Almost everyday, new energy-intensive technologies, systems and applications for various sectors, including transportation sector, are entering into our daily routines, which drastically increase the fuel and power requirements for economic activities and societal developments. Such needs have greatly impacted the energy equation with various constraints related to the environment, health and sustainable development will continue to do so in the next decades. It is now terribly necessary to set up the energy equation without hydrocarbon fuels which is fully recognized by many researchers, scientists, organizations, companies, etc. A new era is now on where it is really time to move for renewables and carbon-free fuels (particularly with hydrogen and ammonia). Today, hydrogen energy technologies are offering a promising option to solve energy and environmental-related issues. Therefore, governments, agencies, companies and local authorities have been developing new strategies and policies particularly for hydrogen economy and implementing them accordingly. However, such a transition brings high costs for investment, research, development and innovation as well as technology developments. It is better to go one step ahead of hydrogen to consider ammonia as a unique option to better achieve the hydrogen economy, which will serve as a shortcut solution in this transition to hydrogen economy.

This book is an outcome of a comprehensive project on ammonia utilization and aims to cover all necessary information regarding ammonia from production to utilization, particularly including the entire life cycle. Firstly, some introductory information regarding hydrogen and ammonia and their productions along with numerous energetic, environmental and economic dimensions are provided and discussed thoroughly. Secondly, there is a strong focus on ammonia production methods along with ammonia storage techniques. Thirdly, there are case studies presented on the methods of ammonia utilization with various scenarios. The methods for analysis of ammonia energy systems will be introduced with the illustrative examples. Furthermore, the book covers detailed modeling and analysis studies, experimental investigations and life cycle assessment studies. Finally, the concluding remarks and future directions of ammonia energy technologies, in addition to the potential developments in the area, are discussed.

Oshawa, Canada — Ibrahim Dincer
Oshawa, Canada — Dogan Erdemir
Oshawa, Canada — Muhammed Iberia Aydin
Oshawa, Canada — Huseyin Karasu
Mississauga, Canada — Greg Vezina

Contents

Chapter 1
Ammonia

Nomenclature

CCUS	Carbon capture utilization and storage
CNG	Compressed natural gas
CO	Carbon monoxide
DME	Dimethyl ether
GJ	Gigajoule
ICE	Internal combustion engine
IEA	International Energy Agency
kJ	Kilojoule
LHV	Lower heating value
LNG	Liquified natural gas
LPG	Liquefied petroleum gas
MJ	Megajoule
NO_x	Nitrate oxides
PM	Particulate matter
SCR	Selective catalytic reduction
SO_2	Sulfur dioxide
VOC	Volatile organic compounds

1.1 Introduction

The need for fossil fuels has risen drastically with the increases of population and human activities as well as technological developments. Although there were some concerns as we will run out of these sources, (as they have been depleted so fast), increased levels of greenhouse gas emissions, particularly CO_2 emissions, have brought global warming and climate change issues forefront, due to the excessive

© The Author(s), under exclusive license to Springer Nature Switzerland AG 2022

I. Dincer et al., *Ammonia Energy Technologies*, Lecture Notes in Energy 91,
https://doi.org/10.1007/978-3-031-13532-3_1

consumption of fossil fuels. Although the unconventional fossil fuel sources are four times higher compared to conventional fossil fuels, those sources are located at extreme geographic formations and may require advanced technologies and their extraction costs are much higher [1].

Not only the depletion of fossil fuels, but also the environmental effects of using these fuels are one of the main reasons to turn onto alternatives due to the fact that the emissions resulting from the extensive consumption of fossil fuels have also causing harms on the environment. The CO, NO_x, VOC, SO_2 and PM emissions caused by the transportation sector in Canada were 53.9, 51.7, 15.9, 1.6, and 2% in 2017, respectively [2]. Depletion of ozone layer, photochemical smog formation, acid rains, and chronic respiratory diseases are caused by those gases.

The systems should be evaluated in detail from source to end use to reduce environmental problems caused by energy production and to develop a truly sustainable approach. For this reason, a 3S concept (Source-System-Service) which was wisely introduced by Prof. Ibrahim Dincer in various conferences and formally published in Dincer and Acar [3]. This 3S concept covers the energy spectrum from source to system and system to service. Here, it is important to emphasize for any energy application, a source is needed, and for the targeted application we need commodities which are treated under services. Once we identify the source(s) and services, we need to design the systems to cover the needs by using the sources. This is illustrated in Fig. 1.1, by adopting this methodology, there are now three steps that should be assessed accordingly:

- Source: Identifying a clean, abundant, economically feasible, and available energy source.
- System: Developing a right system based on what sources are available and what useful outputs are needed by the users and/or communities.
- Service: Determining the needs, in terms of power, heat, cooling, freshwater, fuel, etc., for the targeted application.

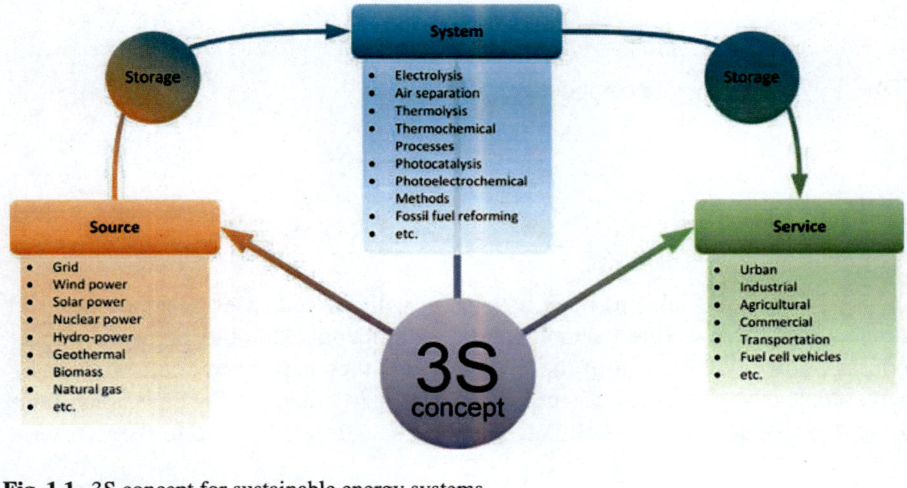

Fig. 1.1 3S concept for sustainable energy systems

For this reason, fuel alternatives such as natural gas, hydrogen, methanol, and ammonia has become one of the main topics of researchers. There are studies for various types of compounds that can be an alternative to fossil fuels or to reduce fossil fuel consumption. In addition to the use of fuels such as methanol, hydrogen, ammonia alone in the engines, there are studies in which these fuels are used by mixing with conventional fuels such as diesel and gasoline.

Ammonia is recognized as a colorless gaseous molecule under normal conditions and has a distinctive odour. Three hydrogen and one nitrogen atoms form a polar structure to create ammonia. Ammonia was discovered for the first time in history as ammonia salt near the temple of Jupiter Ammon in a region of Libya. The name ammonia derives from the name of the temple, Ammon [4]. Ammonium chloride salt, *sal ammoniac*, is a chemical known and used since ancient times. Salt derivatives of ammonia are found in nature, especially in regions with volcanic activities. These substances are formed with the decomposition of nitrogen-containing organic materials in volcanic regions. Also, ammonium salts can be found near coal deposits.

Ammonia and ammonium ions are generally released to nature because of biological activities in every water source where there is life, and they function as an important component of the nitrogen cycle. Figure 1.2 summarizes the nitrogen cycle in nature. The nitrogen cycle is essential for the continuation of life. Ammonia is formed by the decomposition of nitrogen-containing organic materials, and then converted to nitrate derivatives through bacterial activities to be used as nutrients by plants [5].

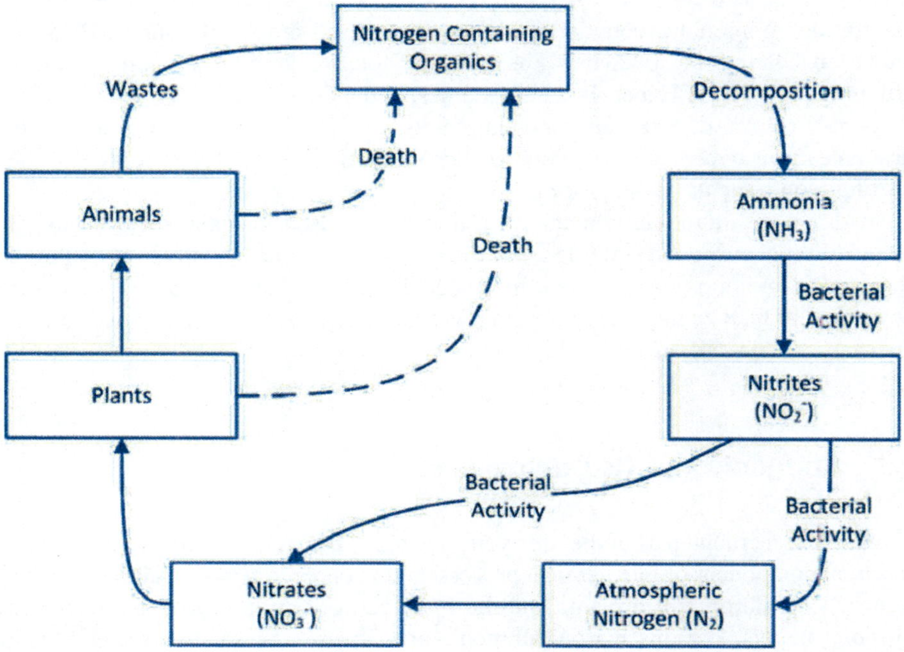

Fig. 1.2 The nitrogen cycle in the nature (modified from [5])

The industrial production of ammonia essentially began with the discovery of the Haber–Bosch reactor, which produces ammonia catalytically under high pressure and temperature, by the German chemists Fritz Haber and Carl Bosch. About 235 million metric tonnes of ammonia was manufactured by the industry in 2019 [6]. The energy spent for the production of ammonia in 2021 corresponds to 2% of the global energy consumption [7]. Ammonia is utilized as a feedstock in chemical and fertilizer production. Only a minor fraction is used in direct applications. A simplified life cycle of ammonia describing its production sources, methods and application areas is given in Fig. 1.3. Today, industrial production of ammonia is carried out with modified versions of Haber–Bosch process. Nitrogen is provided from the air using various technologies. Carbon-free nitrogen production can be achieved through renewable energy sources. However, carbon emissions from the Haber-Bosch process are mainly caused by hydrogen production. Industrial hydrogen production relies on fossil-fuel based technologies such as steam methane reforming and coal gasification. Although near-zero carbon emissions can be achieved with carbon capture utilization and storage (CCUS) technologies, electrolysis, and other clean hydrogen production technologies, those technologies increase the ammonia production costs.

Figure 1.4 shows the market share of exports for ammonia in 2019. China, Russia, Middle East and the US are the largest producers of ammonia [8]. However, most of the ammonia produced in those countries utilized by domestic sources. On the other hand, Saudi Arabia, Russia, and Trinidad and Tobago are the largest ammonia exporting countries. The export price of ammonia per ton varied between $230 and $470 in 2019, and the world average was $286 [9]. However, energy and ammonia production costs have increased drastically due to Russia's attack on Ukraine in 2022. It is known that a total of 96% of ammonia is globally produced by the Haber–Bosch processes, and CO_2 emissions are estimated to account for 1.2% of total anthropogenic emissions [10]. These emissions mainly originate from hydrogen production technologies that utilize fossil fuels and the heating requirements of the process.

In summary, ammonia is a chemical that can be used in many sectors and for many different purposes. Although the utilization of ammonia is vastly in agricultural sector, it is an important feedstock for many chemical products. And also, it could be the key to finalize the ever-increasing search for an alternative clean fuel, since it does not contain carbon.

1.2 Ammonia and Its Importance

Ammonia is a crucial part of the life cycle, due to its nitrogen content. Nitrogen is an essential component of biochemical processes. It is one of the basic building blocks to produce chlorophyll in plants and the production of amino acids and enzymes in living beings. For this reason, ammonia and fertilizer products produced from ammonia are needed for the growth of plants in the agriculture sector. Aside from

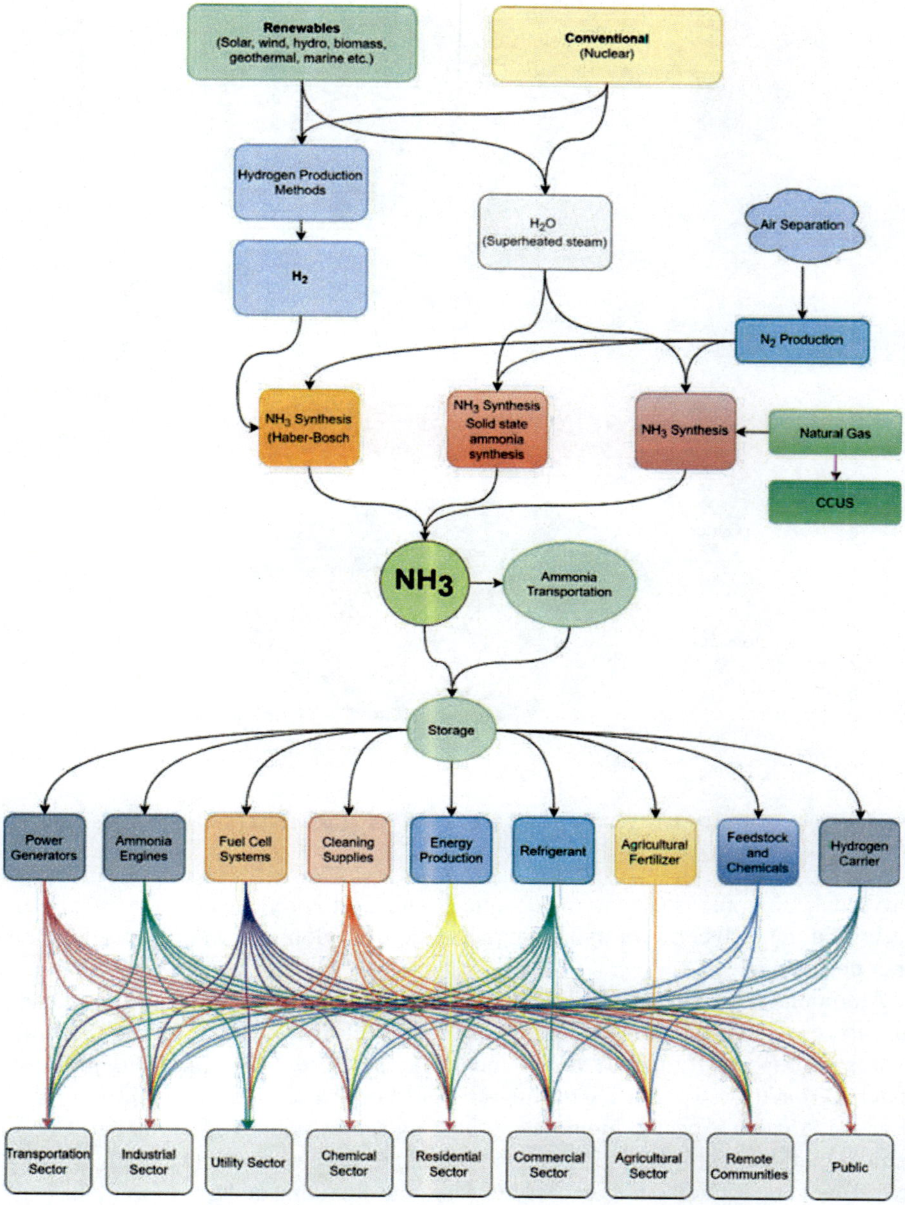

Fig. 1.3 Cradle-to-grave life cycle of ammonia

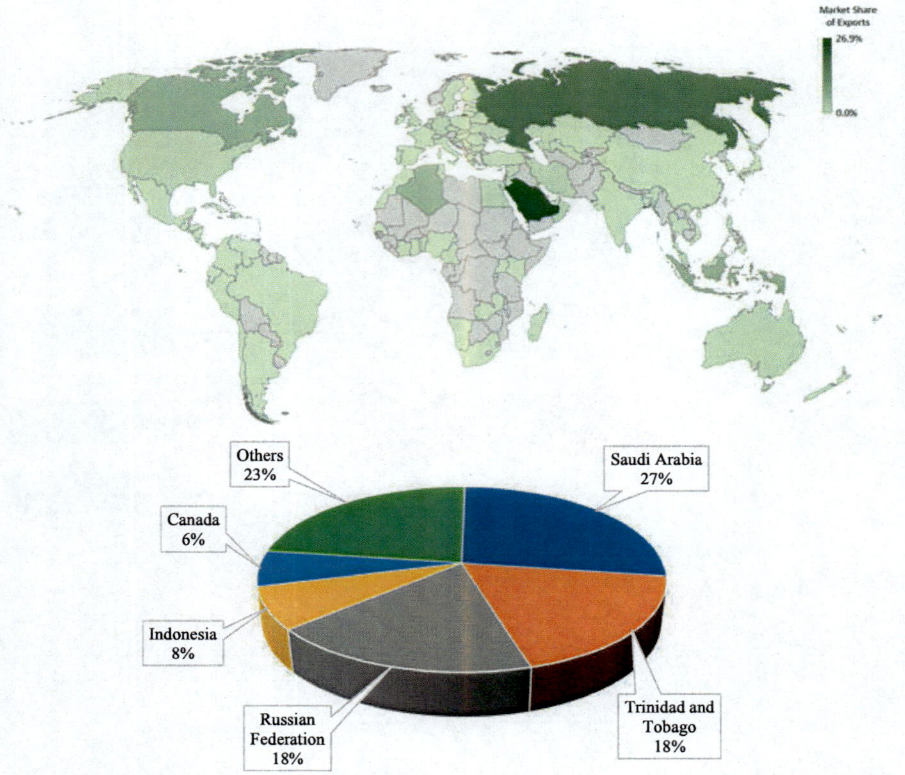

Fig. 1.4 Ammonia export market share by countries in 2019 (data from [8])

fertilizers, ammonia is used in the chemical industry to produce derivatives such as nitric acid, nitroglycerin for dynamite production, acrylonitrile and caprolactam for fiber production.

Ammonia is also considered as an excellent hydrogen carrier, with the recent shift to zero-emission vehicles in the fight against climate change. To transport the hydrogen properly, it must be stored under high pressure. Hydrogen is compressed under 200 bar to transport using the tube-trailers. For fuel cell devices, hydrogen refueling stations are expected to operate at high pressures such as 350–700 bar. Cryogenic liquefaction at −123.15 °C and below should be performed to store hydrogen at 700 bar [11]. Liquefaction process increases the operating and transportation costs of hydrogen production and storage systems. Therefore, ammonia emerges as a good alternative as a hydrogen carrier. Ammonia turns into liquid under atmospheric conditions at −33 °C. Also, its auto ignition temperature is higher than hydrogen, and it can be easily detected in the environment with its pungent, sharp odor.

In addition, ammonia has the potential to be a working fluid for thermodynamic cycles, including heat engines (such as Rankine cycles). Note that cooling/refrigeration, heating, electricity production, and combination of those cycles can easily be done in an integrated fashion by utilizing ammonia. Furthermore, it can be utilized as fuel in internal combustion engines, power generators, gas turbines, etc. Some key facts about ammonia can be listed as follows:

- Ammonia molecule has one nitrogen and three hydrogen atoms in its structure. Nitrogen can be obtained with air separation, and the hydrogen can be obtained either from conventional fossil fuel based production methods or from clean hydrogen production technologies with renewable resources.
- Ammonia is the second major chemical that is produced globally.
- The physical properties of ammonia (and three hydrogen atoms in its structure) make it an excellent hydrogen carrier. Also, it has no carbon atom in its structure thus makes it a zero-carbon and sustainable option.
- Making ammonia life cycle carbon free requires clean hydrogen production where renewable sources are deployed.
- Ammonia can be utilized in gas or liquid phases, also some of the derivatives of ammonia can be stored in solid phase.
- Ammonia can be stored utilizing relatively low pressures compared to hydrogen, thus making it easier to transport.
- Ammonia can be transported with existing stainless-steel pipelines with minimum maintenance requirements.
- Ammonia can be utilized in ammonia fuel cells directly, or in combustion engines with modifications.
- Ammonia can be used as a refrigerant in various applications.

Compared to hydrogen, the transportation costs of ammonia are lower. Special cooling systems and tank designs are required to transport liquid hydrogen due to safety concerns. While the density of liquefied hydrogen is 70.85 kg/m^3, this value for liquefied ammonia is 609 kg/m^3 at 20 °C under 10 bar. Figure 1.5 shows the seaborne transport costs for ammonia and hydrogen. While the transportation price for liquefied hydrogen is around 0.12 US$/t km, this value may become about 0.03 US$/t km for ammonia.

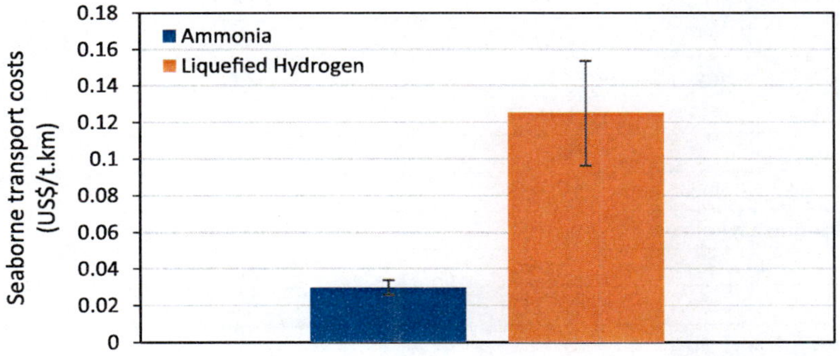

Fig. 1.5 Seaborn transport costs for ammonia and liquefied hydrogen (data from [12])

1.3 Ammonia and Its Properties

Ammonia consists of 17.6% hydrogen and 82.4% nitrogen. Approximately, one liter of ammonia contains 0.137 kg of hydrogen, which is higher than most of the chemicals. A detailed information on various chemicals and their hydrogen content was given in Table 1.1. The energy density of ammonia is 13.6 GJ/m^3 at 10 bar. Table 1.2 shows the comparison of various properties of ammonia with other types of fuels. Its energy density is very close to methanol; however, its specific energetic cost is low compared to methanol. At the same time, the hydrogen content makes the ammonia a suitable hydrogen carrier. Hydrogen is an excellent clean fuel, albeit utilizing hydrogen as an energy carrier has its difficulties. Liquefaction of hydrogen requires high energy. Also, providing adequate safety standards for transportation and storage of hydrogen is harder than ammonia. Hydrogen's liquefaction temperature is −253 °C. However, ammonia can be liquified at −33 °C and can be stored under room temperature without additional cooling unlike hydrogen. Although hydrogen has the highest higher heating value among other fuels, its energy density is very low. The autoignition temperature of ammonia is 651 °C. The flammability limit is in the range of 16–25% in air, making ammonia a safer hydrogen carrier [13].

Ammonia can be produced using fossil fuels such as heavy fuel oil, coal, coke oven gas, refinery gas, and natural gas. Note that a total of 72% of ammonia is produced worldwide by steam reforming of natural gas, and 22% of ammonia is produced by coal gasification as shown in Fig. 1.6. The rest 6% of ammonia is produced using other feedstock such as fuel oil, naphtha etc. Among the several methods of ammonia production, steam methane reforming is currently the least energy intensive technique. Coal is widely employed in China and is often associated with high energy intensities. The cost of natural gas is 70–90% of the cost of producing ammonia. Because ammonia production is based on natural gas in the SMR process, as natural gas prices rise, so the do ammonia production costs [15].

Table 1.1 Hydrogen content of various fluids (Modified from [16])

Carrier	H_2 in one liter of liquid (kg)	Energy to release one kg of H_2 (kWh)
Ammonia	0.137	6.3
Dodecahydro-N-ethylcarbazole	0.055	7.6
Formic Acid	0.053	4.3
Methanol	0.149	6.7
Perhydro-benzyltoluene	0.055	9
Perhydro-dibenzyltoluene	0.057	12.7
Toluene	0.047	11.2
Water	0.112	50.4

Source Davies et al. [16]

Table 1.2 Comparison of ammonia with other fuel types

Properties	Gasoline	Diesel	LPG	CNG	Gaseous hydrogen	Liquid hydrogen	Ammonia
Formula	C_8H_{18}	$C_{12}H_{23}$	C_3H_8	CH_4	H_2	H_2	NH_3
Lower heating value (MJ/kg)	44.5	43.5	45.7	38.1	120.1	120.1	18.8
Flammability limits, gas in air (vol.%)	1.4–7.6	0.6–5.5	1.81–8.86	5.0–15.0	4–75	4–75	16–25
Flame speed (m/s)	0.58	0.87	0.83	8.45	3.51	3.51	0.15
Autoignition temperature (°C)	300	230	470	450	571	571	651
Minimum ignition energy (MJ)	0.14	N/A	N/A	N/A	0.018	N/A	8
Flash point (°C)	− 42.7	73.8	− 87.7	− 184.4	N/A	N/A	− 33.4
Octane	90–98	N/A	112	107	> 130	> 130	110
Fuel density (kg/m^3)	698.3	838.8	1898	187.2	17.5	71.1	602.8
Energy density (MJ/m^3)	31,074	36,403	86,487	7132	2101	8539	11,333
Latent heat of vaporization (kJ/kg)	71.78	47.86	44.4	104.8	0	N/A	1369
Storage method	Liquid	Liquid	Comp. liquid	Comp. gas	Comp. gas	Comp. liquid	Comp. liquid
Storage temperature (°C)	25	25	25	25	25	− 253	25
Storage pressure (kPa)	101.3	101.3	850	24,821	24,821	102	1030
Cost (US$/l)	1.19	1.49	0.79	0.57	1.16	1.18	0.85
Cost (US$/MJ)	0.038	0.041	0.009	0.080	0.552	0.138	0.075

Source Erdemir and Dincer [14]

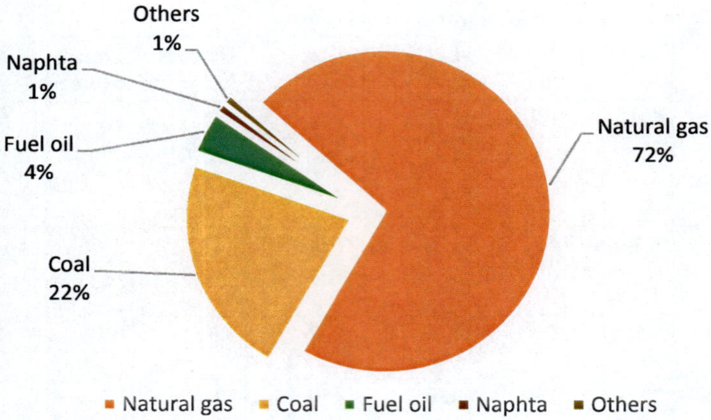

Fig. 1.6 Sources of global ammonia production based on feedstock use (data from [17])

1.4 Sectoral Ammonia Utilization

Ammonia has a wide range of possible applications in many sectors, including agriculture, transportation, power generation, refrigeration, food, chemical, petrochemical, etc. for various useful commodities. It is been using as a refrigerant, fertilizer, cleaning agent, hydrogen carrier, nitrogen source for fermentation and many more, as depicted in Fig. 1.7. In 2019, global ammonia production was around 235 M tonnes, thus making ammonia the second most-produced chemical after sulfuric acid [6]. Figure 1.8 shows the market share of ammonia. Today, almost a total of 80% of ammonia production is supplied to the agricultural sector by fertilizer use. Also, ammonia is used in many production pathways such as polyamide, nitric acid, nylon, explosives, pharmaceutical production. Ammonia can be used directly as fertilizer, or it can be used to produce ammonium nitrate, ammonium sulfate, ammonium hydrogen phosphate, and urea by using different processes. Although it has the highest nitrogen content as a percentage, ammonia is not preferred as fertilizer, except in the United States. The reason for this can be listed as the presence of ammonia as a gas under normal conditions, storage conditions, environmental and health risks. For this reason, ammonia derivatives containing ammonium are preferred in the rest of the world [18].

As noted earlier, ammonia is a zero-carbon fuel that can be used in transportation sector. It has no global warming potential throughout the usage phase. Besides to its desirable features as a fuel, ammonia is frequently employed as a reducing agent for nitrogen oxides in exhaust emmision gases. Renewable ammonia is a carbon-free fuel, refrigerant, and working fluid, as well as a hydrogen storage medium. After electric and hydrogen powered cars, ammonia as a clean and sustainable fuel for road vehicles has the lowest global warming potential. Consequently, the use of ammonia in the transportation industry will result in considerable financial and environmental advantages, as well as public satisfaction.

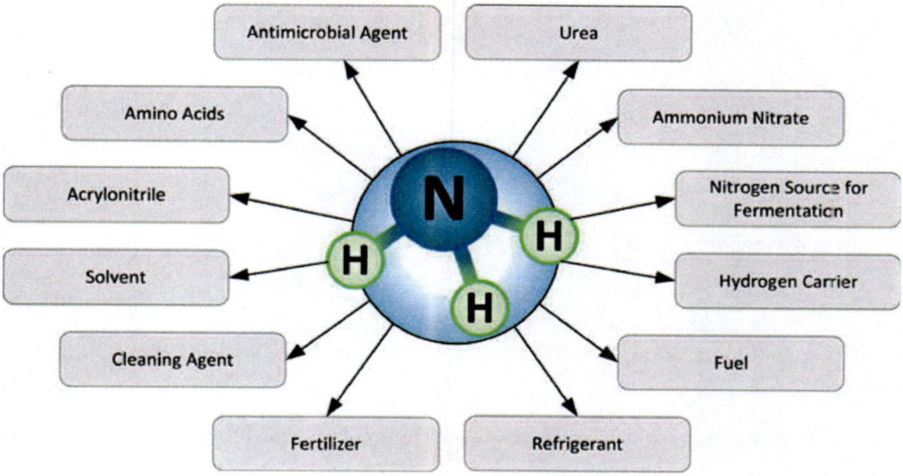

Fig. 1.7 Uses of ammonia in various sectors

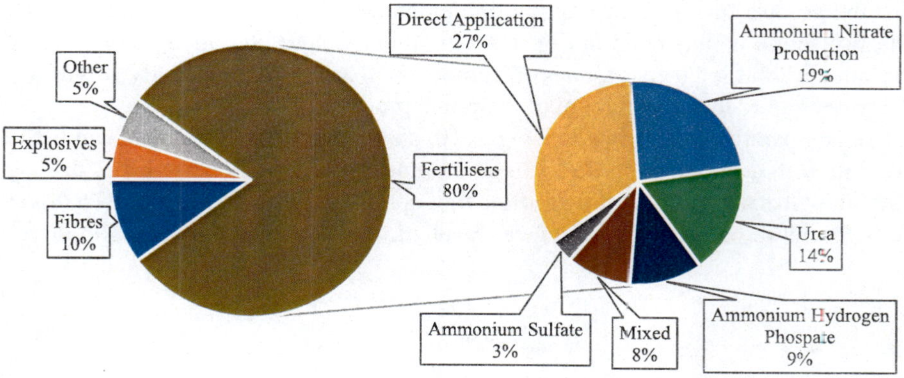

Fig. 1.8 Market share of ammonia related products (data from [8])

It is expected that the use of ammonia will increase since ammonia is used in many sectors for various purposes. Figure 1.9 depicts the ammonia production changes over the years. The data between 1940 and 2017 is recorded from the U.S. Geological Survey [14]. Global capacity of ammonia production was about 150 million metric tonnes in 2019 [6]. Less then 3% of ammonia is used in direct applications in agricultural sector. Those applications mostly limited to United States. Remaining portion of the production is utilized in chemical processing and fertilizer production [7]. Because of widespread application in agriculture, and being a good energy carrier, the production is expected to rise in the future. Utilization of ammonia as a carbon-free fuel is researched globally due to environmental concerns and endeavors to reduce CO_2 emissions. Countries have shifted some of their attention to ammonia-related projects for off-grid applications, internal combustion engines, and power

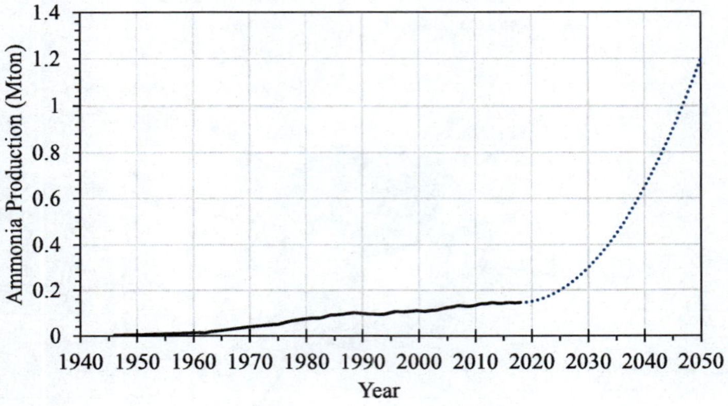

Fig. 1.9 The amount of ammonia produced around the world (modified from [14])

generation, etc. In addition, studies on the use of ammonia as a hydrogen carrier will increase in accordance with the research on clean hydrogen technologies. We foresee that the production and consumption of ammonia will be increasing exponentially. Our projection is illustrated in Fig. 1.9 with a blue-dotted line. In 2050, we expect that about 1.2 billion metric tonnes of ammonia will be produced globally. It is nearly 8.2 times higher than the amount of ammonia produced in 2019.

As comparatively illustrated in Fig. 1.10, an ammonia driven vehicle can travel 100 km with a fuel cost of US$6.30 as of June 2021. On the other hand, driving 100 km with same car but with gasoline fuel option costs US$8.11. It costs a driver US$18.1 more per 1000 km distance travelled. If we assume a driver who drives

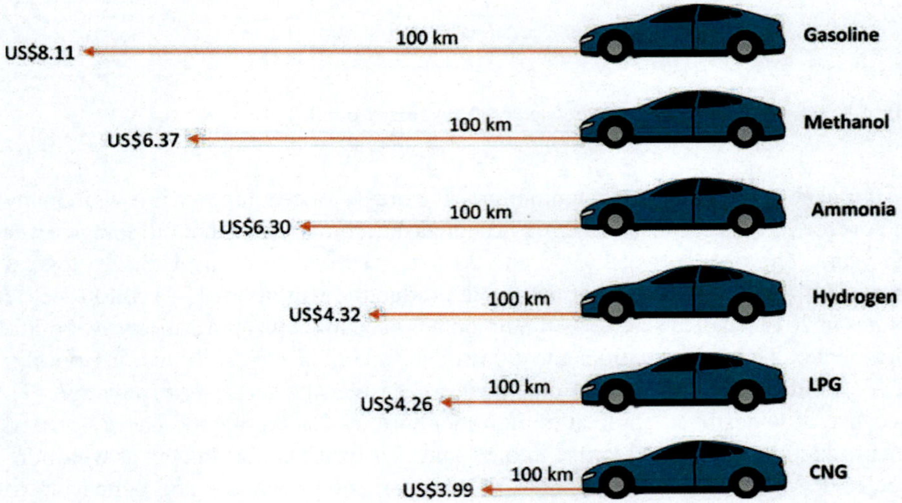

Fig. 1.10 Comparison of driving cost per 100 km for various fueled vehicles

40,000 km per year, then this driver must pay US$724 more per year. In terms of cost and on an energy content basis, ammonia has traditionally been priced competitively with gasoline and diesel. The ability of ammonia fuel to be utilized in diesel engines, gasoline engines, fuel cells, and gas turbines is a huge benefits. Ammonia is a promising transportation fuel. By-product refrigeration has the benefit of lowering costs and requiring less maintenance during vehicle service. Economic availability and feasibility, a worldwide distribution network, and ease of handling are all benefits of ammonia. In comparison with gasoline and methanol, ammonia is a cost-effective fuel per unit energy stored onboard.

Both hydrogen and ammonia are recognized as clean fuels, as they contain no carbon and result in no carbon footprint when used. Due to the increased levels of environmental concerns, there has been increasing attention to these two clean fuels. The researchers, scientists and technologists have been intensely working on systems using ammonia and hydrogen in order to increase sustainability indexes of systems. Many recent studies in the open literature have focused on either developing novel systems that using ammonia or hydrogen or converting present conventional systems into systems that using ammonia or hydrogen as a fuel [19–24].

It is anticipated that the usage of ammonia and hydrogen will expand significantly in the near future due to energy supply/demand challenges and environmental concerns. The MAN Energy Solutions, for instance, intends to provide a retrofit kit for their ammonia engine by 2025 [25]. In recent times, the Japanese shipping company NYK line has launched a number of prominent technology development agreements targeted at the building of ammonia-fueled ships and the delivery of fuel. Among its partners in these enterprises are the classification society Class NK, the engine producer IHI Power Systems, and the shipbuilding Japan Marine United Corporation. Three kinds of ships, including an ammonia-powered ammonia gas carrier, an offshore ammonia barge, and an ammonia-powered tugboat, have been disclosed so far in the open literature. These partnerships seek to commercialize and bring these vessels into practical use, moving beyond the initial research process [26]. Japan is anticipated to be a major ammonia and/or hydrogen importer within the next decade, having already shown a keen interest in Australia's shipping of environmentally friendly products. JERA, the biggest power company in Japan, said recently that it wants to shut down all of its old coal-fired power plants by 2030 and replace them with green fuels like ammonia and hydrogen [27].

As indicated in Fig. 1.11, it is predicted that the total worldwide ammonia production will increase from around 235 million metric tonnes in 2019 to roughly 290 million metric tonnes by 2030. This expected increase in capacity is the result of roughly 107 planned and announced ammonia facilities, most of which are situated in Asia and the Middle East. In 2019, Canada generated over 5 million metric tonnes of ammonia [29]. Approximately 80% of the ammonia produced by industry is used as a fertilizer. Ammonia is being used in the manufacture of textiles, explosives, insecticides, as well as other chemicals. It is utilised in several home and commercial cleaning products. By combining ammonia gas with water, 5–10% ammonia cleaning

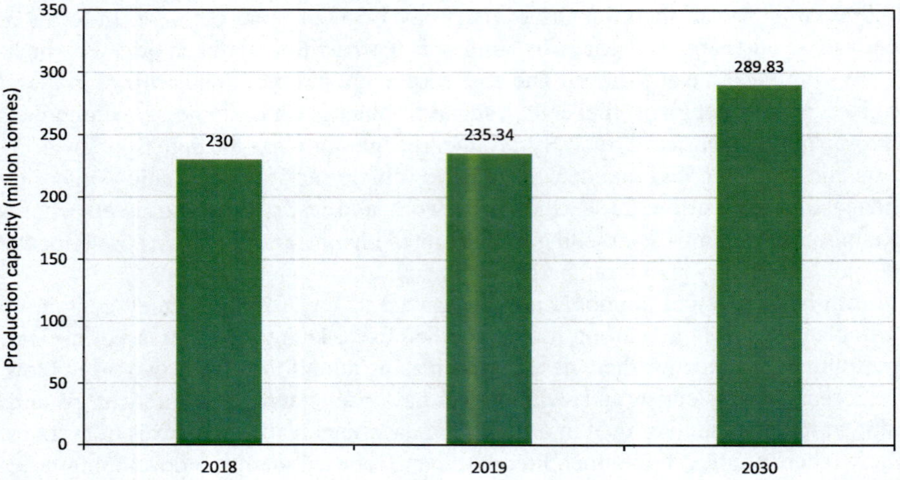

Fig. 1.11 Global production capacity of ammonia from 2018 to 2030 (data from [27])

solutions for the home may be produced. Ammonia solutions for industrial usage may have concentrations of at least 25% and are highly corrosive [30].

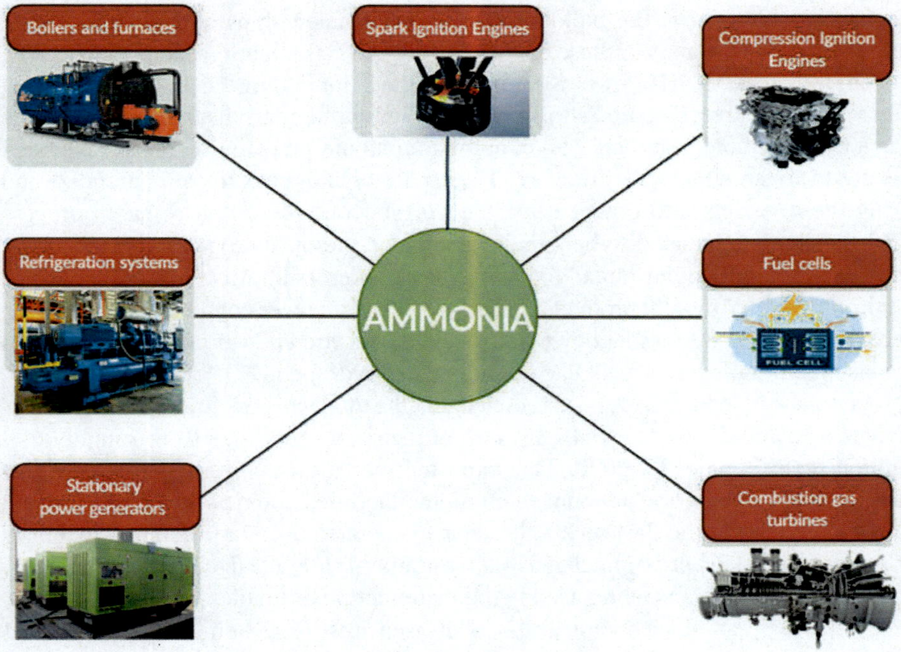

Fig. 1.12 Ammonia utilization in various applications

Ammonia can be used in many applications as illustrated in Fig. 1.12. On the present market, ammonia is mostly used as a fertiliser and a refrigerant. In addition, there are prototypes of cars powered by ammonia that use either engines or fuel cells. As an alternative fuel, ammonia may be used in all kinds of internal combustion engines, gas turbines, and burners, as well as directly in fuel cells, with only modest changes. This is a significant advantage compared to other fuel types. In an ammonia-based economy, fuel cells, stationary power generators, furnaces/boilers, and even automobiles will provide decentralised power production and smart grid applications. Due to very low energy needs, ammonia may be converted into hydrogen for almost any use [31].

Figure 1.13 depicts an illustrative cost comparison of various vehicle fuels in terms of energy cost per gigajoule. In terms of current energy market prices, ammonia is the cheapest fuel, and it costs US$10.27 per gigajoule of energy while gasoline costs US$22.48 per gigajoule, methanol costs US$47.05 per gigajoule, compressed natural gas (CNG) costs US$29.59, hydrogen costs US$27.19 per gigajoule, and liquefied petroleum gas (LPG) cost US$22.02 per gigajoule of energy. This demonstrates that ammonia is a cost-effective transportation fuel. By-product refrigeration has the advantage of lowering costs and requiring less maintenance during vehicle operation. Commercial availability and viability, a global distribution network, and ease of handling are all advantages of ammonia. As illustrated in Fig. 1.13, ammonia is a more cost-effective fuel per unit energy stored onboard than methanol, CNG, hydrogen, gasoline, and LPG.

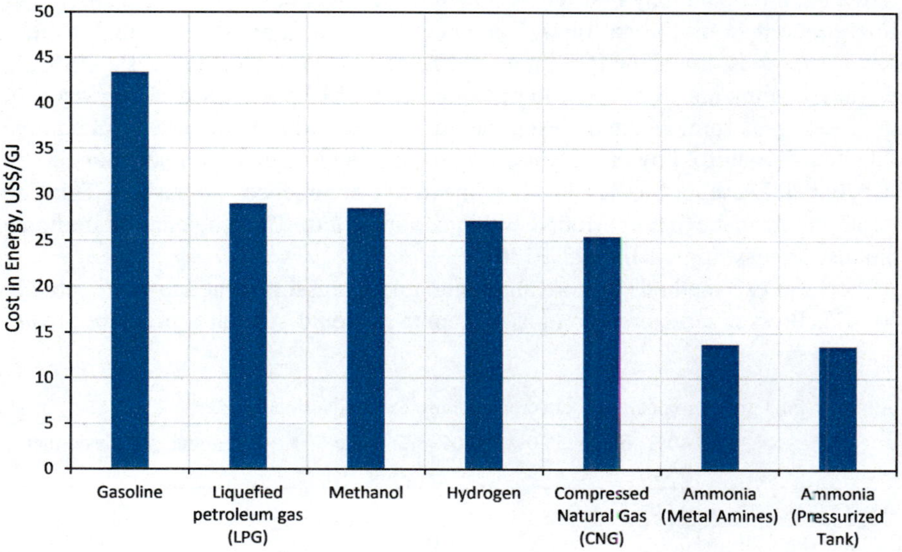

Fig. 1.13 Comparison of the cost per gigajoule of energy for different vehicle fuels (Modified from [33–35])

In typical internal combustion engines (ICE), ammonia and a combustion promoter will be injected in tandem. Combustion of ammonia is characterised by low flame temperature, poor laminar burning velocity, high ignition energy, and low flammability limitations, which would substantially degrade engine performance. Combustion and emission characteristics of a compression-ignition engine that consumes ammonia and dimethyl ether (DME) mixtures were studied in a research [32]. The research advocated employing direct liquid injection to restrict combustion within the cylinder bowl so as to limit ammonia emissions. DME, which has a high cetane value, is combined with ammonia. Due to its great resistance to autoignition, ammonia cannot spontaneously ignite. DME, like ammonia, has a high vapour pressure and must be compressed to stay a liquid. DME and ammonia are mixable, and their polarity allows the combination to stay stable, making it more convenient. DME is an effective diesel engine fuel that contains little fossil fuels. The availability of certain ammonia/DME combinations as refrigerants increases its viability as an alternative fuel.

When ammonia is utilised, the operational capability of the engine is lowered. Possible causes include the following: ammonia has a high latent heat of 18.6 MJ/kg, but a low heating value of 42 MJ/kg compared to diesel fuel. To accomplish the same engine load with the same amount of fuel energy, roughly 2.26 times as much ammonia as diesel fuel is required. Assuming an equivalency ratio of 0.5 in a diesel engine running, this quantity of ammonia and its greater latent heat may lower the in-cylinder air temperature by about 100 °C. This chilling will slow down chemical reactions [32].

In cars, ammonia may be used as a mixed fuel. Ammonia is less expensive per kilogramme than traditional fuels. The energy costs of ammonia and diesel fuels, including blends, are shown in Table 1.3. As indicated, diesel fuel costs US$3.21 per gallon, ammonia costs US$680 per tonne, and DME costs US$555 per tonne. In dual fuelling of ammonia and diesel, the initial step is to provide gaseous ammonia to the intake manifold by producing a premixed combination of ammonia and air in the cylinder. Diesel or biodiesel is then injected to commence combustion. Thus, no alterations are made to the current diesel injection system. The efficiency of ammonia combustion may approach to 95% [36].

There are two methods to store ammonia: under pressure or at a low temperature [40, 41]. Pressurized storage retains ammonia in liquid state at a pressure greater

Table 1.3 Fuel prices comparison delivered to compression ignition engine

	40% ammonia/60% diesel	40% ammonia/ 60% dimethyl ether	Ammonia	Diesel fuel
LHV (MJ/kg)	32.6	24.5	18.6	42
Fuel rate (kg/kWh)	0.316	0.42	0.554	0.245
Fuel price (US$/kg)	0.78	0.61	0.68	0.85
Fuel energy cost (US$/kWh)	0.25	0.26	0.38	0.21

Source [37–39]

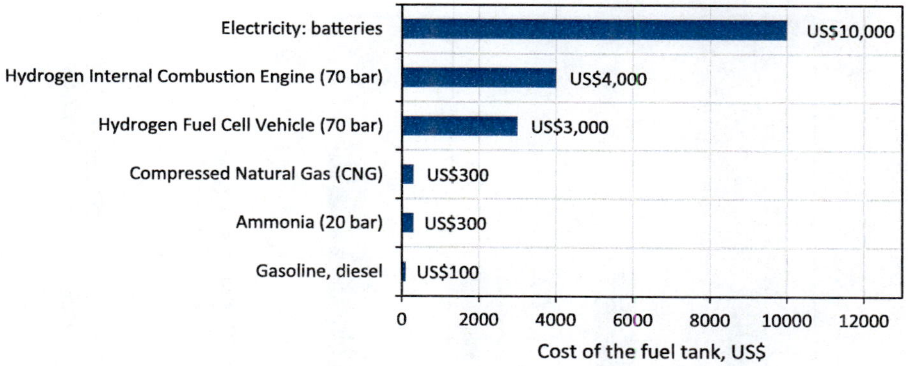

Fig. 1.14 Costs of on-board storage tanks for different types of fuelled vehicles

than 8.6 bar at ambient temperature (20 °C). However, ammonia is often kept above 17 bar to maintain liquid phase if ambient temperature rises. The energy density of pressurised liquid ammonia storage is 13.77 MJ/L. As a general rule, 2.8 tonnes of ammonia may be kept every tonne of steel. This storing requires no energy to maintain its compressed condition. Typically, storage at low temperatures is employed for large-scale storage.

This form of storage needs energy to maintain a low temperature and prevent boil-off caused by ambient temperature. For extended storage, low temperature is favoured due to its lower initial cost. The energy density of ammonia stored in this manner is 15.37 MJ/L, as opposed to 13.77 MJ/L for pressured storage. If 183 days of storage are considered, reflecting the interval between winter and summer, the ammonia storage cost will be US$4.03 per gigajoule. This price is much less than the cost of hydrogen storage, which is US$98.74/GJ.

As demonstrated in Fig. 1.14, the projected cost of a fuel tank for a personal car with a 300-mile range is smallest for ammonia after standard and traditional gasoline/diesel tanks. Figure 1.14 demonstrates that on-board ammonia storage is priced similarly to compressed natural gas and gasoline cars. As expected, since hydrogen storage requires more complicated systems, the cost of the storage tank is higher then gasoline, diesel, CNG and ammonia. The highest cost is seen in batteries.

Figure 1.15 depicts the driving costs of various fuels. At the current market prices, gasoline is most costly fuel per 1 km distance travelled and compressed natural gas (CNG) is the most low-cost option due low demand in the world for natural gas. However, when pre-pandemic price of fuels taken into consideration ammonia was most cost-effective fuel in various fuels. During COVID-19 pandemic, higher demands and lower supplies have caused ammonia prices to soar. The prices of ammonia shot up nearly 60% since fall 2020 and are now at US$680 per tonne as of June 2021 [42]. As a result, worldwide values have skyrocketed, as companies empty the storage, demand spikes. It is expected that ammonia prices will drop nearly pre-pandemic levels when ammonia production increase with shutdowns lifted [39]. On

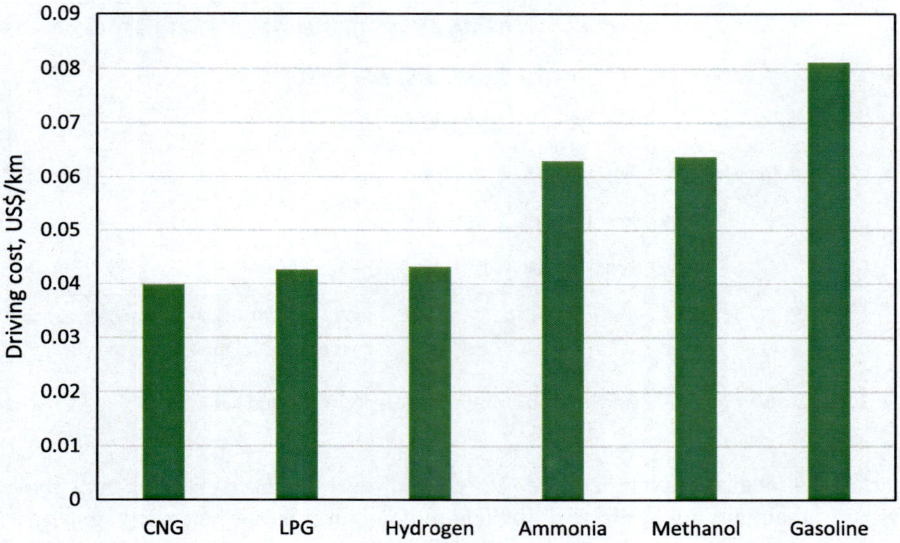

Fig. 1.15 Driving cost of various fuels

the other hand, the analysts expect coronavirus-related shutdowns to ease throughout the 2021, when the COVID-19 vaccine begins to reach the world.

According to the studies carried out by the analysts [43], the global oil demand will rise 6.6 million barrels per day in 2021, reversing nearly two-thirds of the nearly 10 million barrels per day decline seen in 2020. China's oil demand has already risen this quarter, surpassing levels observed in the same quarter of 2019. The reversal in China's oil consumption is the first indicator of what will soon become a reality: rapid global demand rise year over year in 2021. By the second half of 2021, this tendency will have tightened the supply–demand balance, supporting oil prices. Figures 1.16 and 1.17 illustrate comparative the energy and exergy efficiencies for several ammonia manufacturing strategies. The highest energy and exergy efficiencies are achieved when using hydropower as energy source. Also, solar PV has the lowest values. Although fossil fuel pathways, such as gasification, have high energy and exergy efficiencies, renewable energy pathways using tidal and wave and hydropower also yield high results in terms of energy and exergy efficiency. It is therefore possible to produce ammonia with high efficiency using renewable technologies. Fossil fuel-based pathways have mediocre efficiency ratings. However, they are more favored by industry due to lower costs, matured installed infrastructure and easy access to fossil fuels.

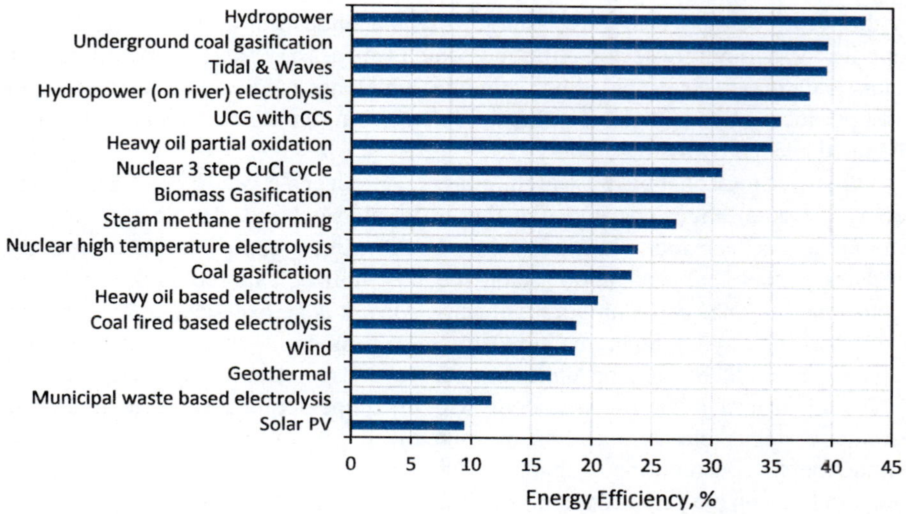

Fig. 1.16 Comparison of the energy efficiencies of diverse ammonia production techniques

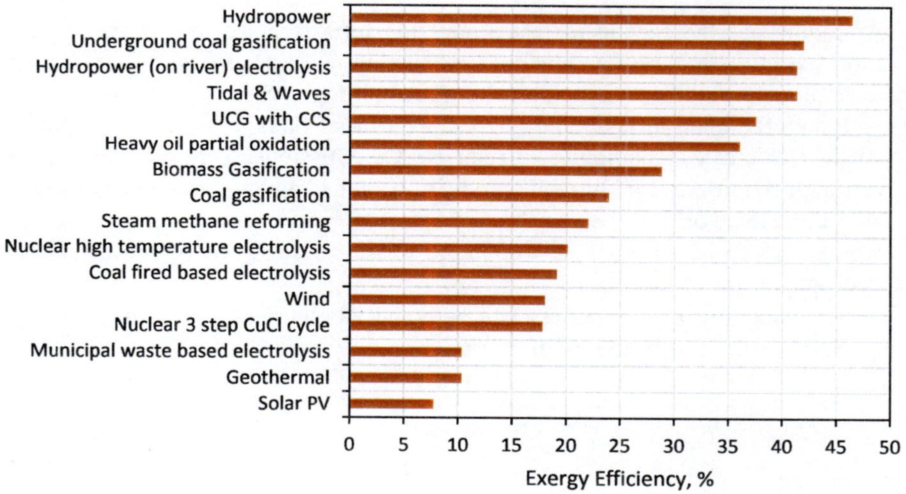

Fig. 1.17 Comparison of the exergy efficiencies of diverse ammonia production techniques

1.5 Closing Remarks

Ammonia is formed as a result of biological activities in nature and is an important part of the nitrogen cycle. The nitrogen cycle is an endless loop and is vital for the continuity of life. In nature, ammonia occurs as a result of decaying dead bodies of living beings and converted to nutrients by microorganisms to be used again by the living beings. Therefore, a large extent of industrial ammonia production supplies the agriculture sector. Nitrogen, which is necessary for the development of plants and animals, is provided by ammonia, which has made ammonia the second most produced chemical globally. In addition to the agriculture sector, ammonia is a chemical needed by many sectors.

Although a total of about 80% of the ammonia produced is used by the agricultural sector, it has an enormous potential due to its physical and chemical properties. Ammonia is a feedstock for nitric acid, plastics, fibres, explosives, and cleaning products in chemical production sector. It can be utilized as a fuel both in internal combustion engines and ammonia fuel cells. Also, it is an excellent energy carrier, due to the three hydrogen atoms in its structure. Ammonia serves as a viable, safe, cost-effective, and hassle-free alternative for the transportation of hydrogen.

The importance of ammonia emerges with the scope of climate change measures. Ammonia stands out as a carbon-free chemical, fuel, and energy carrier from production to consumption with the utilization of clean hydrogen technologies. Reducing carbon emissions depends on minimizing fossil fuel consumption. Therefore, ammonia utilization is important for a sustainable future to find carbon-free alternatives to fossil fuels in both production technologies and human activities. The key properties of ammonia can be summarized as follows:

- Ammonia has the second largest production infrastructure worldwide. Therefore, the infrastructure for ammonia production and transportation is a mature technology providing an excellent alternative for fossil-fuels.
- Although the main infrastructure depends on fossil-fuels today, it can be easily adapted to clean technologies to reduce CO_2 emissions. Zero-carbon or low-carbon technologies for ammonia production studied and researched extensively all around the world.
- Ammonia can be stored in gas or liquid phase with less energy compared to hydrogen.
- Existing stainless steel pipelines can potentially be utilized for ammonia transportation with little or no modification.

References

1. Erias A, Karaka C, Grajetzki C, Carton J, Paulos M, Jantunen P, Baral P, Samal Bex JMV (2016) World energy resources. World Energy Council. https://www.worldenergy.org/publications/entry/world-energy-resources-2013-survey
2. Statistics Canada (2019) Canadian passenger bus and urban transit industries, fuel consumption, by industry. https://doi.org/10.25318/2310008401-eng
3. Acar C, Dincer I (2019) Review and evaluation of hydrogen production options for better environment. J Clean Prod 218:835–849. https://doi.org/10.1016/j.jclepro.2019.02.046
4. Hoad TF (2008) The concise Oxford dictionary of English etymology. Paw Prints
5. Vesilind PA, Morgan SM, Heine LG (2009) Introduction to environmental engineering. Cengage Learn

6. Ghavam S, Vahdati M, Wilson IAG, Styring P (2021) Sustainable ammonia production processes. Front Energy Res 9:34
7. International Energy Agency (2021) Ammonia technology roadmap—towards more sustainable nitrogen fertiliser production
8. International Fertilizer Association (2022) World NH_3 statistics by country
9. Trade Map (2021) List of exporters for the selected product in 2020. Product: 281410 Anhydrous ammonia. https://www.trademap.org/Country_SelProduct.aspx?nvpm=1%7C%7C%7C%7C%7C%7C281410%7C%7C%7C6%7C1%7C1%7C2%7C1%7C%7C2%7C1%7C1%7C1. Accessed 23 Mar 2021
10. Smith C, Hill AK, Torrente-Murciano L (2020) Current and future role of Haber-Bosch ammonia in a carbon-free energy landscape. Energy Environ Sci 13:331–344. https://doi.org/10.1039/C9EE02873K
11. Aydin MI, Dincer I, Ha H (2021) Development of Oshawa hydrogen hub in Canada: a case study. Int J Hydrogen Energy 46:23997–24010. https://doi.org/10.1016/j.ijhydene.2021.05.011
12. Valentini A (2020) Argus white paper—green ammonia
13. Zamfirescu C, Dincer I (2009) Ammonia as a green fuel and hydrogen source for vehicular applications. Fuel Process Technol 90:729–737. https://doi.org/10.1016/j.fuproc.2009.02.004
14. Erdemir D, Dincer I (2021) A perspective on the use of ammonia as a clean fuel: challenges and solutions. Int J Energy Res 45:4827–4834. https://doi.org/10.1002/er.6232
15. Bicer Y, Dincer I, Zamfirescu C, Vezina G, Raso F (2016) Comparative life cycle assessment of various ammonia production methods. J Clean Prod 135:1379–1395. https://doi.org/10.1016/j.jclepro.2016.07.023
16. Davies J, Dolci F, Klassek-Bajorek D, Cebolla Ortiz R, Weidner E. 2020. JRC Science for Policy Report: Current status of Chemical Energy Storage Tecnologies. Petten.
17. International Energy Agency (2012) Energy technology perspectives 2012: pathways to a clean energy system
18. Mosier A, Syers JK, Freney JR (2013) Agriculture and the nitrogen cycle: assessing the impacts of fertilizer use on food production and the environment. Island Press
19. Ishaq H, Dincer I (2021) Biomass and bioenergy a novel biomass gasification based cascaded hydrogen and ammonia synthesis system using Stoichiometric and Gibbs reactors. Biomass Bioenerg 145:105929. https://doi.org/10.1016/j.biombioe.2020.105929
20. Ghavam S, Garcia-Garcia G, Styring P (2021) A novel approach to ammonia synthesis from hydrogen sulfide. Int J Hydrogen Energy 46:4072–4086. https://doi.org/10.1016/j.ijhydene.2020.10.192
21. Bicer Y, Dincer I (2018) Clean fuel options with hydrogen for sea transportation: a life cycle approach. Int J Hydrogen Energy 43:1179–1193. https://doi.org/10.1016/j.ijhydene.2017.10.157
22. Siddiqui O, Dincer I (2020) A new solar energy system for ammonia production and utilization in fuel cells. Energy Convers Manag 208:112590. https://doi.org/10.1016/j.enconman.2020.112590
23. Cesaro Z, Ives M, Nayak-Luke R, Mason M, Bañares-Alcántara R (2021) Ammonia to power: forecasting the levelized cost of electricity from green ammonia in large-scale power plants. Appl Energy 282. https://doi.org/10.1016/j.apenergy.2020.116009
24. Yapicioglu A, Dincer I (2019) A review on clean ammonia as a potential fuel for power generators. Renew Sustain Energy Rev 103:96–108. https://doi.org/10.1016/j.rser.2018.12.023
25. The Motorship (n.d.) MAN ES unveils 2025 ammonia retrofit target
26. Brown T (2020) Japan's NYK and partners to develop ammonia fueled and fueling vessels—Ammonia Energy Association. Ammon. Energy Assoc. Available from: https://www.ammoniaenergy.org/articles/japans-nyk-and-partners-to-develop-ammonia-fueled-and-fueling-vessels/, (Accessed at 29/11/2022).
27. Comodity Independed Intelligence Services (2020) Japan's government embracing ammonia as fuel of the future in zero-carbon emissions drive. Available from: https://www.icis.com/explore/resources/news/2020/10/28/10568460/japan-s-government-embracing-ammonia-as-fuel-of-the-future-in-zero-carbon-emissions-drive/, Accessed at 29.11.2022.

28. Fernández, L. (2021) Ammonia annual production capacity globally 2030, Accessed 07.10.2021. *Statista.*
29. Statistics Canada (2022) Canadian fertilizer production, by product type and fertilizer year, cumulative data. (x1000) https://www150.statcan.gc.ca/t1/tbl1/en/tv.action?pid=3210003701&pickMembers%5B0%5D=1.1&cubeTimeFrame.startYear=2006+%2F+2007&cubeTimeFrame.endYear=2021+%2F+2022&referencePeriods=20060101%2C20210101
30. New York State Department of Health (2004) The facts about ammonia, Accessed 29.11.2022
31. Bicer Y (2017) Investigation of novel ammonia production options using photoelectrochemical hydrogen, Ontario Tech University
32. Gross CW, Kong SC (2013) Performance characteristics of a compression-ignition engine using direct-injection ammonia-DME mixtures. Fuel 103:1069–1079. https://doi.org/10.1016/j.fuel.2012.08.026
33. Schorn F, Breuer JL, Samsun RC, Schnorbus T, Heuser B, Peters R, and Stolten D (2021) Methanol as a renewable energy carrier: An assessment of production and transportation costs for selected global locations. Adv. Appl. Energy, 3, 100050.
34. HAFNIA (2020) Ammonfuel - An industrial view of ammonia as a marine fuel.
35. U.S. Department of Energy (DOE) (2022) Clean Cities - Alternative Fuel Price Report.
36. Ryu K, Zacharakis-Jutz GE, Kong SC (2014) Effects of gaseous ammonia direct injection on performance characteristics of a spark-ignition engine. Appl Energy 116:206–215. https://doi.org/10.1016/j.apenergy.2013.11.067
37. ECHEMI (2022) Methyl ether Market Price & Analysis, https://www.echemi.com/productsInformation/pid_Seven1867-dimethylether.html, Accessed 29.11.2022
38. GlobalPetrolPrices.com (2021) Gasoline prices around the world, https://www.globalpetrolprices.com/gasoline_prices/, Accessed 07.06.2021
39. Echemi (2022) Ammonia Market Price & Analysis, https://www.echemi.com/productsInformation/pid_Rock19411-ammonia.html, Accessed 29.11.2022
40. Dimitriou P, Javaid R (2020) A review of ammonia as a compression ignition engine fuel. Int J Hydrogen Energy 45:7098–7118. https://doi.org/10.1016/j.ijhydene.2019.12.209
41. Sánchez A, Castellano E, Martín M, Vega P (2021) Evaluating ammonia as green fuel for power generation: a thermo-chemical perspective. Appl Energy 293:116956. https://doi.org/10.1016/j.apenergy.2021.116956
42. Morgan, T (2021) Price Gouging or Price Reality? Anhydrous Ammonia Prices Climb 60% Since Fall | The Scoop, https://www.thedailyscoop.com/news/retail-business/price-gouging-or-price-reality-anhydrous-ammonia-prices-climb-60-fall, Accessed 29.11.2022
43. Crooks, E (2020) Ten predictions for energy in 2021, https://www.woodmac.com/news/opinion/ten-predictions-for-energy-in-2021/, Accessed 29.11.2022

Chapter 2
Ammonia Production

Nomenclature

Symbols

A	Area (cm^2)
A_a	Estimated equipment capacity
A_b	Base capacity for the equipment
Ag	Silver
Au	Gold
C	Carbon
C_0	Cost of operating labor (US$)
C_a	Estimated equipment cost (US$)
C_b	Base price of the equipment (US$)
C_{BM}	Bare module cost (US$)
C_C	Fixed capital investment (US$)
C_{CM}	Manufacturing cost (US$)
C_D	Direct costs (US$)
C_F	Cost of feedstocks (US$)
C_G	General expenses (US$)
C_M	Fixed costs (US$)
c_p	Specific heat capacity at constant pressure (kJ/kg K)
$C_p{}^0$	Purchased module cost for base condition (US$)
C_U	Cost of utilities (US$)

© The Author(s), under exclusive license to Springer Nature Switzerland AG 2022

I. Dincer et al., *Ammonia Energy Technologies*, Lecture Notes in Energy 91,
https://doi.org/10.1007/978-3-031-13532-3_2

C_W	Cost of waste treatment (US$)
CH_4	Methane
CO	Carbon monoxide
CO_2	Carbon dioxide
F	Faradays constant (Coulomb/mol)
F_{BM}	Bare module cost factor
H_2	Hydrogen
i	Current (A)
i_0	Exchange current density (A/cm^2)
kWh	Kilowatt-hour
NH_3	Ammonia
NH_4	Ammonium ion
R	Gas constant (kJ/kmol K)
T	Temperature (°C, K)
V	Voltage (V)

Subscripts

act	Activation potential
ael	Alkaline electrolysis
an	Anode
cat	Cathode
diff	Diffusion
el	Electrode
elc	Electrolyte
is	Isentropic
mem	Membrane
ohm	Ohmic
suct	Suction

Greek Letters

α	Charge transfer activity
ϵ	Volume fraction
η	Efficiency
ρ	Density (kg/m^3)
σ	Membrane resistance (1/ohm-cm)
ΔG^0	Gibbs free energy variation (kJ/kmol)

Acronyms

AEL	Alkaline electrolysis
BEV	Battery electric vehicle
BG	Biomass gasification
CEPCI	Chemical engineering plant cost index
CNG	Compressed natural gas
Cu-Cl	Copper chlorine
DGE	Diesel gallon equivalent
DLE	Diesel liter equivalent
DME	Dimethyl ether
HHV	Higher heating value
ICE	Internal combustion engine
LNG	Liquefied natural gas
LPG	Liquefied petroleum gas
MAPS	Micro ammonia production system
Mg-Cl	Magnesium chlorine
NG	Natural gas
PEM	Proton exchange membrane electrolysis
PSA	Pressure swing adsorption
PV	Photovoltaic
SI	Sulfur iodine
SMR	Steam methane reforming

2.1 Introduction

Ammonia and hydrogen are called clean fuels as they have no carbon footprint. Due to environmental concerns, attention to them and their usage is emerging. Scientists and technologists have been intensely working on systems using ammonia and hydrogen to increase the sustainability of systems. Current studies have focused on either developing novel systems using ammonia or hydrogen or converting present conventional systems into systems using clean ammonia or hydrogen as fuel. Soon, it is expected that the use of ammonia and hydrogen will increase with energy supply/demand issues and environmental challenges as we have been facing.

In this chapter, the production methods of ammonia are discussed. Detailed information about conventional ammonia production is presented, which includes the Haber–Bosch reactor and hydrogen production by steam methane reforming. In addition to conventional methods, electrolytic ammonia production systems, where hydrogen is produced by electrolysis technology, are also examined. Carbon-free ammonia production can be achieved with electrolysis by utilizing renewable energy. However, the downside of electrolysis technology is the capital and operating costs. Since the technology is relatively new and requires further development to catch up the mature technologies such as steam reforming and coal gasification. Hence, the costs required to manufacture electrolysis systems have been estimated using cost assessment methodologies. The costs to manufacture a facility that produces ammonia on-site for small settlements and communities were examined. Accordingly, the expenses required for the establishment and operation of the ammonia plant and the production cost of ammonia have been calculated. Studies on ammonia costs have been conducted by investigating the required infrastructure for the generation of ammonia with surplus electricity that occurs in microgrid systems. In addition, energy and exergy efficiencies were investigated by modeling the sub-components required for ammonia production using the Aspen Plus software package.

2.2 Hydrogen Production Methods

Ammonia is produced using hydrogen and nitrogen, but high feedstock costs come from hydrogen production. Therefore, the role of hydrogen production to reduce ammonia costs becomes significant. Today, the required hydrogen for ammonia production is predominantly made using fossil fuels. However, it is vital to use clean hydrogen produced by renewable resources for clean ammonia production. Therefore, to provide a sustainable ammonia economy, hydrogen production should be freed from fossil fuels. Figure 2.1 shows various hydrogen production methods. Today, most of the hydrogen production is done with fossil fuels. Although there are studies on cleaner production technologies, most of these methods are still in the research and development and pilot-scale studies. In this section, fossil fuel-based hydrogen production and electrolysis, which are the most used methods, are discussed.

2.2.1 Fossil Fuel Based Hydrogen Production

Steam methane reforming and coal gasification are the most utilized fossil fuel-based hydrogen production methods. The production of hydrogen from fossil fuels is based on producing syngas containing carbon monoxide (CO) and hydrogen. In the next step, syngas containing carbon dioxide (CO_2) and H_2 is obtained by reacting CO with water. Also, additional processes are applied for the separation of contaminants such as CO_2, sulfur, and nitrogen compounds from the syngas. The main reactions for the steam methane reforming can be summarized as follows:

$$\text{Steam methane reforming: } CH_4 + H_2O \rightarrow CO + 3H_2 \tag{2.1}$$

$$\text{CO shift reaction: } CO + H_2O \rightarrow CO_2 + H_2 \tag{2.2}$$

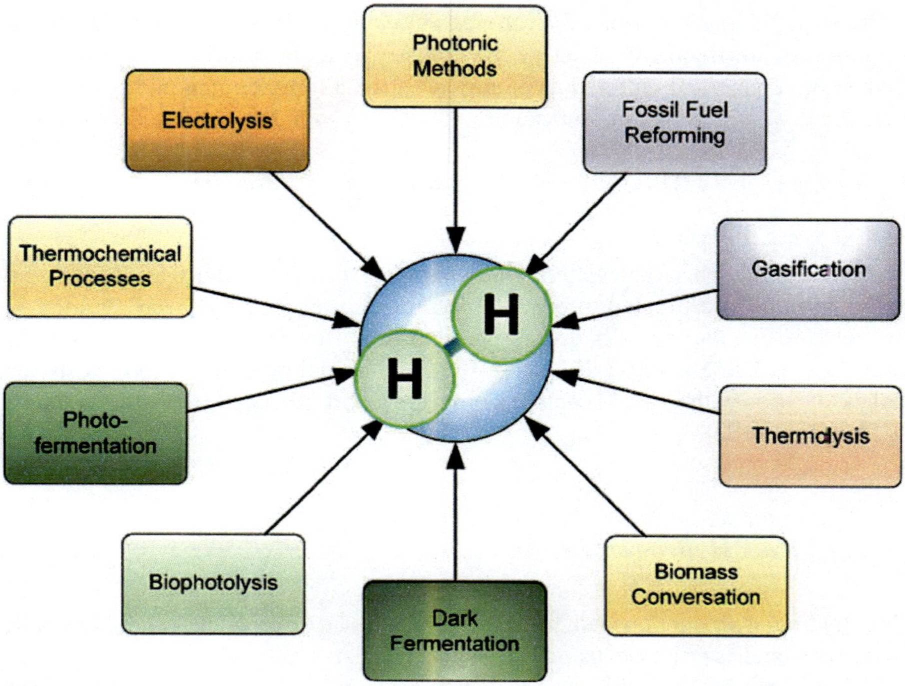

Fig. 2.1 Hydrogen production methods (modified from [1])

Gasification is required to use coal for the hydrogen source. In this case, carbon is gasified to produce synthetic methane, followed by steam-methane reforming [2]:

$$C + O_2 \rightarrow CO_2 \tag{2.3}$$

$$C + 1/2O_2 \rightarrow CO \tag{2.4}$$

$$C + CO_2 \leftrightarrow 2CO \tag{2.5}$$

$$C + H_2O \leftrightarrow CO + H_2 \tag{2.6}$$

$$C + 2H_2O \leftrightarrow CO_2 + 2H_2 \tag{2.7}$$

$$C + 2H_2 \leftrightarrow CH_4 \tag{2.8}$$

$$CO + H_2O \leftrightarrow CO_2 + H_2 \tag{2.9}$$

Price of the fossil-fuel based hydrogen varies in the literature. However, those technologies are frequently used by the industry as a cheap solution to hydrogen production. Penner [3] reported an empirical formula for obtaining hydrogen using natural gas without using carbon capture:

$$\textit{Hydrogen Price}\ (\text{US\$/kg}) = 0.359 \cdot \textit{Natural Gas Price}\ (\text{US\$/m}^3) + 0.15 \tag{2.10}$$

Accordingly, when calculated with current Ontario natural gas prices, the hydrogen price is estimated to be US$1305/m^3. However, it has been reported in the literature that the cost of hydrogen production by steam methane reforming from natural gas is US$1.46–3/kg H_2 [4, 5]. In addition, hydrogen production costs for coal gasification with a carbon capture system are reported between US$1.73–1.98/kg H_2 [5].

2.2.2 *Clean Hydrogen Production*

Clean hydrogen production methods are generally based on producing hydrogen from water. Hydrogen is produced using electricity and heat or both. The carbon emissions are non-existent for the hydrogen production systems utilizing clean electricity and heat sources. Clean hydrogen production can be achieved by utilizing nuclear or renewable power sources such as hydroelectric, wind, and solar.

Electrolysis is a water-splitting process that utilizes membranes or diaphragms to dissociate water into hydrogen and oxygen by applying an external current. Currently, electrolysis is the most frequently used industrial process to produce clean hydrogen. Hydrogen production cost for electrolysis is reported to be as low as US$2.9/kg H_2 [6]. Hydrogen price from electrolysis is heavily dependent on electricity consumption. Electricity consumption is reported to be 50–78, 50–83, and 40–50 kWh/kg H_2 for alkaline electrolysis, proton electrolyte membrane electrolysis, and solid oxide electrolysis, respectively [7]. It is estimated that the hydrogen prices for distributed and centralized systems will be as low as US$2.30 and 2.00/kg H_2 by 2030 [8].

In addition to electrolysis, thermochemical cycles such as Cu–Cl (Copper-chlorine cycle), Mg–Cl (Magnesium-chlorine cycle), and SI (sulfur-iodine cycle) can also be used for clean hydrogen production. Thermochemical cycles adopt a series of chemical reactions to dissociate water, and the other chemicals in the system are recycled and reused. These systems are not energy extensive like electrolysis. However, investment costs are high, especially since the equipment of the processes must be durable to corrosive chemicals are used in the process. Razi and Dincer [9] estimated the capital costs for the Cu–Cl cycle as US$23.9 M for a capacity up to 1619.3 kg/h. The cost of hydrogen production is between US$5.54–7.08/kg. Similarly, preliminary costing studies for the Mg–Cl cycle show that the price would be around US$3.36/kg [10].

2.3 Air Separation Methods

The stoichiometric nitrogen to hydrogen ratio should be 1:3 for ammonia production. Therefore, nitrogen production is as vital as hydrogen production. Different production pathways can be followed for nitrogen production. These production methods have their advantages and disadvantages and should be evaluated carefully. Delivery and production options for nitrogen are given in Fig. 2.2. Depending on the plant capacity, outsourcing or on-site production is possible. On-site production can be realized by pressure swing adsorption, membrane separation and cryogenic separation.

Note that the largest nitrogen source is air. Figure 2.3 shows the chemical composition of atmospheric gases. 78.1% of the atmosphere consists of nitrogen. The remaining percent of the air consists of oxygen and inert gases. Oxygen and inert gases should be separated from the air to produce nitrogen.

Nitrogen can be produced by various methods such as membrane separation, pressure swing adsorption, or cryogenic distillation. Although gas or liquid delivery in cylinders is also possible, this solution is usually adopted by small-scale plants to avoid increasing capital costs. Today, large-scale facilities prefer cryogenic distillation, and 90% of the nitrogen requirement of the industry is met by this technology [12].

Different technologies nitrogen production technologies can be used for various production capacities. An overview of the technologies used according to nitrogen requirement and nitrogen purity is given in Fig. 2.4. Accordingly, in cases where the nitrogen requirement is low, gas and liquid nitrogen can be obtained by outsourcing.

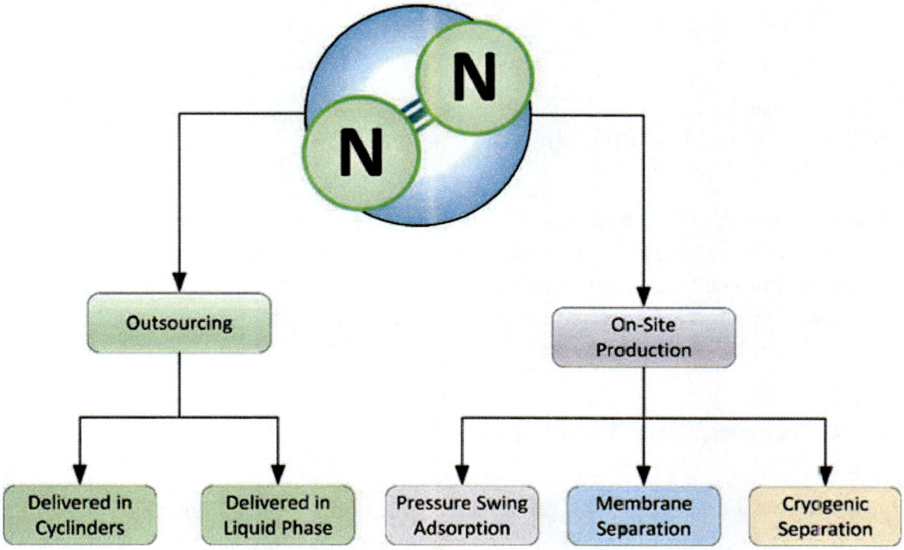

Fig. 2.2 Production and delivery options for nitrogen as feedstock for ammonia production

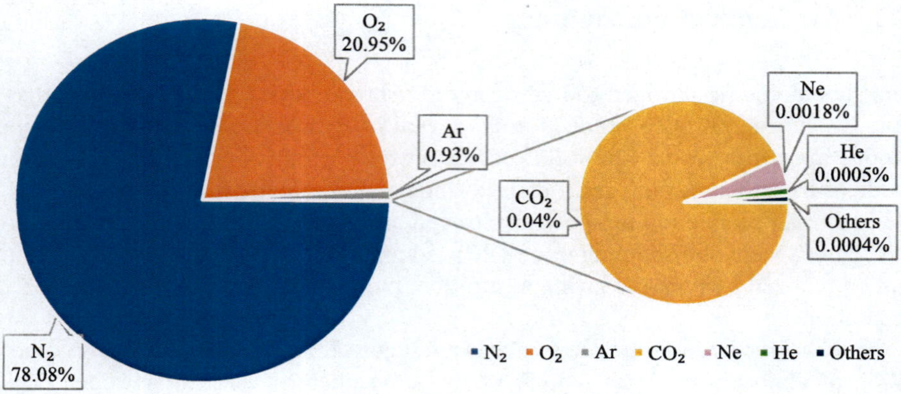

Fig. 2.3 Chemical composition of the atmosphere by volume (data from [11])

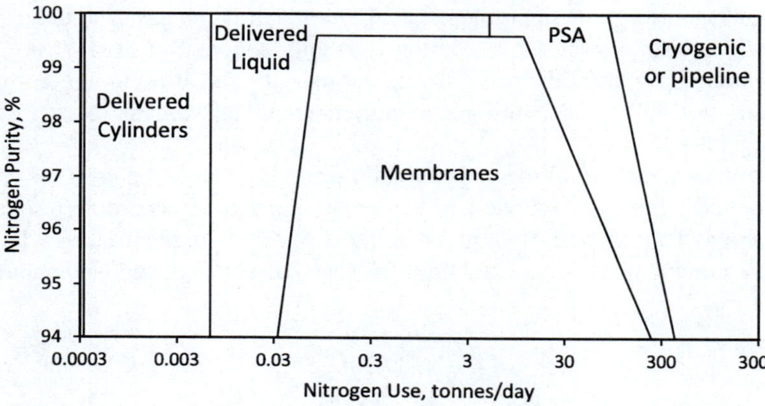

Fig. 2.4 Comparison of different nitrogen production technologies according to production capacity and purity (modified from [13])

However, when the daily requirement exceeds 30 kg, membrane processes can be used if there is no need for high purity. In cases where the daily need exceeds five tonnes, outsourcing is not feasible, and usually, pressure swing adsorption and cryogenic separation are preferred to produce nitrogen with high purity [13].

2.3.1 Cryogenic Air Separation

Cryogenic air separation is based on separating the components of the air by bringing them to their boiling points. The boiling points of oxygen, nitrogen, and argon, which have the highest proportions in air, are −182.95, −195.75, and −185.85 °C, respectively [12]. The cryogenic air separation process consists of two parts, hot and cold. Accordingly, the hot part consists of compression, dehumidification, and purification; the cold part consists of heat exchangers and distillation columns. The moisture and CO_2 in the air can be removed from the flow by adsorption with silica gel. Mass production of nitrogen in high volumes can be achieved with cryogenic air separation efficiently and economically [14]. The main components of the cryogenic air separation unit are listed below:

- Air compressor
- Compressed air tank
- Chiller
- Air purification unit
- Heat exchangers
- Distillation column
- Liquid nitrogen tank.

2.3.2 *Pressure Swing Adsorption*

The pressure swing adsorption (PSA) process separates the least adsorbed component from a gas mixture under high pressure. Therefore, oxygen-adsorbing adsorbents such as carbon molecular sieves separate nitrogen [16]. Figure 2.5 summarizes nitrogen generation with the pressure swing adsorption process. The filtered air is taken to the adsorbent tanks with the help of a blower. Oxygen molecules are smaller than nitrogen molecules. Therefore, oxygen can be fixed on the adsorbent bed by using carbon molecular sieves, which has the same pore size as air molecules. The main components of the pressure swing adsorption can be summarized as follows:

- Air compressor
- Compressed air tank
- Two or more parallel adsorption tanks
- Buffer tank.

2.3.3 *Membrane Separation*

Membrane separation is the process of separating the air by permeation using polymeric membranes. Standard membrane nitrogen separation systems can vary between 0.5 and 200 N m^3/h. However, there are also systems with higher flow rates. Figure 2.6 represents the flow chart of a membrane separation unit. The air is filtered before being taken into the membrane units, and the gases in the air are purged by passing through the membranes. Nitrogen gas is obtained from the end of the line [17]. The purity of the nitrogen gas varies between 94 and 99.5%. There is a trace amount of oxygen and argon in the produced nitrogen gas. In cases where higher purity nitrogen

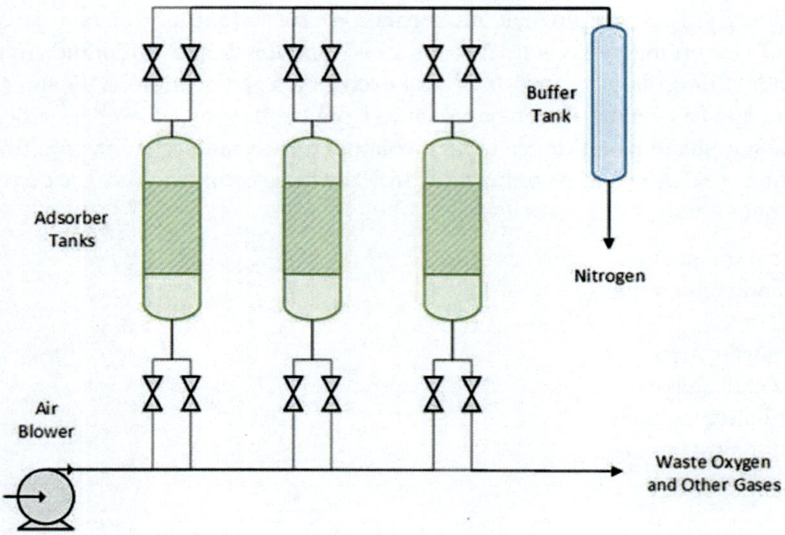

Fig. 2.5 Pressure swing adsorption process (modified from [15])

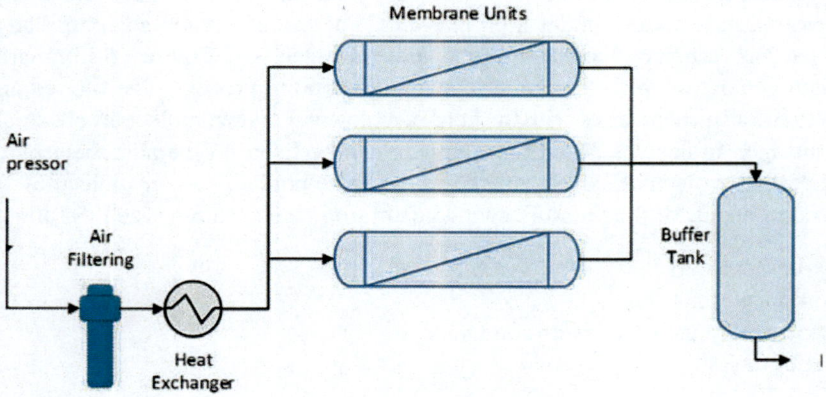

Fig. 2.6 Air separation and nitrogen production with membranes

gas is required, PSA systems are more suitable. However, membrane processes are more suitable for low-volume applications, especially if purity is not essential [13]. The main components of traditional membrane separation systems can be listed as follows:

- Air compressor
- Particulate filter
- Heat exchangers
- Parallel membrane units
- Buffer tank.

2.4 Ammonia Production Methods

Ammonia production by the Haber–Bosch process accounts for 1.4% of global fossil fuel consumption [18]. For each tonne of ammonia, 2.7–3.4 tonnes of carbon dioxide emissions occur. Figure 2.7 summarizes some of the industrial or lab-scale ammonia production methods. Although there are different ammonia production processes, the systems used by the industry are modified versions of the Haber–Bosch process. 96% of the ammonia produced today is produced in this way. The reaction that takes place in the processes can be defined as follows:

$$N_2 + 3H_2 \rightarrow 2NH_3 \quad \Delta H^0 = -92.44\ \text{kJ/mol} \tag{2.11}$$

Although ammonia consists of nitrogen and hydrogen, the reaction does not occur at standard temperature and pressure. For this reason, reactions are carried out in the presence of high temperature, pressure, and catalyst. The formation reaction of ammonia is exothermic, and approximately 2.7 GJ of energy is released per tonne. The reaction takes place partially in the reactor. Only a part of the hydrogen and nitrogen can react simultaneously. Ammonia, hydrogen, and nitrogen at the outlet should be separated in modern production systems. Then the separated hydrogen and nitrogen gases are looped in the reactor. The reaction rate of hydrogen in one cycle is between 25 and 34% [19].

Hydrogen and nitrogen production form the basis of industrial ammonia production. Pressure swing adsorption systems are generally used for nitrogen production. These systems are very straightforward and can produce 97–99.9% pure nitrogen. For hydrogen production, natural gas and coal-based reformer systems are generally adopted by the industry. These systems consume a massive amount of fossil fuels and have the highest emissions in the ammonia production process. Carbon capture systems are used to reduce carbon dioxide emissions. Ammonia produced

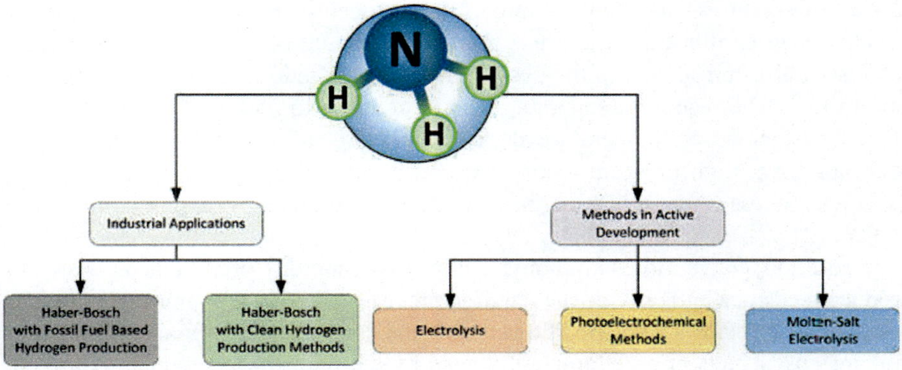

Fig. 2.7 Various ammonia production methods

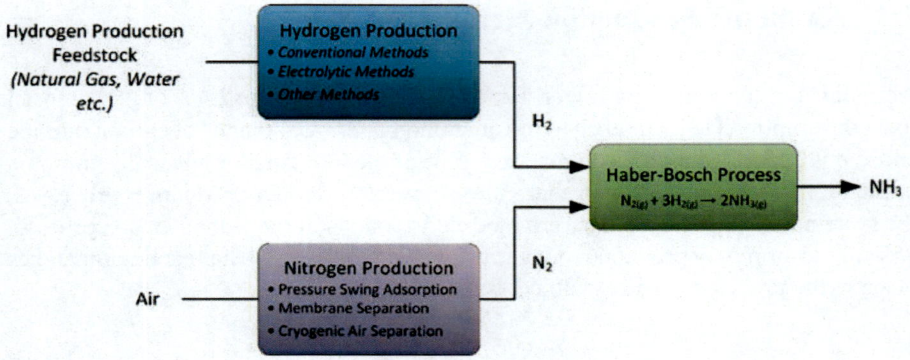

Fig. 2.8 Conventional ammonia production pathway with Haber–Bosch process

with carbon capture systems is called blue ammonia. However, green ammonia can be obtained using clean hydrogen production systems such as electrolysis, and thermochemical cycles.

The conventional reaction of nitrogen and hydrogen to produce ammonia with the presence of a catalyst requires high temperature and pressure. Also, the reaction is exothermic and reversible. Conversion of NH_3 is limited by the chemical equilibrium of the species in the reactor. Figure 2.8 shows the conventional ammonia production pathway. Designing ammonia production facilities with adequate efficiency requires a base knowledge of thermodynamic conditions inside the reactor [20]:

- Pressure, volume, and temperature of the reactor.
- Thermodynamic properties of the reactor and species.
- Phase equilibrium between the species including condensed ammonia and gaseous ammonia.
- Transport properties of the species.

Due to the thermodynamic restrictions, the reaction shifts to the reactants' side at low temperatures. The stable nature of the nitrogen increases the temperature of the chemical equilibrium to produce a tangible amount of ammonia. Therefore, a catalyst is also introduced to the system to lower the temperature of the chemical equilibrium. The reaction temperature can be reduced to 350–500 °C with the use of catalysts. However, temperature alone is not enough to provide thermodynamical conditions. So, high pressure is also required to complete the reaction. Although there are processes that utilize 65 bar, the generally preferred pressure is minimum 140 bar [21].

In addition, conventional ammonia synthesis is a mixture of various processes to produce feedstock and provide the required process conditions for ammonia production. Figure 2.9 shows a detailed flowchart of the conventional ammonia production system. Fossil fuels are predominantly used by the industry to produce hydrogen. A typical ammonia synthesis plant constitutes N_2, O_2, H_2, CH_4, and Argon gases. Also, small fractions of oxidized compounds can be found in the gas mixture. The

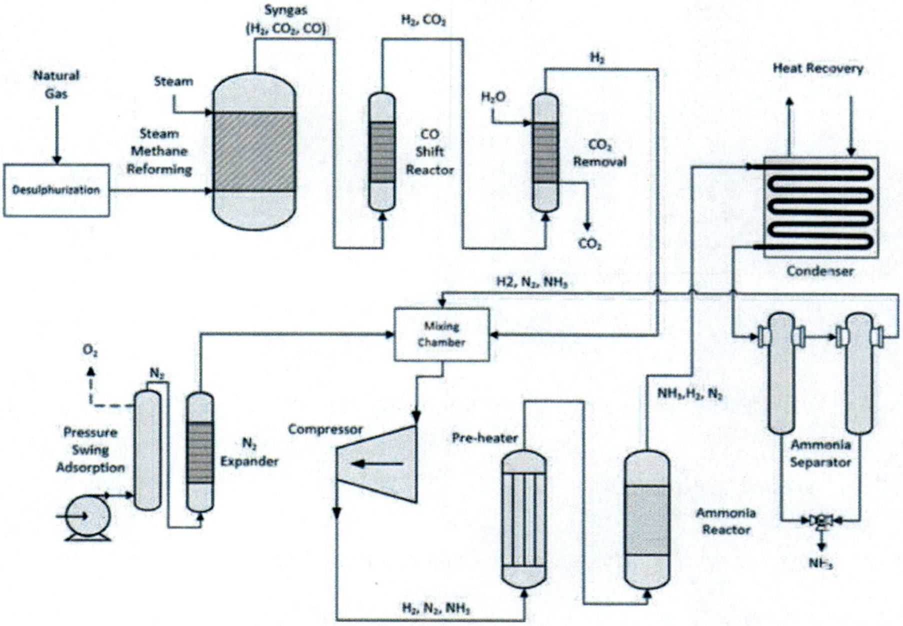

Fig. 2.9 Conventional ammonia production system

presence of oxygen and oxidized species in the Haber-Bosch reactor is undesirable as poisoning of the catalyst causes performance losses in the reaction. However, a 100% separation of oxygen-containing gases is not possible, hence this is one of the main reasons that shorten the life of catalysts.

Nitrogen can be obtained from the air via various processes such as pressure swing adsorption, membrane separation, etc. Hydrogen is obtained from hydrocarbons in industrial applications. Coal and natural gas are the most common feedstocks for the hydrogen production process. The carbon to hydrogen ratio of those varies for each process. However, natural gas or light hydrocarbons have a minimal environmental impact and higher efficiency among these processes [1]. Commonly, steam reforming, autothermal reforming, and partial oxidation are used to produce hydrogen from natural gas. A flow chart of a typical steam methane reforming process is shown in Fig. 2.10. The process starts with the desulphurization process where the sulfur is separated from natural gas. The syngas is then formed in a steam methane reforming reactor. Note that syngas is a mixture of hydrogen, carbon monoxide and carbon dioxide gases, and the CO shift and CO_2 removal reactors are then used to separate these gases.

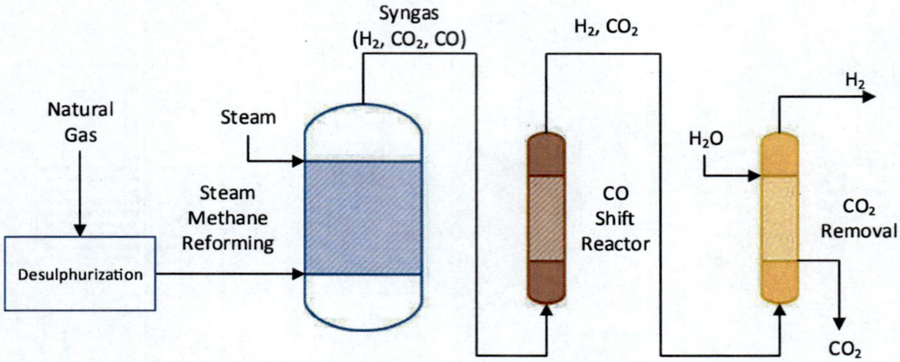

Fig. 2.10 Hydrogen production via steam methane reforming process

2.4.1 *Steam Reforming*

Steam reforming with natural gas occurs in a reaction as follows:

$$CH_{4(g)} + H_2O_{(g)} \rightarrow CO_{(g)} + 3H_{2(g)} \quad \Delta H^0 = 206 \text{ kJ/mol} \tag{2.12}$$

This steam reforming process is combined with a water gas shift reaction. Water gas shift reactions are exothermic while the steam reforming reaction is endothermic. Heat is supplied for the reaction in tubes placed within furnaces. Also, some part of the hydrocarbon feed is placed under combustion to provide heat. Efficient heating systems provide better hydrogen yields. Also, some of the processes include a secondary reformer with an air inlet for the conversion of residual methane. The primary and secondary reformer can be optimized for various applications. The stoichiometric ratio between hydrogen and nitrogen is also considered depending on the air amount. So, the nitrogen yield of the production process is adjusted accordingly. The operational conditions of a primary reformer are between 750 and 820 °C. Usually, a nickel-based catalyst is used for efficient conversion. However, nickel-based catalysts are sensitive to sulfur-containing chemicals. Therefore, the desulfurization process should be carried out before the steam reforming process [21].

2.4.2 *Water Gas Shift Process*

A gas mixture containing H_2, CO, and CO_2 is obtained after the steam reforming process. However, oxidized species cannot be allowed inside the Haber–Bosch reactor to prevent the poisoning of the catalyst. Then, a water gas shift process can be used to efficiently remove CO from the mixture. Water–gas shift reaction can be expressed as follows:

$$CO + H_2O \leftrightarrow CO_2 + H_2 \tag{2.13}$$

The most common applications of the water gas shift process consist of two phases. High-temperature gas shift process operates between 350 and 500 °C, followed by a low-temperature gas shift process at 200–250 °C. CO concentration in the gas mix drops down to 3% after the high-temperature gas shift process, and to 0.2–0.4% after the low-temperature gas shift [22].

2.4.3 CO_2 Removal and Methanation

The syngas from the water gas shift process has 20% CO_2 in it [22]. A huge portion of this carbon dioxide can be removed from the stream by using chemical absorption or adsorption. However, the remaining part of the CO_2 is still unwanted for the Haber–Bosch process. The methanation process is used for further removal of CO_2 from the gas mixture. This process nearly removes all the CO_2 inside the gas mixture. After CO_2 removal, the gas is cooled down to remove water, and pressurized before passing to the Haber–Bosch reactor.

2.4.4 Other Ammonia Production Technologies

Today, industrial ammonia production technologies are mostly carried out with Haber–Bosch systems. Although higher yields are obtained in catalyst systems using rare earth elements, these materials increase the production cost. Ammonia production is carried out in cycles as the efficiency of iron catalysts is low. This enlarges the installed facility volume and increases the investment costs. As shown in modeling studies, ammonia production cost in small-scale or decentralized systems exceeds the feasible level for Haber–Bosch systems. For this reason, research and development of on-demand and on-site systems, which can produce the needs for small-scale applications, continues actively.

Recently, electron or proton-driven processes have started to emerge as an alternative to the Haber–Bosch process. Instead of reacting nitrogen and hydrogen inside a reactor, electron/proton-driven processes utilize electrochemical reactions to produce ammonia. Here, water is oxidized to provide hydrogen to the reaction, and the nitrogen is reduced in the reactor [23]:

$$2N_2 + 6H_2O \rightarrow 4NH_3 + 3O_2$$

Nazemi et al. [24] developed an electrosynthesis unit to produce ammonia by using Pd–Ag nanoparticles. In the liquid phase, the unit achieved $45.6 \pm 3.7\ \mu g/cm^2$ h at a cell voltage of −0.6 V (vs. RHE) and the current density was reported as 1.1 mA/cm^2. The faradaic efficiency was found to be 9.9% and the energy efficiency was 19.6%.

In the gas phase, the ammonia yield rate was reported as 19.4 ± 2.1 μg/cm^2 h at −0.07 V (vs. RHE). The faradaic efficiency of the gas phase electrosynthesis unit was 7.9% and the energy efficiency was 27.1%. The highest energy efficiency was reported at −0.01 V as 72%. In this particular system, there is a need for 7.7 MWh of energy to produce one tonne of ammonia to achieve this efficiency, this value was reported as approximately 7.8 MWh in the Haber–Bosch process [24].

Although electron or proton-driven processes offer an alternative method for ammonia production, there are problems with efficiency. In particular, very low solubility of nitrogen in water and strong nitrogen bonds affect productivity. For this reason, Zhou et al. [25] conducted a study on electrolytic ammonia production by using ionic liquids with high nitrogen solubility. By utilizing an ionic liquid and iron-based electrocatalyst, researchers achieved 60% faradaic efficiency. However, the ammonia yield rate varies between 2.5 and 12.5 mg/m^3 h, and the application is in the early development stage.

Various catalysts may be utilized in order to achieve electrosynthesis of ammonia such as Ag_2O–Au, Ag–Au, Ag_2O, AuNCs, AuNSs [23]. Nazemi and El-Sayed [23] studied various electrosynthesis ammonia production methods and reported that the production capacity of the Micro Ammonia Production System (MAPS) could reach 165 t per year with the most efficient plasmonic nanocatalyst by using a large-scale electrode. They also predict that this amount may even go up to 207 t/year.

As discussed in Chap. 1, ammonia is one of the most hydrogen-containing substances compared to other liquids. This makes ammonia a good hydrogen and energy carrier. Decentralized ammonia production would become feasible by combining this technology with a technology that can efficiently convert ammonia to hydrogen. Especially in using renewable energy sources where surplus electricity production is high, decentralized options may become useful by directing surplus electricity to ammonia production or by producing ammonia in cases where electricity costs are low. Thus, ammonia could become an effective agent for energy storage. It would be possible to store the energy with zero carbon emissions by converting the ammonia to hydrogen and utilizing hydrogen in fuel cells to efficiently convert it to electricity when needed.

There are studies and patents on releasing hydrogen from ammonia by using inductive heating and Ru-Rb catalysts [26]. Those systems can efficiently convert ammonia to hydrogen with induction. An example of a hydrogen release reactor is given in Fig. 2.11. The system consists of a catalyst-coated multichannel reactor and enables exothermic reactions by induction. Also, the sizing of this type of reactor is very flexible, the size can be increased according to the need or the desired production amounts can be reached by utilizing more than one reactor in parallel.

Compared to hydrogen, ammonia is easier to liquefy and store. The energy density of liquid hydrogen is 1.209 kWh/kg. High pressures and cryogenic applications are required to reach the liquid state, which is the highest energy-containing state of hydrogen. However, ammonia can be liquified at significantly higher temperatures and lower pressure.

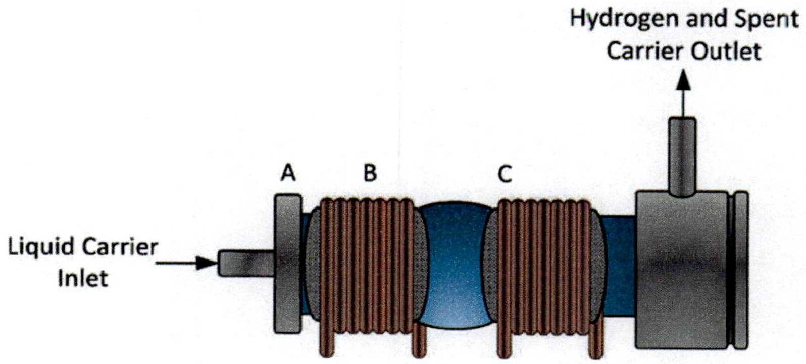

Fig. 2.11 A hydrogen release reactor module. **a** Diffuser, **b** vaporizer, **c** hydrogen release section (adapted from [26])

The diesel liter equivalent of ammonia is 1.94 kg/DLE. The cost of ammonia in the small scale ammonia plant constructed in this chapter varies between 4.24 and 11.46 kg/DGE. Nazemi and El-Sayed [24] reported that the capital cost of a MAPS system that produces approximately 500 kg of ammonia per day was estimated as US$25,000. This value is quite low when compared to the Haber–Bosch reactor. Capital costs have a great impact on reducing ammonia costs. However, the major cost contribution comes from hydrogen production. MAPS reactor seems to be feasible since it is decreasing the capital costs on a larger extent. However, one should note that the comparison between a lab-scale study to a mature technology should be made very carefully since operational problems are creating a huge uncertainty.

2.5 Ammonia Storage

Ammonia can be stored pressurized or under low temperatures. Anhydrous ammonia is usually stored in cylindrical or spherical vessels. Pressurized ammonia above 8.6 bar is in liquid form and no energy is required. Usually, the pressure is kept around 17 bar if there is a possibility of an increase in ambient temperature [27]. Carbon steel vessels are sufficient for storage at this pressure. Figure 2.12 shows a typical ammonia storage tank. Since ammonia has a corrosive effect on stainless steel, the inner layer is built with carbon steel. The outer layer is made of concrete or steel; however, there is always an insulating layer between these two layers.

The choice of high pressure or low-temperature ammonia storage methods is generally determined by the amount of ammonia to be stored. Ammonia liquefied at low temperatures takes up 15 times less space than pressurized ammonia. This is important, especially in facilities with very high capacity. The smaller volume of liquid ammonia primarily reduces the use of carbon steel amount for vessels and capital costs. However, with the need for continuous cooling, the operating costs increase. In this case, a feasibility study about which process will be preferred must carefully be done.

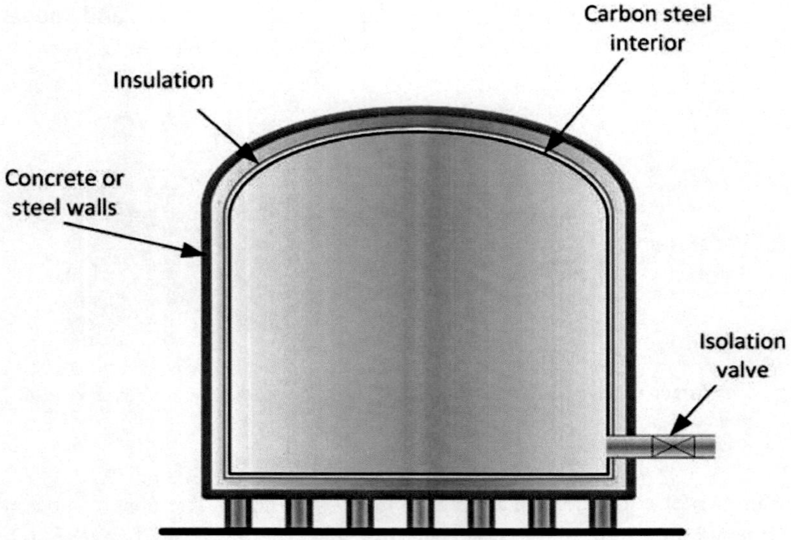

Fig. 2.12 Typical ammonia storage tank (adapted from [28])

2.6 Economics of Ammonia Production Plants

One has to note that the capital and operating costs directly affect ammonia prices. Therefore, these cost variables should be evaluated in a comprehensive manner when estimating the ammonia prices. This section presents an assessment of methodologies used to determine the ammonia price and discuss the cost analyses for various scenarios.

2.6.1 Cost Assessment Methodology

Capital cost estimation approaches can be summarized in five classifications. Each of these classifications has definitions in different details and accuracy ranges. Table 2.1 shows the classification of price estimates. According to this table, class 5 is the least, and class 1 is the most accurate estimation method.

In this study, the stochastic method (class 3) was adopted for price estimations. This approach has an accuracy of +40 to 25%. Accordingly, price estimations were made by examining the already installed plant costs and working tables of common equipment. The equation most commonly used in price estimates can be expressed as follows [29]:

Table 2.1 Classification of cost estimates

Class	Level of project definition	The typical purpose of the estimate	Methodology
5	0–2	Screening of feasibility	Stochastic or judgment
4	1–15	Concept study	Primarily stochastic
3	10–40	Budget, authorization, control	Mixed but primarily stochastic
2	30–70	Control or bid/tender	Primarily deterministic
1	50–100	Check estimate or bid/tender	Deterministic

Source Turton et al. [29]

$$\frac{C_a}{C_b} = \left(\frac{A_a}{A_b}\right)^n \tag{2.14}$$

where

C_a: Estimated equipment cost, US$
C_b: Base price of the equipment, US$
A_a: Estimated equipment capacity
A_b: Base capacity for the equipment
n: Cost exponent.

The price of existing equipment is taken as a base price while estimating the price for each piece of equipment. The proportional relationship between capacity and price increase is expressed with the cost exponent. This value varies for different equipment and the required values for equipment commonly used in processes can be obtained from Perry's Chemical Engineers' Handbook [30]. With a similar approach, the cost exponent for the entire facility is 0.6 on average, and this approach is referred to as the six-tenths rule [29].

Inflation should also be considered in price estimations based on current equipment prices. For this reason, various indexes are used in the industry to make pricing according to inflation. The main indexes are Chemical Engineering Plant Cost Index (CEPCI), Engineering News-Record Construction Index, Marshall, and Swift Process Industry Index. CEPCI is the most frequently used index among them. In this study, the CEPCI index was used to adjust the inflation for today. Price adjustment can be made with the following formula:

$$\frac{C_2}{C_1} = \frac{I_2}{I_1} \tag{2.15}$$

where

C_2: Adjusted cost, US$
C_1: Base cost, US$
I_1: Cost index for the adjusted cost
I_2: Cost index for the base cost.

By combining Eqs. 2.14 and 2.15, capital cost estimates can be made with the following equation:

$$C_a = C_b \left(\frac{A_a}{A_b} \right)^n \frac{I_2}{I_1} \tag{2.16}$$

where

C_a: Estimated equipment cost, US$
C_b: Base price of the equipment, US$
A_a: Estimated equipment capacity
A_b: Base capacity for the equipment
n: Cost exponent.
I_1: Cost index for the adjusted cost
I_2: Cost index for the base cost.

Although equipment costs have an important place in feasibility studies, they are not enough on their own. In determining the costs of a facility, in addition to the equipment cost, costs such as materials required for installation, labor costs, project design costs, contingency costs, and construction costs should also be considered. One of the methods commonly used in estimating the costs that may occur directly and indirectly is the module costing technique. It is one of the most convenient methods for making preliminary costing estimates. According to this technique, the estimation of direct and indirect costs is based on the basic price of the equipment. Equation 2.17 shows the relation between the bare module cost (C_{BM}) and purchased cost for base conditions (C_p^0). Bare module cost factor (F_{BM}) is used to define the relationships between these variables based on operating conditions such as materials and pressure.

$$C_{BM} = C_p^0 F_{BM} \tag{2.17}$$

where

C_{BM}: Bare module cost, US$
C_p^0: Purchased cost for base conditions, US$
FBM: Bare module cost factor (FBM)

2.6.2 *Operational and Maintenance Costs*

Operational and maintenance costs are estimated for ammonia production at various capacities. Those calculations are used in the estimation of manufacturing costs and profitability analysis.

Overall costs to produce a product can be described in three major categories: (i) direct manufacturing costs, (ii) fixed manufacturing costs, and (iii) general expenses [29]. Direct costs include raw material costs, waste treatment costs, utility costs, labor, maintenance, operating supplies, etc. After the initial construction of the facility, the production rate may change to supply the demand. Hence, direct costs are directly dependent on the production rate. On the other hand, fixed costs are not directly related to the rate of production. Instead, those costs include depreciation, local taxes, insurance, and plant overhead costs. Administration costs, distribution and selling costs, and R&D costs make up the general expenses. Those expenses rather include management-level and administrational activities, and they are not directly related to manufacturing costs. The cost of manufacturing ammonia plants can be calculated with the following equation:

$$C_{CM} = C_D + C_M + C_G \tag{2.18}$$

where

C_{CM}: Cost of manufacturing, US$
C_D: Direct costs, US$
C_M: Fixed costs, US$
C_G: General expenses, US$

A basic calculation of manufacturing costs generally requires the following costs to be known:

1. Fixed capital investment (C_C)
2. Cost of operating labor (C_O)
3. Cost of utilities (C_U)
4. Cost of waste treatment (C_W)
5. Cost of feedstocks (C_F).

For maintenance and repairs, it is accepted in the literature that the costs will be between 2 and 10% of the fixed capital cost. In average cases, this value will be around 6%. Also, the costs required for consultancy and official procedures may vary between 1 and 25% of operating labor costs. Table 2.2 summarizes all the costs affecting manufacturing costs with their relations. Costing calculations were done for each unit in the process, based on their technological advancement level. Pressure swing adsorption and Haber–Bosch have matured technologies; however, electrolysis requires more attention when it comes to cost calculations. Although the alkaline electrolysis unit is a matured technology, PEM electrolyzers are still in the research and development phase. In particular, the replacement time of membrane stacks should be examined, and cost calculations should be made accordingly. Similarly, for AEL, part changes that may occur due to corrosion must be considered. For other units, unexpected costs are not expected to be high. Therefore, the rates of maintenance and repairs are assumed as given in Table 2.2 for the units other than electrolysis.

Table 2.2 Manufacturing costs and their typical ranges and relations to other cost types

Cost type	Notation
Direct costs	
Feedstock	C_F
Waste treatment	C_W
Utilities	C_U
Operating labor	C_O
Supervision and official procedures	$C_S = (0.1 - 0.25)C_O$
Maintenance and repairs	$C_M = (0.02 - 0.1)C_C$
Operating supplies	$(0.1 - 0.2)C_M$
Laboratory charges	$(0.1 - 0.2)C_O$
Fixed costs	
Depreciation	$(0.1)C_C$
Taxes and insurance	$(0.014 - 0.05)C_C$
Plant overhead	$(0.50 - 0.7) * (C_O + C_S + C_M)$
General expenses	
Administration costs	$(0.15) * (C_O + C_S + C_M)$
Marketing costs	$(0.02 - 0.2)C_{CM}$
R&D costs	$(0.05)C_{CM}$

Source Turton et al. [29]

2.6.3 System Descriptions of Ammonia Plant Case Studies

To understand the effect of different pathways on ammonia production, the conventional ammonia production system was evaluated. Accordingly, an economic review of conventional and electrolytic ammonia production was made. While conducting the economic analysis, the investment and operating costs of plants and the production cost of ammonia were calculated. Facilities with different small-scale capacities were selected as case studies. The formation reaction of ammonia can be expressed as follows:

$$N_2 + 3H_2 \rightarrow 2NH_3 \quad \Delta H^0 = -92.44\ \text{kJ/mol} \tag{2.19}$$

Accordingly, the nitrogen and hydrogen amounts needed by the ammonia plants of different capacities stoichiometrically were calculated and presented in Table 2.3.

The case studies presented in this section were made according to the stoichiometric ratios specified in Table 2.3. However, the ammonia formation reaction is reversible. The reaction may shift backward under different pressures and temperatures. Therefore, it is crucial to minimize the temperature and pressure changes in the ammonia reactor. Even under the most favorable conditions, some of the hydrogen and nitrogen leave the reactor without reacting. For this reason, gases and ammonia are separated at the reactor outlet, and the remaining gases are looped back to the ammonia reactor.

Table 2.3 Amount of nitrogen and hydrogen feedstock required for ammonia production at different capacities

NH_3, mt	N_2, mt	H_2, mt
0.5	0.41	0.09
1	0.82	0.18
2.5	2.06	0.44
5	4.11	0.89

Conventional Ammonia Production

The flowchart of the system designed for conventional ammonia production is presented in Fig. 2.9. Accordingly, the hydrogen required for the reaction is produced by steam methane reforming, and the pressure swing adsorption process produces nitrogen. First, feedstock gases are mixed with an H_2:N_2 ratio of 1:3 and then transferred to the pre-heater with a compressor. Next, pre-heated gases are transferred to the ammonia reactor. After the catalytic ammonia production, unreacted compounds and ammonia pass to a condenser for separation. Finally, the separated H_2 and N_2 gasses are sent back into the mixing chamber and cycled through the process to react again.

Electrolytic Ammonia Production

Electrolytic ammonia production systems are similar in structure to conventional ammonia production systems. However, instead of the steam methane reforming system, which has high carbon dioxide emissions and fossil fuel consumption, hydrogen production is provided by electrolytic methods. Hydrogen production can be done using alkaline, proton exchange membrane, or solid oxide electrolysis units. Figure 2.13 represents the flow chart of an electrolytic ammonia production process. The hydrogen requirement in conventional ammonia production, provided by steam methane reforming, is studied using electrolysis in this case study. The effects of electrolysis on energy costs and capital costs is also investigated.

Electrolytic Ammonia Production with Renewable Options

In this case study, ammonia production cost and its effects on the process economy are investigated if a part of the electricity and heat need is used with a renewable (solar and wind) source. Oshawa, Ontario is selected as the center of this case study. Therefore, the sun exposure and wind speed data are based on the data of the city of Oshawa. However, it is not always possible to benefit from sunlight. For this reason, wind energy is utilized when sunlight is insufficient, and grid electricity is used during on-peak times. Figure 2.14 shows the current diagram of the proposed system. This system uses the Haber-Bosch reactor to produce ammonia. Hydrogen is produced by electrolysis instead of conventional methods with high carbon emissions. This makes it possible to eliminate carbon emissions from ammonia production. In this scenario, carbon emissions occur as upstream emissions from the production of the raw materials used to manufacture the systems.

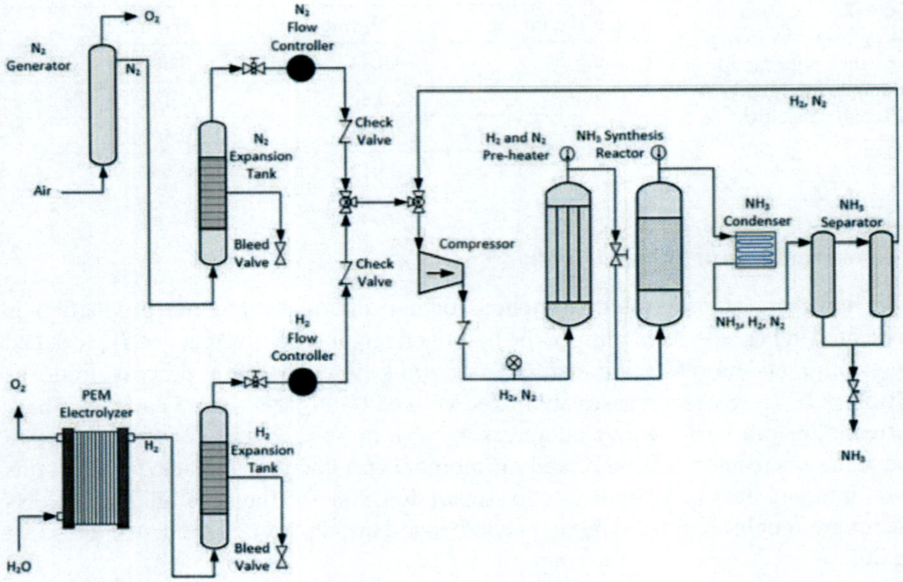

Fig. 2.13 Electrolytic ammonia production

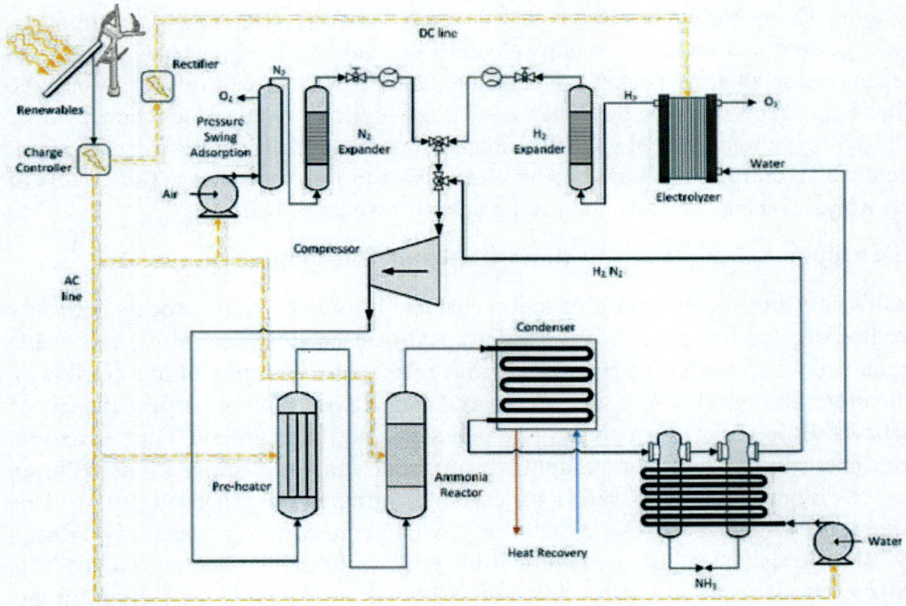

Fig. 2.14 Electrolytic ammonia production with renewable sources

Economics of Hydrogen Production

Proton exchange membrane electrolyzer (PEM) and alkaline electrolyzer (AEL) options have been investigated for hydrogen production. Both units have a similar structure. However, membranes or separators and feedstock requirements are different. PEM electrolyzer utilizes proton exchange membranes and requires distilled water to operate. Operational and maintenance costs of a PEM unit include electricity costs, regular stack change due to wearing of the membranes, pumping needs of the water, and treatment of the water. On the other hand, an alkaline electrolyzer needs an alkaline solution to operate. Potassium hydroxide or sodium hydroxide is the most common alkaline solutions used in an AEL operation. Operational and maintenance costs of an AEL unit include electricity consumption, replacement of the stack because of corrosion, KOH feedstock consumption, pumping needs, and regular maintenance costs. Due to the nature of the alkaline solutions, fittings and piping of the unit should be maintained regularly.

The system is modeled to calculate the energy required for hydrogen production by electrolysis. The following formula can be used to determine the potential required for the electrolysis process:

$$V_{cell} = V_{ideal} + \Delta V_{act} + \Delta V_{diff} + \Delta V_{ohm} \tag{2.20}$$

where

V_{cell}: Cell voltage, V
V_{ideal}: Minimum thermodynamic voltage for water dissociation, V
V_{act}: Potential required for the activation of electrodes, V
V_{diff}: Diffusion overvoltage, V
V_{ohm}: Ohmic resistance of the cell, V

Accordingly, the potential (V_{cell}) to be applied to the electrolysis cell for hydrogen production is the sum of the minimum potential to be applied between the two electrodes for water dissociation (V_{ideal}), the potential required for the activation of the electrodes (ΔV_{act}), the diffusion overvoltage caused by the concentration difference in the system components (ΔV_{diff}), the voltage drop caused by the internal resistance of the cell (ΔV_{ohm}). The ideal voltage can be calculated with the following equation [31]:

$$V_{ideal} = \left(\frac{\Delta G^0}{2F}\right) + \left(\frac{RT}{2F}\right)\ln\left(\frac{P_{H_2}P_{O_2}^{0.5}}{\alpha_{H_2O}}\right) \tag{2.21}$$

where

V_{ideal}: Minimum thermodynamic voltage for water dissociation, V
ΔG^0: Gibbs free energy variation, kJ/kmol
F: Faraday's constant, sA/mol
R: Gas constant, kJ/kmol-K
T: Temperature, K
P_{H2}: Partial pressure of hydrogen, kPa

P_{O2} Partial pressure of oxygen, kPa
α_{H2O} Charge transfer activity of water

The activation potential of the electrodes can be calculated with a similar approach for both anode and cathode with the following equation with the exchange current density of electrodes ($i_{0,el}$) [32]:

$$\Delta V_{act,el} = \frac{RT}{\alpha_{el} n F} \ln\left(\frac{i}{i_{0,el}}\right) \tag{2.22}$$

where

$V_{act,el}$: Activation potential of electrode, V
F: Faraday's constant, sA/mol
R: Gas constant, kJ/kmol-K
T: Temperature, K
n: Number of moles of electrons involved in the overall water splitting reaction
$i0_{,el}$: Activation overvoltage, A/cm^2
i: Current passing through the electrode, A/cm^2

At high current densities, the dissociation reaction of water occurs not electrically, but by substance transport. This influences polarization and increases the potential that must be applied to the system. The following formula is used to calculate the diffusion overvoltage [33]:

$$\Delta V_{diff} = \frac{RT}{\alpha_{el} n F} \ln\left(1 + \left(\frac{i}{i_L}\right)\right) \tag{2.23}$$

where

V_{diff}: Diffusion overvoltage, V
i: Current passing through the cell, A/cm^2
i_L: Limiting current density, A/cm^2
F: Faraday's constant, sA/mol
R: Gas constant, kJ/kmol-K
T: Temperature, K
n: Number of moles of electrons involved in the overall water splitting reaction

The internal resistance of the PEM electrolyzer cell depends on the resistances of the membrane electrode assembly, the surface areas of these components, and the applied current:

$$\Delta V_{ohm} = (R_{mem} + R_{an} + R_{cat}) i A \tag{2.24}$$

where

V_{ohm}: Voltage drop due to cell resistance, V
R_{mem}: Resistance of the membrane, ohm

R_{an}: Resistance of the anode, ohm
R_{cat}: Resistance of the cathode, ohm
i: Current passing through the cell, A/cm^2
A: Area of the cell, cm^2

Resistance of the membrane can be calculated as follows [34]:

$$\sigma_{mem} = (0.005139\lambda - 0.00326)e^{1268\left(\frac{1}{303}-\frac{1}{T}\right)} \quad (2.25)$$

where

σ_m: Membrane conductivity, 1/ohm-cm
λ: Humidification degree of the membrane
T: Temperature, K

However, the resistance of the electrolyte must also be considered for alkaline electrolysis. Electrolyte conductivity can be calculated with an empirical approach [35]:

$$\begin{aligned}\sigma_{ael} = &-2.96396 - 0.02371T + 0.12269w_{el} + (5.7 \times 10^{-5})T^2 \\ &+ 0.00173w_{el}^2 + (4.7 \times 10^{-4})w_{el} - (3.6 \times 10^{-8})T^3 \\ &+ (2.7 \times 10^{-6})w_{el}^3 - (8.9 \times 10^{-6})Tw_{el}^2 + (2.4 \times 10^{-7})T^2w_{el} \quad (2.26)\end{aligned}$$

where

σ_{ael}: Specific conductivity of the electrolyte, S/cm
w_{el}: Mass percent of alkaline solution, %
T: Temperature, K

Table 2.4 shows the selected parameters for modeling of alkaline and PEM electrolysis units. These values were selected based on the units available commercially. Figure 2.15 shows the modeling results for the PEM electrolysis unit. The energy requirement of the electrolysis system decreases with increasing temperature. At low current density values, the effect of temperature is small. However, as

Table 2.4 Selected variables for modeling of alkaline and PEM electrolysis

	Alkaline electrolysis (AEL)	PEM electrolysis
Stack size, cm^2	20,000	3600
Cell amount	Variable	Variable
Membrane	Diaphragm	PtB coated Nafion
Electrode	Nickel electrodes	Pt-coated titanium
Catalyst load	–	3 gr/cm^2
KOH%	30%	–
Current density	0.6	2.4

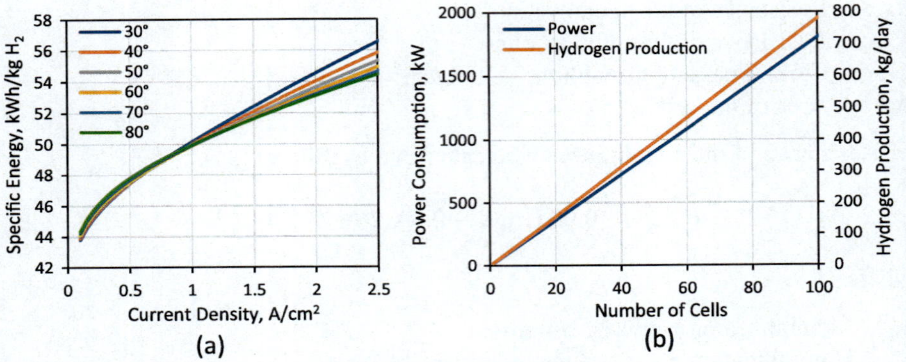

Fig. 2.15 Modeling results for PEM electrolysis. **a** Effect of temperature change on hydrogen production. **b** Effect of cell number on hydrogen production rate and power consumption

the current density increases, the effect on the energy requirement becomes larger. In a PEM unit with an active area of 3600 cm^2, daily hydrogen production can reach 700 kg. However, it has been calculated that the power consumption is approximately 1800 kW. Therefore, according to the data obtained, the power consumed per kilogram of hydrogen production is 55.7 kWh.

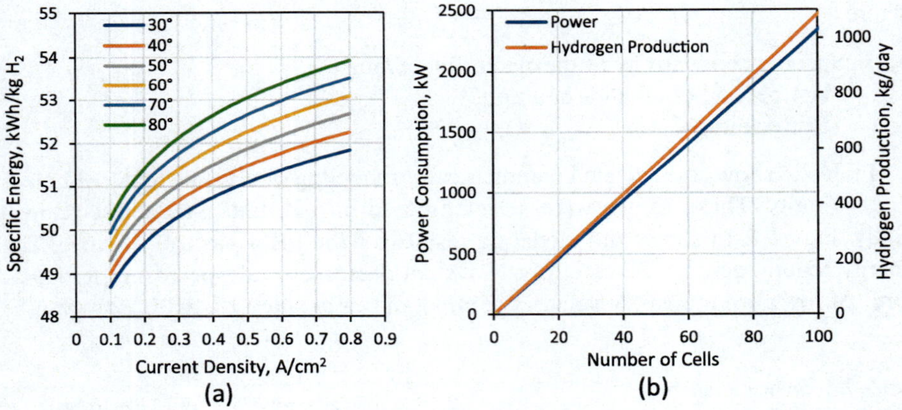

Fig. 2.16 Modeling results for AEL electrolysis. **a** Effect of temperature change on hydrogen production. **b** Effect of cell number on hydrogen production rate and power consumption

Figure 2.16 presents the modeling data for the AEL electrolysis system. The increase in temperature has a positive effect on power consumption as in PEM electrolysis. However, the operating range of AEL is more limited compared to PEM. Also, the effect of current density on temperature is not as apparent as in PEM. To prevent the formation of chlorine in alkaline electrolysis systems, low current densities are used. Therefore, the surface area must be large. In an alkaline electrolysis unit with an active area of 2 m^2, daily hydrogen production can exceed 1000 kg. The amount of energy required for unit hydrogen production was calculated as 51.7 kWh. AEL systems are more mature than PEM systems. Although PEM systems can produce high hydrogen in low surface areas, investment and maintenance costs can be higher due to the need for catalyst-coated membranes and electrodes. However, while PEM systems can work with pure water, a certain mass percentage of NaOH or KOH solutions is used in AEL systems. By their nature, these chemicals are corrosive, and it is essential to design equipment accordingly.

Average capital costs for PEM and AEL are reported as US$1212 and 948/kW in 2020, respectively [36]. Although those data do not give precise results about the stack prices, these values can be used to roughly determine the stack prices. Using this data would be appropriate according to the stochastic cost assessment method if an appropriate margin of error was considered in the calculations. The calculated investment costs for PEM and AEL stacks are presented in Fig. 2.17. The rare earth elements on the electrodes and membrane electrode assemblies for efficient water dissociation are the main reason for the high cost of PEM reactors. As the capacity increases, the amount of rare earth elements per unit area increases, so the costs also increase.

The price breakdowns for AEL and PEM units are given in Fig. 2.18. Considering the price breakdowns of AEL and PEM units, the highest percentage belongs to the electrolyzer stacks. The reason for this is the high production costs of the electrode and membrane materials used in both systems. Also, catalysts using both membranes and electrodes in PEM stacks lead to higher costs. This is a situation that should be evaluated mainly in terms of maintenance costs. The lifetime of the Electrolyser stacks varies between 5 and 7 years [37]. While calculating capital costs, the effect of indirect costs should also be considered. Contribution of the indirect costs such as engineering and supervision is between 4 and 20%, construction costs are between 4 and 17%, and the contingency is between 5 and 15% [38].

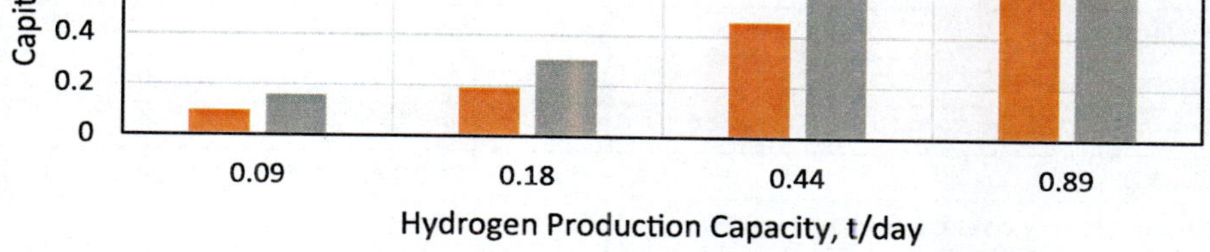

Fig. 2.17 Capital costs for AEL and PEM electrolysis stacks

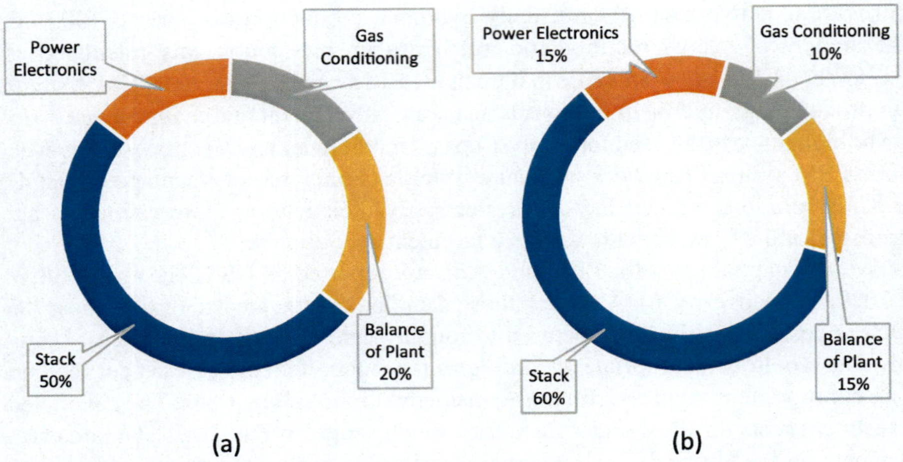

Fig. 2.18 Cost breakdowns for **a** AEL and **b** PEM electrolyzer units

The design parameters for the PEM unit are given in Table 2.5. Considering the lifetime of the membrane stacks, the replacement interval selected as 7.5 years. As stated in the modeling studies, the specific energy consumption was calculated as 55.7 kg H_2/kWh. Water consumption was calculated stoichiometrically.

Table 2.6 gives the estimates of annual costs for a PEM unit. The biggest contributor to the annual costs of the electrolysis unit is electricity consumption. Also, stack replacement costs are spread over 25 years. As stated in the literature, the amount of electricity required to separate water with electrolysis is high [39]. Today, hydrogen production with this method varies between US$2–12/kg H_2 [1]. In utilities, after electricity, the highest contribution comes from stack replacement. The electrolysis stack has highest cost in an electrolysis unit for both AEL and PEM. In a plant with 25 years of operational time, a stack needs to be replaced on average three times depending on the operational conditions and usage. Therefore, the electrolysis stack with high-cost components such as rare earth elements and membranes increases costs.

Table 2.5 Design parameters for PEM unit

	Case I	Case II	Case III	Case IV
Stack size, m^2	3600			
Stack amount	1	2	5	10
Power consumption of BoP equipment, kWh	20	40	100	200
Stack price	158,000	302,000	750,000	1,500,000
Specific energy consumption, kg H_2/kWh	55.7	55.7	55.7	55.7
Stack replacement interval, years	7.5	7.5	7.5	7.5
Water consumption, m^3/day	800	1600	4000	8000

Table 2.6 Manufacturing costs for PEM unit

Direct costs	Case I	Case II	Case III	Case IV
Feedstock				
Water	US$14,600.00	US$29,200.00	US$73,000.00	US$146,000.00
Utilities				
Electricity	US$203,960.37	US$407,920.74	US$1,019,801.86	US$2,039,603.72
Operating labor	US$6000.00	US$9000.00	US$15,000.00	US$15,000 00
Supervision and official procedures	US$1050.00	US$1575.00	US$2625.00	US$2625.00
Maintenance and repairs	US$5580.76	US$10,697.31	US$26,511.19	US$52,867 64
Stack replacement	US$22,571.43	US$43,142.86	US$107,142.86	US$214,285.71
Operating supplies	US$837.11	US$1604.60	US$3976.68	US$7930.15
Laboratory charges	US$900.00	US$1350.00	US$2250.00	US$2250.00
Fixed costs				
Depreciation	US$55,807.59	US$106,973.15	US$265,111.85	US$528,675.36
Taxes and insurance	US$17,858.43	US$34,231.41	US$84,835.79	US$169,175.44
Plant overhead	US$7578.46	US$12,763.39	US$26,481.71	US$42,295.58
Total	US$339,041.50	US$660,755.81	US$1,629,034.29	US$3,223,007.94

The design parameters for the AEL unit are given in Table 2.7. The lifetime of AEL stacks is approximately 50,000 h. Therefore, the stack lifetime was selected as 7 years. The energy consumption of the unit is calculated as 51.7 kg H_2/kWh. Also, the specific power consumption of the unit was calculated as 51.7 kg H_2/kWh. Calculated costs for manufacturing an AEL unit was given in Table 2.8. The electricity consumption for the electrolysis process accounts for a large part of the operational costs. In addition, the stacks where the water dissociation process takes place also contribute to the initial investment and subsequent maintenance costs.

Table 2.7 Design parameters for AEL unit

	Case I	Case II	Case III	Case IV
Stack size, m^2	20,000			
Stack amount	1	2	5	10
Power consumption of BoP equipment, kWh	20	40	100	200
Stack price, US$	94,800	189,000	455,000	909,000
Specific energy consumption, kg H_2/kWh	51.7	51.7	51.7	51.7
Stack replacement interval, years	7	7	7	7
Water consumption, m^3/day	800	1600	4000	8000

Table 2.8 Manufacturing costs for AEL unit

	Case I	Case II	Case III	Case IV
Direct costs				
Feedstock				
Water	US$14,600.00	US$29,200.00	US$73,000.00	US$146,000.00
Utilities				
Electricity	US$189,313.31	US$378,626.62	US$946,566.54	US$1,893,133.08
Operating labor	US$6000.00	US$9000.00	US$15,000.00	US$15,000.00
Supervision and official procedures	US$1050.00	US$1575.00	US$2625.00	US$2625.00
Maintenance and repairs	US$4311.27	US$8562.19	US$20,651.00	US$41,301.99
Stack replacement	US$13,542.86	US$27,000.00	US$65,000.00	US$129,857.14
Operating supplies	US$646.69	US$1284.33	US$3097.65	US$6195.30
Laboratory charges	US$900.00	US$1350.00	US$2250.00	US$2250.00
Fixed costs				
Depreciation	US$43,112.73	US$85,621.88	US$206,509.96	US$413,019.93
Taxes and insurance	US$13,796.07	US$27,399.00	US$66,083.19	US$132,166.38
Plant overhead	US$6816.76	US$11,482.31	US$22,965.60	US$35,356.20
Total	US$296,222.06	US$583,233.69	US$1,425,881.30	US$2,819,037.38

Calculated direct and indirect costs for both systems are given in Fig. 2.19. The highest proportion of costs for all cases is due to equipment costs. Engineering and supervision costs make the second largest contribution. The cost of electrodes, electrode coatings and membranes used in the system drives up equipment prices. Also, operational difficulties for AEL systems increase the engineering and supervision costs compared to PEM electrolysis. Direct and indirect cost calculations are made by considering the best, worst, and average scenarios. The cost of capital investment per kW of energy consumption reported for PEM in literature is broader than AEL [36]. Because of the technological maturity level of PEM electrolysis, it is predictable that the uncertainty will be higher than AEL. However, when the best-case scenarios are compared, the costs calculated for PEM are more appropriate. Again, this is due to the wide uncertainty range given for PEM. However, AEL performs better in mid and worst-case scenarios. Therefore, the cost of an electrolysis plant with a daily hydrogen production capacity of 0.89 t is calculated as US$2.04–5.35 M and US$1.47–8.02 M for AEL and PEM, respectively.

As AEL is a mature technology, it has a more extensive production network. However, it operates at a lower current density compared to PEM. In addition, the gas crossover is higher due to the diaphragm or membranes used, affecting gas purity. At the same time, corrosive electrolyte requirement is one of its disadvantages. However, pure water is used in PEM electrolysis, and higher current densities can be obtained.

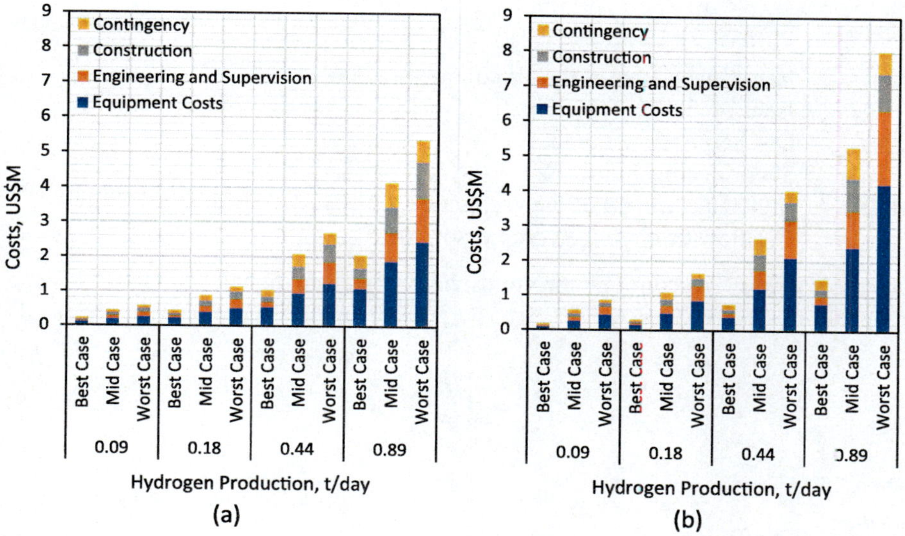

Fig. 2.19 Direct and indirect costs for **a** AEL plant, and **b** PEM plant

Economics of Nitrogen Production

The chemical stoichiometry required for a small-scale ammonia production facility was calculated and presented in Table 2.3. Accordingly, the nitrogen production costs are calculated for capacities of 0.41, 0.82, 2.06, and 4.11 t/day. The air requirement for nitrogen production can be calculated using the formula below.

$$V_{air} = \frac{V_{N_2}}{\eta} \tag{2.27}$$

where

V_{air}: Volume of the air, m^3
V_{N_2}: Volume of the nitrogen gas, m^3
η: Ratio of the nitrogen in air, 0.78

Accordingly, the required air is calculated as 470, 940, 2350, and 4700 N m^3/day for each capacity, respectively. According to Fig. 2.4, pressure swing adsorption processes are the most suitable for small-scale ammonia production facilities considered in this study. Therefore, the capital costs for these processes are evaluated accordingly.

The efficiency of the air compressor is accepted as 0.72. The compressor power can be calculated with the following formula:

$$P_{is} = 2.31\frac{k}{k-1}\frac{T_{discharge} - T_{suct}}{M_{air}}Q_M \tag{2.28}$$

where

P_{is}: Isentropic power of the air compressor, kW
k: Isentropic gas coefficient
$T_{discharge}$: Discharge temperature, K
T_{suct}: Suction temperature, K
M_{air}: Molecular weight of the gas, g/mol
Q_M: Gas flow through the compressor, t/h

The air is then sucked through the atmosphere; therefore, to calculate the discharge temperature:

$$\frac{T_2}{T_1} = \left(\frac{P_2}{P_1}\right)^{\frac{\gamma-1}{\gamma}} \tag{2.29}$$

where

T: Temperature, K
P: Pressure, kPa
γ: Ratio of the specific heat of the air

Finally, to calculate the actual power of the compressor:

$$P_{act} = P_{is}\eta_{compressor} \tag{2.30}$$

where

P_{act}: Actual power of the compressor, kW
P_{is}: Isentropic power of the compressor, kW
$\eta_{compressor}$: Efficiency of the compressor, 0.72

The power requirement of the air compressor for 0.41, 0.82, 2.06, and 4.11 t/day is 2.7, 5.4, 13.6, and 27.1 kW, respectively. The base cost of the compressor is US$133,000 for a 224 kW compressor, and the scaling exponent is 0.84 [30]. Figure 2.20 shows the installed capital costs for air compressors for various cases.

The calculations of compressed air and nitrogen storage are done with stainless steel horizontal pressure vessels. Also, the pressure swing adsorption process requires similarly structured tanks. The base price for these tanks is US$6300 for 3.8 m^3, and the scaling exponent is 0.62 [30]. The cost calculation results for the best, mid and worst cases for nitrogen production are given in Fig. 2.21. Accordingly, most of the costs are related to equipment purchase and installation. The contribution of other expenditures decreases with increasing capacity.

The investment costs for a system to be installed with pressure swing adsorption are given in Fig. 2.22. Accordingly, the costs for systems with a nitrogen production capacity of 0.41, 0.82, 2.06, and 4.11 t/day were determined as US$200,000, US$252,000, US$373,000, US$530,000, respectively.

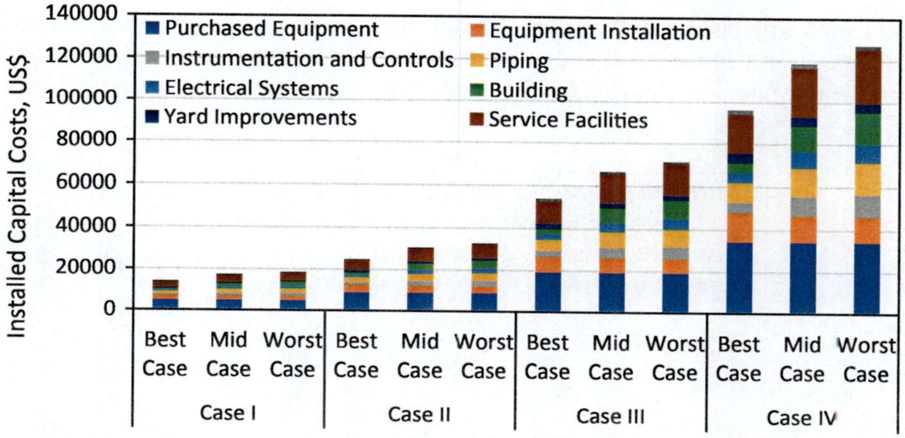

Fig. 2.20 Installed capital costs for air compressor for various cases and scenarios

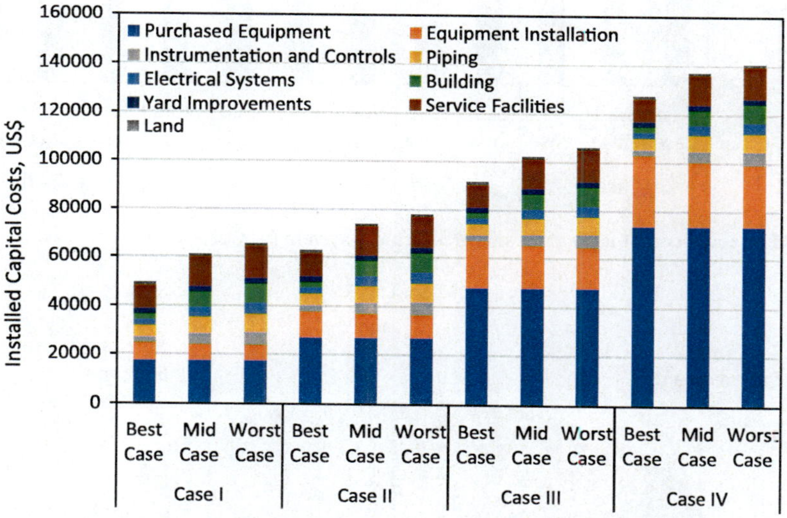

Fig. 2.21 Installed capital costs for one pressure vessel for various cases

The cost distribution for the systems to be installed is shown in Fig. 2.23. Accordingly, while the equipment prices constitute approximately 30% of the total price, the service facilities are determined as 20%, the equipment installations are 10%, and the building installations are 11%.

A pressure swing adsorption system was chosen for nitrogen production. Accordingly, the air enters the pressure swing adsorption unit with the help of a compressor. In addition, nitrogen is passed through an expander before being sent to the Haber–Bosch unit. The feedstock of the unit will be supplied by the air, with no costs associated with air usage. However, the unit needs an adsorbent to operate. Therefore, electricity, operational, and maintenance costs for the compressor, expander, adsorbent requirement, and the ASU is calculated accordingly. Design parameters for

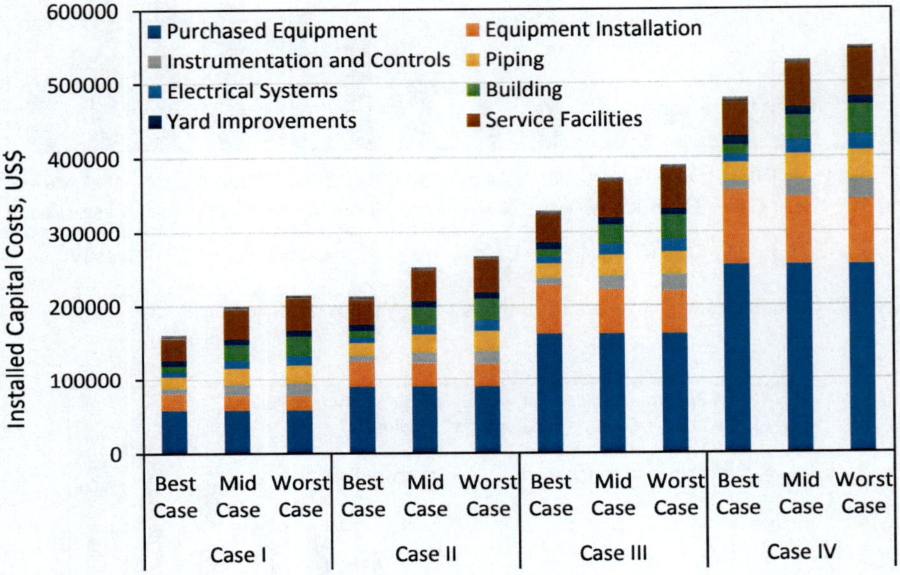

Fig. 2.22 Capital costs for pressure swing adsorption system for various cases

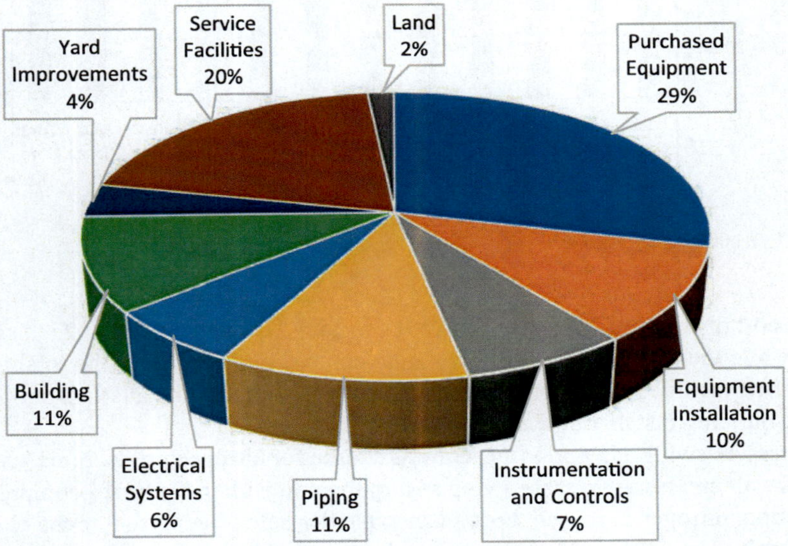

Fig. 2.23 Breakdown of the pressure swing adsorption system capital cost

the air separation unit were given in Table 2.9, and accordingly, manufacturing costs were presented in Table 2.10. Costs for ASU unit were calculated as US$44,743.66, US$61,320.55, US$99,672.17, US$136,554.91 for case I, case II, case III, and case IV, respectively.

Economics of Haber–Bosch Synthesis Unit

The Haber Bosch process is the core of the ammonia production plant. This process consists of a reactor where hydrogen and nitrogen react, a cooling unit where ammonia is shifted to the liquid phase and separated from unreacted gases, and a recycle unit where unreacted gases are recycled through the system. Recycling of unreacted hydrogen and nitrogen gases is important because ammonia production varies between 12 and 16.5% in a single pass [19].

Table 2.9 Design parameters for air separation unit

	Case I	Case II	Case III	Case IV
Air compressor power (kW)	2.7	5.4	13.6	27.1
Reactor volume, m^3	20	40	100	200
Nitrogen expander volume, m^3	20	40	100	200
Adsorbent need, kg	2680	5360	13,400	26,800
Adsorbent price (*Carbon Molecular Sieve*), US$/kg	6.50–9.00			
Adsorbent change interval, years	5	5	5	5
Daily electricity need, kWh/day	64.8	129.6	326.4	650.4

Table 2.10 Manufacturing costs for air separation unit

	Case I	Case II	Case III	Case IV
Direct costs				
Utilities	US$2672.68	US$5345.35	US$13,462.37	US$26,825.75
Operating labor	US$6000.00	US$9000.00	US$15,000.00	US$15,000.00
Supervision and official procedures	US$1050.00	US$1575.00	US$2625.00	US$2625.00
Maintenance and repairs	US$1999.40	US$2522.09	US$3729.75	US$5302.95
Operating supplies	US$299.91	US$378.31	US$559.46	US$795.44
Laboratory charges	US$900.00	US$1350.00	US$2250.00	US$2250.00
Fixed costs				
Depreciation	US$19,993.97	US$25,220.87	US$37,297.53	US$53,029.54
Taxes and insurance	US$6398.07	US$8070.68	US$11,935.21	US$16,969.45
Plant overhead	US$5429.64	US$7858.25	US$12,812.85	US$13,755.77
Total	US$44,743.66	US$61,320.55	US$99,672.17	US$136,554.91

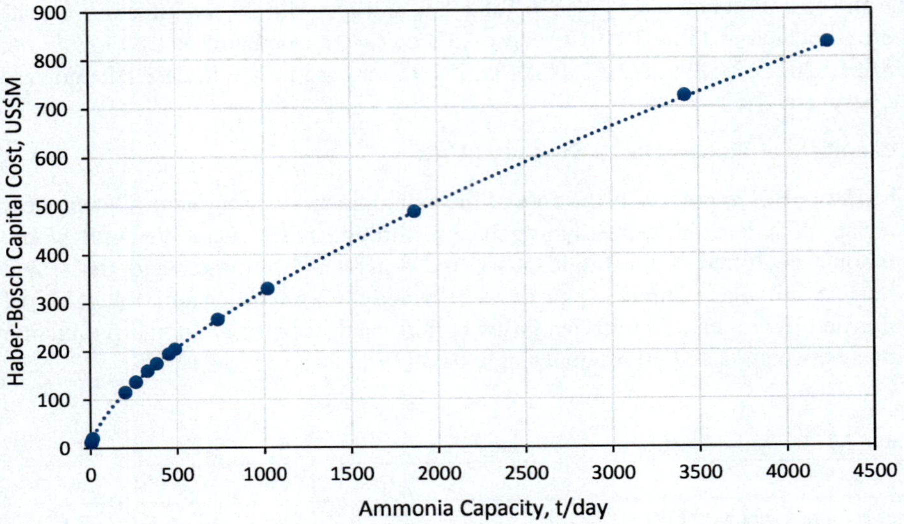

Fig. 2.24 Haber–Bosch capital costs by ammonia production capacity (data from [40])

Iron-based catalysts are used in reactors in most industrial applications. However, there are also studies with rare earth elements such as ruthenium in the literature [41]. Although higher efficiencies can be obtained with different catalysts, catalyst prices are a limiting factor for industrial applications. For this reason, iron catalysts are preferred predominantly [42]. Iron-based catalyst prices vary between US$900 and US$1000/t [43].

Table 2.11 Manufacturing costs for Haber–Bosch unit

Direct costs	Case I	Case II	Case III	Case IV
Operating labor	US$6000.00	US$6000.00	US$12,000.00	US$12,000.00
Supervision and official procedures	US$1050.00	US$1050.00	US$2100.00	US$2100.00
Maintenance and repairs	US$23,800.00	US$37,300.00	US$67,500.00	US$105,700.00
Operating supplies	US$3570.00	US$5595.00	US$10,125.00	US$15,855.00
Laboratory charges	US$900.00	US$900.00	US$1800.00	US$1800.00
Fixed costs				
Depreciation	US$238,000.00	US$373,000.00	US$675,000.00	US$1,057,000.00
Taxes and insurance	US$76,160.00	US$119,360.00	US$216,000.00	US$338,240.00
Plant overhead	US$18,510.00	US$26,610.00	US$48,960.00	US$71,880.00
Total	US$367,990.00	US$569,815.00	US$1,033,485.00	US$1,604,575.00

There are many cost studies on Haber–Bosch systems in the literature. Figure 2.24 shows the results of these cost studies. The relationship between the production capacity and capital costs gives the scale-up exponent, which is 0.63 ± 0.09. The scale-up exponent is given as 0.70 in the literature for ammonia production facilities [30]. This value is given for an entire plant that includes the Haber–Bosch process. Since only the costing of the Haber–Bosch system was studied in this study, the scale-up exponent was taken as 0.63.

The Haber–Bosch unit consists of a pre-heater, reactor, and auxiliary equipment. However, the catalyst used in the reactor must be changed at certain times. Although catalysts such as ruthenium increase efficiency, the costs are high for utilization in commercial processes [44]. For this reason, a magnetite (iron oxide)-based catalyst is used while making the calculations. Catalyst prices vary between US$100–300/t. Table 2.11 summarizes the total annual operational and maintenance costs of a Haber–Bosch unit. A large part of these manufacturing costs consists of depreciation, tax and insurance items.

2.7 Assessment of Ammonia Production Plants

While studying the capital costs of the ammonia plant, the costs of each main equipment need to be estimated. The ammonia production process consists of the following parts [42]:

- Hydrogen production system
- Nitrogen generation system
- Separation of hydrogen and nitrogen gases from other gases
- Mixing, and preheating systems of nitrogen and hydrogen gases
- Reactors containing catalysts and heat control and heat recovery equipment of these reactors
- Condensation of ammonia and separation from process gases
- Recirculation equipment for the recycling of unreacted gases
- Other equipment (heaters, heat exchangers, pumps, etc.).

Table 2.12 shows the estimates for various ammonia production plants. Since investment costs do not increase linearly with increasing capacity, ammonia production costs increase in low-capacity systems. Especially in systems where renewable energy resources are used, the initial investment costs of renewable resources also affect facility investments. Table 2.13 summarizes the selected ammonia production capacities and feedstock requirements. The total operation time of the ammonia production facility is accepted as 25 years. All costs related to manufacturing are estimated for each unit accordingly.

Table 2.12 Ammonia production plant estimates adjusted to 2020 prices

Power source	Hydrogen production process	Air separation	Gas turbine	Ammonia capacity (t/day)	Haber–Bosch capital costs (US$M)	Total capital costs (US$M)	Ammonia production price (US$/t)
Natural gas	Steam methane reforming	No	No	1387	204.5	310.1	629.3
Natural gas	Steam methane reforming	No	No	2004	259.6	486.6	585.4
Natural gas	Steam methane reforming with sequestration	No	No	2004	259.6	545	630.4
Nuclear MHR	Sulfur-iodine thermochemical (low efficiency)	Yes	No	3422	643.9	3012.6	688.8
Nuclear MHR	Sulfur-iodine thermochemical (high efficiency)	Yes	No	4244	740.5	3525	640.5
Solar and NG	Sulfur-iodine thermochemical	Yes	Yes	455	173	1837.2	2062
Solar	Photovoltaic electrolysis (US$5/W)	Yes	Yes	7	12.4	73	6687.1
Solar	Photovoltaic electrolysis (US$0.75/W)	Yes	Yes	7	12.4	25.8	2541.8
Solar	Dish stirling electrolysis	Yes	Yes	7	11.2	36	3670
Solar	Power tower electrolysis	Yes	Yes	331	141.6	842.8	1839.5
Solar	Photovoltaic electrolysis	Yes	Yes	1866	433.7	6684.9	1602.4
Solar	Power tower electrolysis	Yes	Yes	202	102.3	575.3	1576.5
Solar	Sulfuric acid/hybrid thermochemical	Yes	Yes	490	182	616.9	932.7
Wind	Electrolysis (near term)	Yes	Yes	5	10.1	13.5	2631.7
Wind	Electrolysis without electricity coproduction	Yes	Yes	263	121.4	751.8	1940.6
Wind	Electrolysis with electricity coproduction	Yes	Yes	263	121.4	758.5	1902.4

(continued)

Table 2.12 (continued)

Power source	Hydrogen production process	Air separation	Gas turbine	Ammonia capacity (t/day)	Haber–Bosch capital costs (US$M)	Total capital costs (US$M)	Ammonia production price (US$/t)
Coal	Coal gasification with sequestration	No	No	1650	228.1	860.8	361.8
Coal	Coal gasification without sequestration	No	No	1817	242.7	798.9	315.8
Coal	Advanced gasification with sequestration	No	No	2080	265.2	1810.3	165.2
Coal	Coal gasification with sequestration	No	No	4524	439.4	1787.8	356.2
Coal	Coal gasification without sequestration	No	No	4524	439.4	1719.3	310.1
Coal	Coal gasification with sequestration	No	No	1626	225.9	914.7	485.4
Coal	Coal gasification without sequestration	No	No	1499	214.6	764.1	418

Source Bartels [40]

Table 2.13 Selected ammonia production capacities and feedstock requirements per day

Case	NH_3, t	N_2, t	H_2, t	Water need, m^3	Air need, t
I	0.5	0.41	0.09	800	0.83
II	1	0.82	0.18	1600	1.66
III	2.5	2.06	0.44	4000	4.14
IV	5	4.11	0.89	8000	8.28

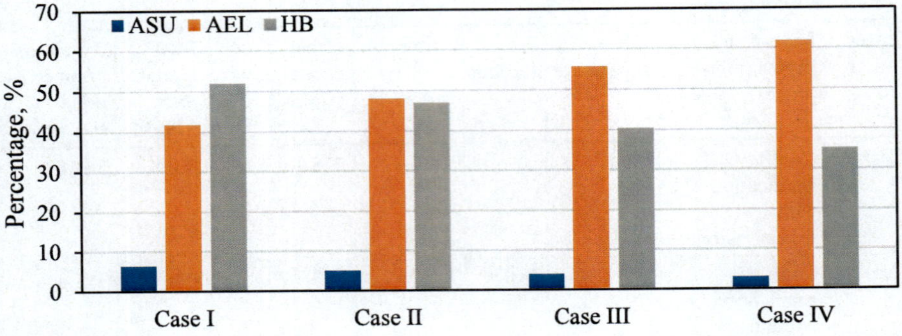

Fig. 2.25 Percentage distribution of annual costs by the unit of a facility using an AEL unit

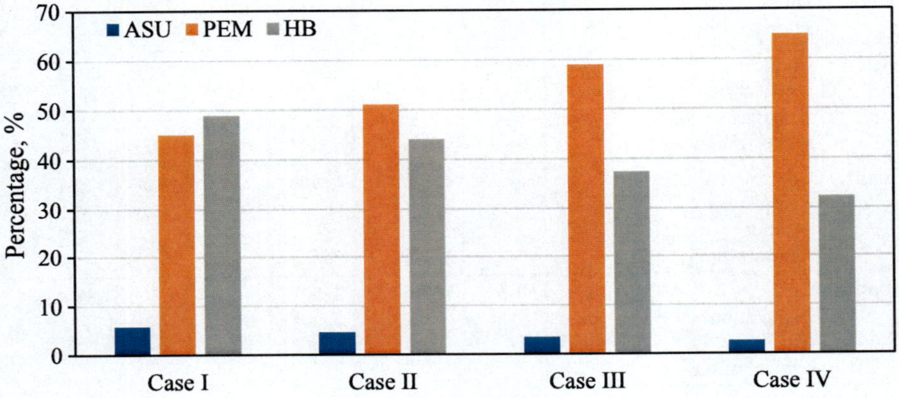

Fig. 2.26 Percentage distribution of annual costs by the unit of a facility using a PEM unit

Figures 2.25 and 2.26 give the percentage of contribution of each unit to annual costs for the system including AEL and PEM units, respectively. In both cases, the electrolysis unit makes the biggest impact on costs. This is due to the high electricity and stack prices. Although it is predicted that stack prices and unit hydrogen prices will decrease in future forecasts, today the costs are high, especially when compared to fossil fuel-based hydrogen production [45, 46]. In addition, the contribution of Haber–Bosch and the air separation unit to the costs decreases with higher capacities. This can be explained by the capacity increase, which reduces raw material costs.

The capital costs of the capacities selected accordingly are presented in Table 2.14. Accordingly, the capital cost required for a Haber–Bosch unit with a daily production capacity of 0.5, 1, 2.5, and 5 t is calculated as US$2.38, 3.73, 6.75, and 10.57 M, respectively.

Table 2.14 Capital costs for various ammonia production capacities

NH_3 production capacity, t/day	Capital costs, US$M
0.5	2.38
1	3.73
2.5	6.75
5	10.57

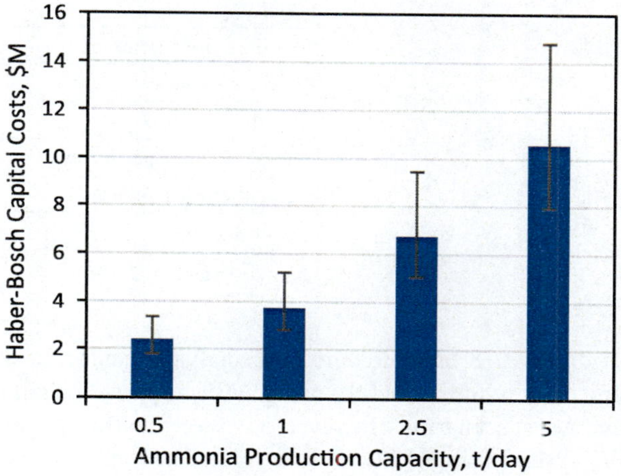

Fig. 2.27 The potential Haber–Bosch capital costs

The margin of error varies between − 25 and + 40%, depending on the sensitivity of the methodology used in determining the capital costs. Accordingly, the values obtained are presented in Fig. 2.27 with margins of error. The rates used to calculate direct and indirect costs are given in Table 2.15. Accordingly, the breakdown of capital costs is calculated according to the best, worst and average scenario for each item.

Table 2.16 shows the capital costs for various ammonia production capacities. The price range for facilities with a capacity between 0.5 and 5 t/day is determined as US$1.79–14.80 M. When these values are compared with the literature data, it is seen that close values are obtained. At the same time, the calculated data are calculated for the current year, considering inflation according to the CEPCI index. the costs obtained from the best- and worst-case scenarios determined according to different scenarios were taken as a basis while calculating these costs. Accordingly, costs have been calculated for ammonia production systems containing both AEL and PEM.

Table 2.17 gives the costs of an ammonia production plant operated using the AEL, in terms of best and worst probabilities. Electricity consumption for hydrogen production and the costs of replacing the electrolysis stack are also added to the calculations. Accordingly, the costs of a facility producing 0.5 t/day are estimated to be between US$590,000 and 915,000. With increasing capacity, costs are increasing predictably. Expenses such as operating labor are taken as constant according to both scenarios.

Table 2.15 Rates for direct and indirect costs

	Best	Mid	Worst
Purchased equipment	15	27.5	40
Equipment installation	6	10	14
Instrumentation and controls	2	7	12
Piping	4	10.5	17
Electrical systems	2	6	10
Building	2	10	18
Yard improvements	2	3.5	5
Service facilities	8	19	30
Land	1	1.5	2
Engineering supervision	4	12	20
Construction expenses	4	10.5	17
Contingency	5	10	10

Source Peters et al. [38]

Table 2.18 tabulates the cost analysis results for ammonia facilities with PEM units. As with the facility using AEL, separate calculations were made for the best and worst scenarios. However, it is seen that the annual costs are slightly higher than the system using AEL. This is related to the fact that PEM is a technology that is still under development. The need for catalysts for membranes and electrodes used in PEM units increases the prices of stacks. Therefore, these costs are higher than AEL in worst-case scenarios.

2.7.1 *Ammonia Production Prices*

While calculating the ammonia production costs, the costs that may occur during the 25-year operating period of the plant and the fixed investment costs have been evaluated. Accordingly, minimum, and maximum ammonia production costs were calculated for each case. Figures 2.28 and 2.29 display the results for the cases where ammonia production is 0.5–1 tonne/day. When the best cases are evaluated, the production costs of AEL and PEM systems are very close to each other. However, in the worst scenarios, the price difference of PEM is noticable. However, it is seen that ammonia costs are mainly due to O&M costs. Especially electrolysis units and the electricity need of these units cause these costs to increase.

Figures 2.30 and 2.31 show the ammonia production cost per kg of facilities with a capacity of 2.5 and 5 t/day. Especially, when compared to facilities with 0.5–1 tonne/day capacity, the costs decrease. In particular, the effect of the fixed cost of the nitrogen production unit decreases considerably. The cost difference between AEL and PEM is also seen here.

Table 2.16 Direct and indirect costs for various ammonia production cases, US$M

Ammonia production capacity, t/day	0.5			1.0			2.5			5.0		
	Best	Mid	Worst	Best	Mid	Worst	Best	Mid	Worst	Best	Mid	Worst
Purchased equipment	0.45	0.60	0.83	0.70	0.93	1.31	1.27	1.69	2.36	1.98	2.64	3.70
Equipment installation	0.13	0.18	0.25	0.21	0.28	0.39	0.38	0.51	0.71	0.59	0.79	1.11
Instrumentation and controls	0.12	0.17	0.23	0.20	0.26	0.37	0.35	0.47	0.66	0.55	0.74	1.04
Piping	0.19	0.25	0.35	0.29	0.39	0.55	0.53	0.71	0.99	0.83	1.11	1.55
Electrical systems	0.11	0.14	0.20	0.17	0.22	0.31	0.30	0.41	0.57	0.48	0.63	0.89
Building	0.07	0.10	0.13	0.11	0.15	0.21	0.20	0.27	0.38	0.32	0.42	0.59
Yard improvements	0.04	0.05	0.07	0.06	0.07	0.10	0.10	0.14	0.19	0.16	0.21	0.30
Service facilities	0.20	0.26	0.37	0.31	0.41	0.57	0.56	0.74	1.04	0.87	1.16	1.63
Land	0.02	0.02	0.03	0.03	0.04	0.05	0.05	0.07	0.09	0.08	0.11	0.15
Engineering supervision	0.18	0.24	0.33	0.28	0.37	0.52	0.51	0.68	0.95	0.79	1.06	1.48
Construction expenses	0.11	0.14	0.20	0.17	0.22	0.31	0.30	0.41	0.57	0.48	0.63	0.89
Contingency	0.18	0.24	0.33	0.28	0.37	0.52	0.51	0.68	0.95	0.79	1.06	1.48
Total	1.79	2.38	3.33	2.80	3.73	5.22	5.06	6.75	9.45	7.93	10.57	14.80

Table 2.17 Fixed costs and operational and maintenance costs of ammonia production plant with AEL

	Case I		Case II		Case III		Case IV	
	Best	Worst	Best	Worst	Best	Worst	Best	Worst
Fixed costs								
Air separation	US$162,056	US$214,146	US$212,796	US$266,988	US$329,017	US$389,460	US$479,244	US$549,440
Haber–Bosch unit	US$1,785,000	US$3,332,000	US$2,797,500	US$5,222,000	US$5,062,500	US$9,450,000	US$7,927,500	US$14,798,000
AEL	US$213,173	US$559,410	US$423,362	US$1,110,988	US$1,021,100	US$2,679,574	US$2,042,201	US$5,359,148
Total	US$2,160,230	US$4,105,556	US$3,433,658	US$6,599,977	US$6,412,617	US$12,519,034	US$10,448,944	US$20,706,588
O&M costs, annual								
Direct costs								
Feedstock								
Water	US$14,600	US$14,600	US$29,200	US$29,200	US$73,000	US$73,000	US$146,000	US$146,000
Utilities								
Electricity	US$189,313	US$189,313	US$378,627	US$378,627	US$946,567	US$946,567	US$1,893,133	US$1,893,133
Operating LAbor	US$6000	US$6000	US$9000	US$9000	US$15,000	US$15,000	US$15,000	US$15,000
Supervision and official procedures	US$1050	US$1050	US$1575	US$1575	US$2625	US$2625	US$2625	US$2625
Maintenance and repairs	US$43,205	US$82,111	US$68,673	US$132,000	US$128,252	US$250,381	US$208,979	US$414,132
Stack replacement	US$13,543	US$13,543	US$27,000	US$27,000	US$65,000	US$65,000	US$129,857	US$129,857
Operating supplies	US$6481	US$12,317	US$10,301	US$19,800	US$19,238	US$37,557	US$31,347	US$62,120
Laboratory charges	US$900	US$900	US$1350	US$1350	US$2250	US$2250	US$2250	US$2250

(continued)

Table 2.17 (continued)

	Case I		Case II		Case III		Case IV	
	Best	Worst	Best	Worst	Best	Worst	Best	Worst
Fixed costs								
Depreciation	US$216,023	US$410,556	US$343,366	US$659,998	US$641,262	US$1,251,903	US$1,044,894	US$2,070,659
Taxes and insurance	US$69,127	US$131,378	US$109,877	US$211,199	US$205,204	US$400,609	US$334,366	US$662,611
Plant overhead	US$30,153	US$53,497	US$47,549	US$85,545	US$87,526	US$160,803	US$135,962	US$259,054
Total	US$590,395	US$915,264	US$1,026,518	US$1,555,293	US$2,185,924	US$3,205,695	US$3,944,414	US$5,657,440

Table 2.18 Fixed costs and operational and maintenance costs of ammonia production plant with PEM

	Case I		Case II		Case III		Case IV	
	Best	Worst	Best	Worst	Best	Worst	Best	Worst
Fixed costs								
Air Separation	US$162,056	US$214,146	US$212,796	US$266,988	US$329,017	US$389,460	US$479,244	US$549,440
Haber–Bosch Unit	US$1,785,000	US$3,332,000	US$2,797,500	US$5,222,000	US$5,062,500	US$9,450,000	US$7,927,500	US$14,798,000
PEM	US$155,519	US$844,883	US$298,103	US$1,631,499	US$738,789	US$4,020,448	US$1,473,267	US$8,022,284
Total	US$2,102,576	US$4,391,029	US$3,308,399	US$7,120,488	US$6,130,306	US$13,859,908	US$9,880,010	US$23,369,724
O&M costs, annual								
Direct costs								
Feedstock								
Water	US$14,600	US$14,600	US$29,200	US$29,200	US$73,000	US$73,000	US$146,000	US$146,000
Utilities								
Electricity	US$189,313	US$189,313	US$378,627	US$378,627	US$946,567	US$946,567	US$1,893,133	US$1,893,133
Operating labor	US$6000	US$6000	US$9000	US$9000	US$15,000	US$15,000	US$15,000	US$15,000
Supervision and official procedures	US$1050	US$1050	US$1575	US$1575	US$2625	US$2625	US$2625	US$2625
Maintenance and repairs	US$42,052	US$87,821	US$66,168	US$142,410	US$122,606	US$277,198	US$197,600	US$467,394
Stack replacement	US$22,571	US$22,571	US$43,143	US$43,143	US$107,143	US$107,143	US$214,286	US$214,286
Operating supplies	US$6308	US$13,173	US$9925	US$21,361	US$18,391	US$41,580	US$29,640	US$70,109
Laboratory charges	US$900	US$900	US$1350	US$1350	US$2250	US$2250	US$2250	US$2250

(continued)

Table 2.18 (continued)

	Case I		Case II		Case III		Case IV	
	Best	Worst	Best	Worst	Best	Worst	Best	Worst
Fixed costs								
Depreciation	US$210,258	US$439,103	US$330,840	US$712,049	US$613,031	US$1,385,991	US$988,001	US$2,336,972
Taxes and insurance	US$67,282	US$140,513	US$105,869	US$227,856	US$196,170	US$443,517	US$316,160	US$747,831
Plant overhead	US$29,461	US$56,922	US$46,046	US$91,791	US$84,139	US$176,894	US$129,135	US$291,012
Total	US$589,795	US$971,967	US$1,021,742	US$1,658,361	US$2,180,921	US$3,471,764	US$3,933,831	US$6,186,613

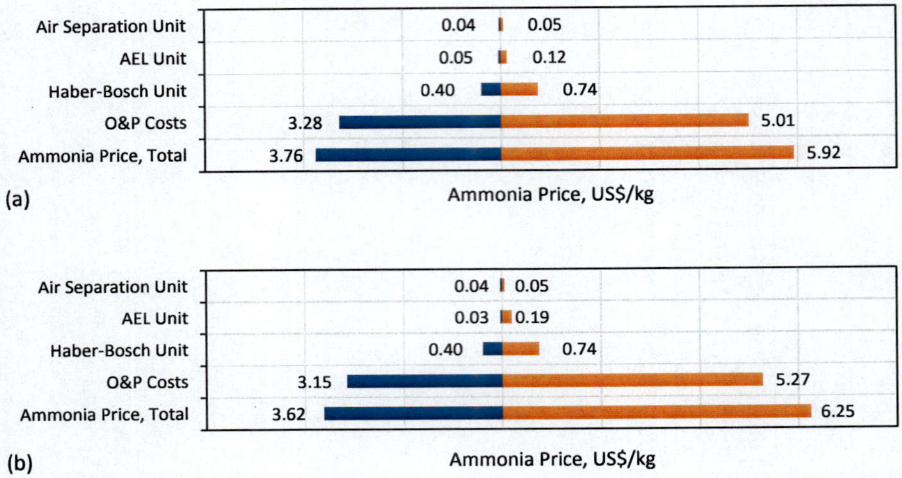

Fig. 2.28 Ammonia costs for a facility producing 0.5 t/day with **a** AEL and **b** PEM unit

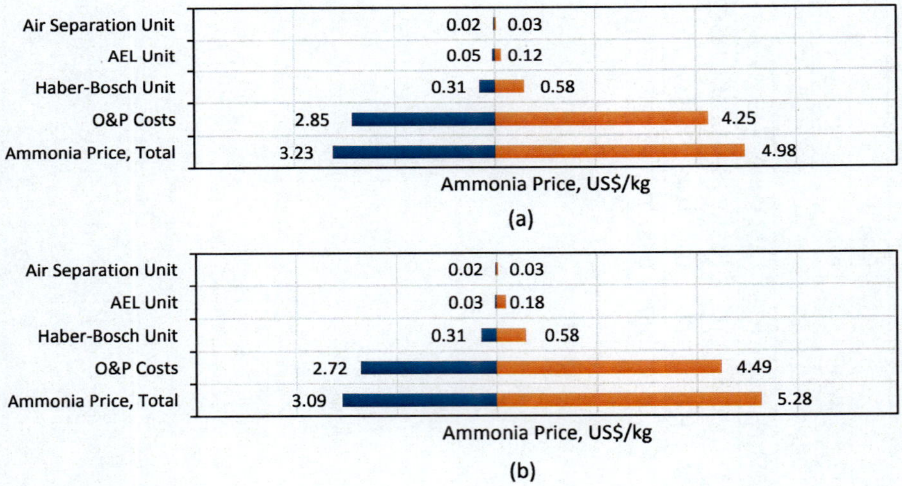

Fig. 2.29 Ammonia costs for a facility producing 1 t/day with **a** AEL and **b** PEM units

The results of the calculated ammonia production costs are compared in Fig. 2.32. Accordingly, with the increase in capacity, the costs also decrease. Despite this, it is seen that the ammonia production cost in the working capacities is quite high compared to the facilities with much higher capacity. This is an expected result when compared with the industry data. In particular, the production prices reported in the plants that meet the hydrogen source from electrolysis vary between US$660

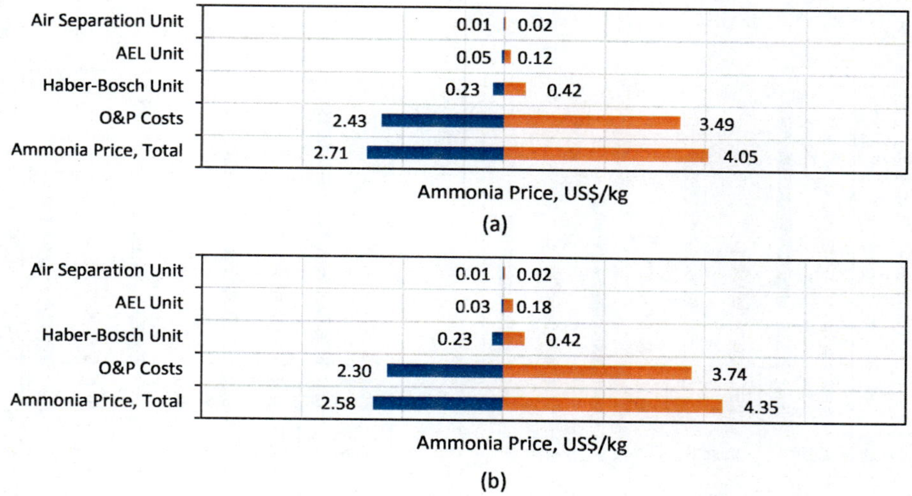

Fig. 2.30 Ammonia costs for a facility producing 2.5 t/day with **a** AEL and **b** PEM units

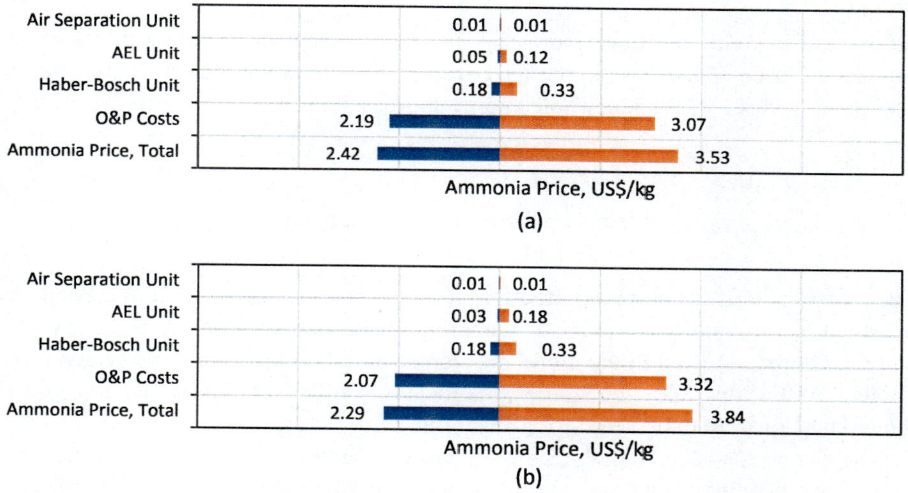

Fig. 2.31 Ammonia costs for a facility producing 5 t/day with **a** AEL and **b** PEM units

and US$5900/t [40]. However, with steam methane reforming and coal gasification, costs can be reduced to US$150 in cases where daily production exceeds 2000 t.

When the results of this comprehensive analysis are compared, it is seen that the effect of the generation of hydrogen by using electricity from the mains and electrolysis on ammonia production is quite high. However, it is seen that the effect of capacity on costs is also great. It is predicted that the current costs may become feasible, especially in remote regions where transportation costs are high, and the

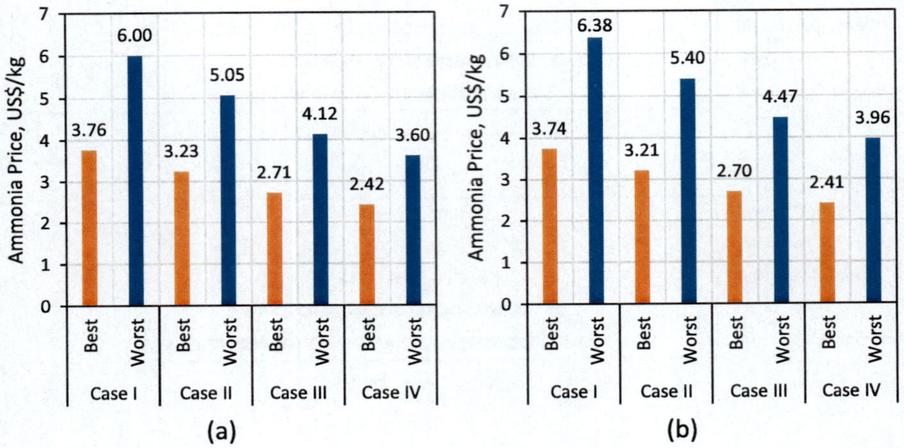

Fig. 2.32 Comparison of ammonia prices for best- and worst-case scenarios. **a** An ammonia facility with AEL. **b** An ammonia facility with PEM

production is continued during off-peak periods or when local renewable energy sources are used. Therefore, in the next section, ammonia production costs will be presented in cases where renewable energy sources are used.

2.7.2 Cost Assessment for Plants with Various Capacities

The electrical infrastructure required for ammonia production at different capacities in a remote community is modeled. Modeling was done using Homer Pro software and the equipment used in the modeling studies and their information was given in Table 2.19. For the modeling studies, the locally available equipment as defined in the HomerPro were considered for utilization.

There is no road access to the Sandy Lake community selected for the study. The most convenient way of transportation has been reported as air. This makes it difficult to transport chemicals such as ammonia in bulk. Ammonia transportation cost was

Table 2.19 System components of the micro grid

Component	Unit name	Capacity	Capital cost
Diesel genset	CAT-20kVA-60Hz-PP	16 kW	US$8000.00
PV panels	CanadianSolar Quintech CS6K-285M	0.285 kW	US$3000.00
Wind turbine	Eocycle EO25 Class IIA	25 kW	US$65,000.00
Battery	Discover 12VRE-3000TF	3.11 kWh	US$410

reported as US$1.14/GJ by Al-Breiki and Bicer [47]. This creates a negative situation, especially for remote communities. For this reason, in this study, a costing analysis has been made when the energy need is provided by local resources together with the local ammonia production facility. The manufacturing cost of the ammonia plant was presented previously. In addition, cost analyzes have been made for situations where the required electricity is met by renewable energy sources and diesel generators.

Case 1: Ammonia Plant with 500 kg/day Capacity

The electrical infrastructure required for a facility with a capacity of 500 kg/day is calculated as US$3,600,000. However, annual operating expenses are US$283,000. The 25-year cash flow chart for these prices is presented in Fig. 2.33. The average capital cost for an ammonia plant with this capacity was calculated as US$3,250,000. The annual operating cost is around US$780,000 on average.

With the inclusion of electrical infrastructure in the process, the price of ammonia rises to an average of US$6.17/kg from US$5.05/kg. Cash flow graphs related to ammonia prices are presented in Fig. 2.34. As can be seen from the graph, when the electrical infrastructure is included, the system can pay for itself in 20 years if the ammonia sales price is US$7500/t. When the electrical infrastructure is not included, this value goes down to about US$5000/t. In October 2021, the price of ammonia for fertilizer purposes was reported as US$1135/t. In this case, it is seen that the system is not feasible under normal conditions. However, the problem of transportation to remote communities plays an important role in the profitability of the price of ammonia. Especially with the COVID-19 pandemic and the conflict between Russia and Ukraine, serious increases in fuel prices are expected. This will further increase the already high transportation costs for remote communities.

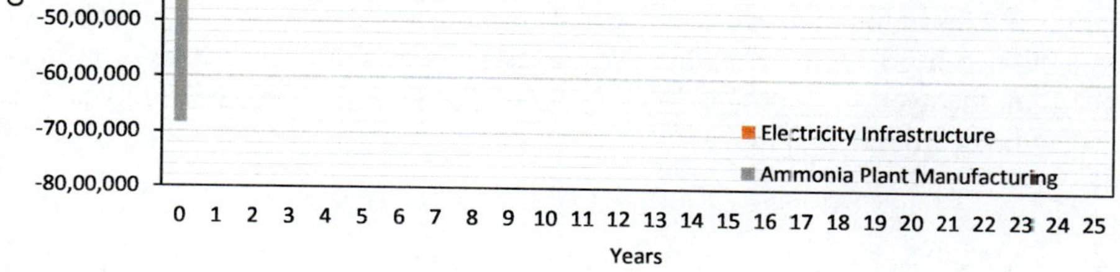

Fig. 2.33 Capital and initial costs for a 500 kg/day ammonia production facility with electricity infrastructure

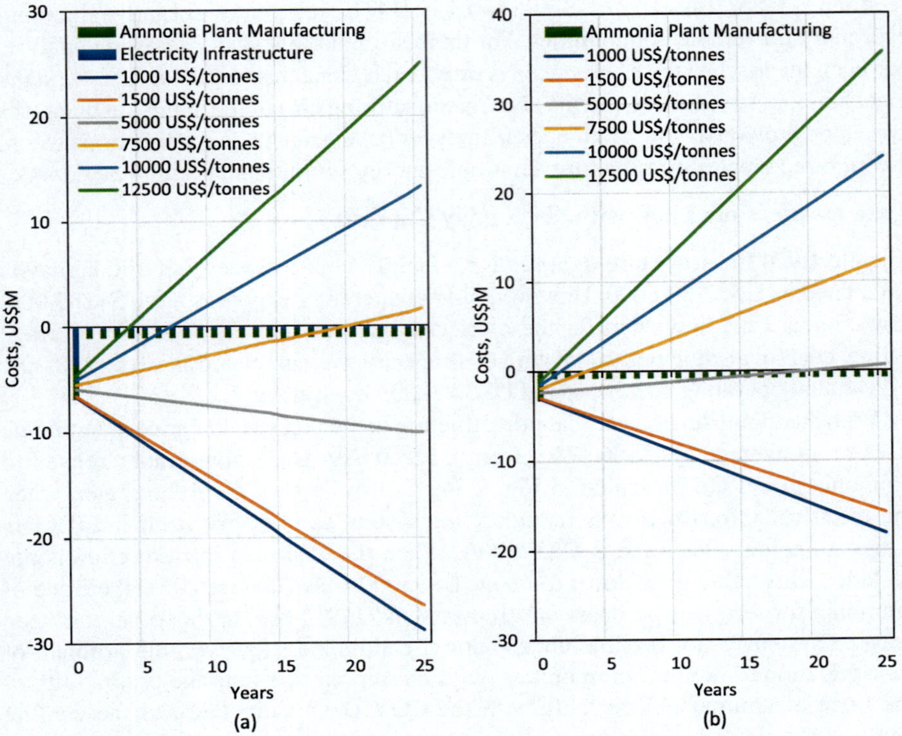

Fig. 2.34 Cash flow of 500 kg/day ammonia production system **a** with and **b** without electrical infrastructure costs

Excess electrical values are presented in Fig. 2.35. Wind turbines used in modeling give high electricity surplus at peak times. Although it is seen that this value rises to 2 MW for peak times, the average value in general remains in the range of 100–200 kW. Excess electricity modeling is utilized by using pumped hydro storage and electrolysis unit. In this way, almost all the hydrogen can be produced with excess electricity.

Figure 2.36 summarizes the average hydrogen load of the electrolyzer. Hydrogen production can go up to 13 kg/h peak values on average. However, on a normal operation, the average hydrogen load is around 3.7 kg/h. In total, yearly hydrogen production is around 32.8 tonnes.

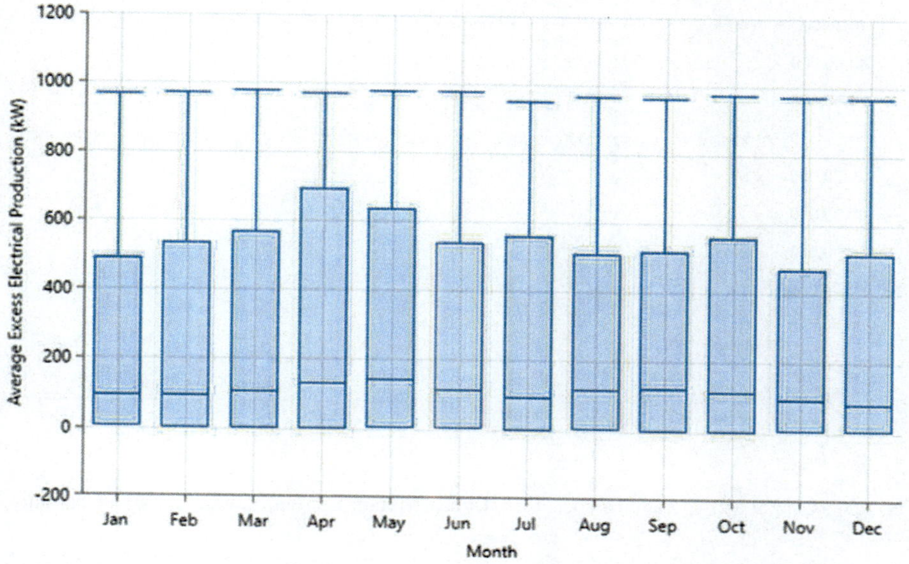

Fig. 2.35 Excess electricity generation of the electrical infrastructure for 500 kg/day ammonia production

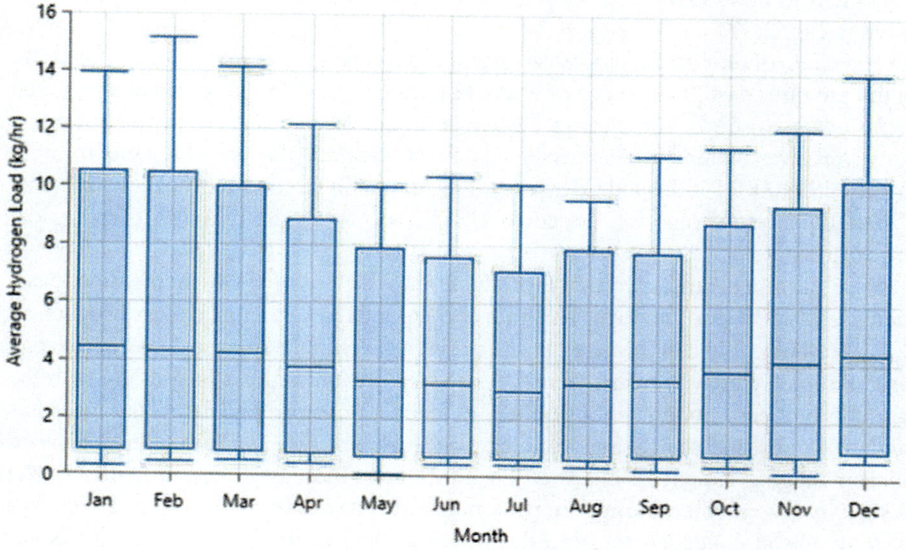

Fig. 2.36 Average hydrogen load of the 500 kg/day ammonia production facility

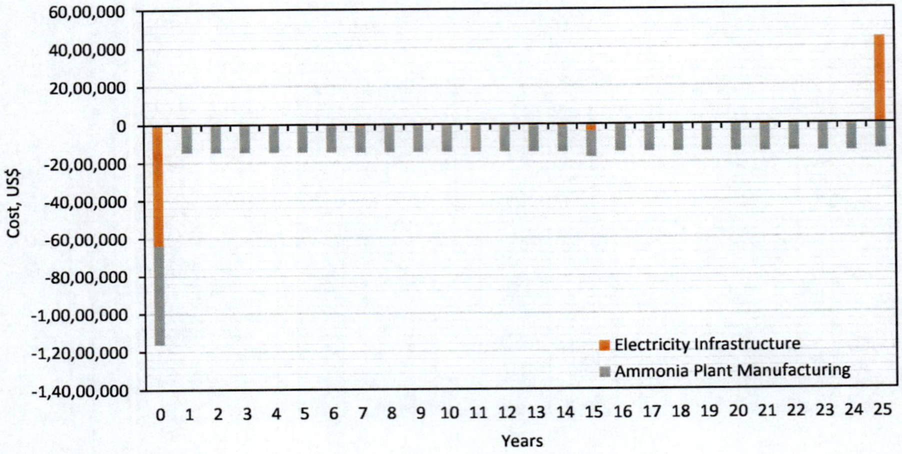

Fig. 2.37 Capital and operational costs for a 1 t/day ammonia production facility with electricity infrastructure

Case 2: Ammonia Plant with 1 t/day Capacity

Similar results were obtained with a system with 1 t/day capacity. The electricity infrastructure cost is estimated to be US$6,394,000, and the operational costs was calculated to be US$100,000. Capital and operational costs for 1 t/day capacity is given in Fig. 2.37.

The effect of the capacity increase on ammonia production cost is seen in Fig. 2.38. In the previous case, in the case of ammonia price of US$7500/t, profit can be made in 20 years, while in the current system, this period has decreased to 7 months. However, even when the electricity cost is subtracted, it does not seem feasible in the US$1000–1500 band. However, the ammonia production cost, which was US$4.30/kg on average, increased to US$5.30/kg with the addition of electrical infrastructure.

In addition, excess electricity production, which was 28% in the first case, increased to 33% for this case. This rate corresponds to an annual excess production of 2000 MWh. It has been stated in the previous sections that excess electricity can be minimized when the diesel Genset is used. However, in this study, along with the increase in electricity consumption due to Haber–Bosch and electrolysis, there has been an increase in the capacity of electricity sources. Even with the diesel Genset, micromanaging the excess electricity is hard due to increased capacity. Most of the electricity is generated during the peak hours of renewable sources. Large-capacity systems are also needed to store the peak surplus electricity produced by large-capacity generators or utilize it for hydrogen production. This situation increases capital costs for storage systems and electrolyzers, thus increasing the electricity costs and indirectly hydrogen and ammonia production costs.

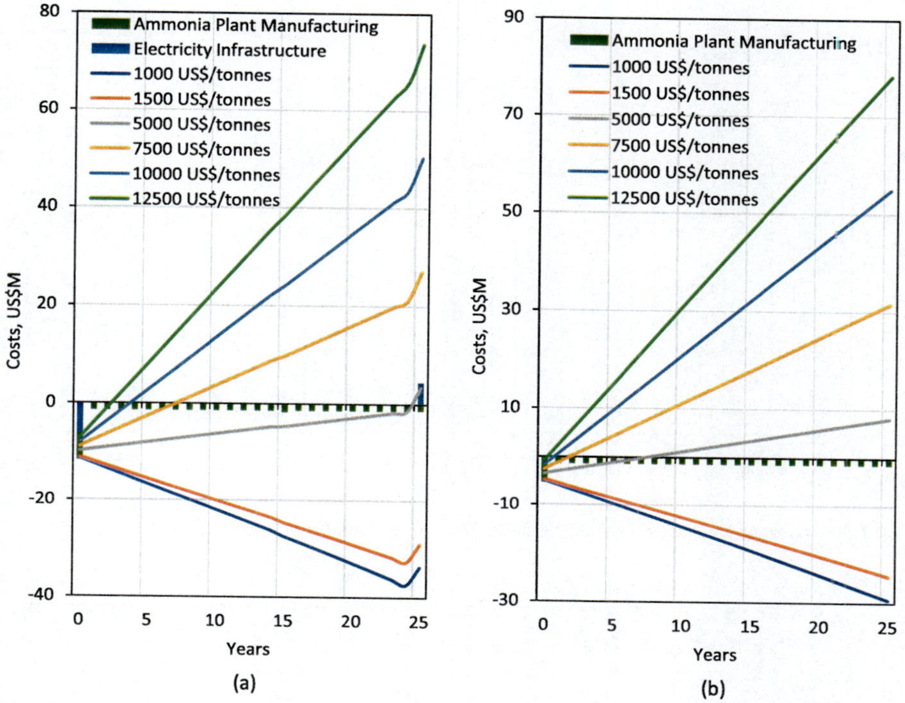

Fig. 2.38 Cash flow of 1 t/day ammonia production system **a** with and **b** without electrical infrastructure costs

Figure 2.39 shows the hydrogen production profile. Like the first case, peak values are obtained when municipal electricity consumption are low. The peak hydrogen production is found to be 22.5 kg/h on average. The average hydrogen load is around 7.5 kg/h on a normal operation. In total, yearly hydrogen production is around 65 tonnes.

Case 3: Ammonia Plant with 2.5 t/day Capacity

The capital costs and operational costs required for an ammonia plant and electricity infrastructure with a daily capacity of 2.5 t are given in Fig. 2.40. Accordingly, the cost required for the electrical infrastructure was calculated as US$13,123,640. The annual expenses are estimated to be US$248,000. As in other cases, a large portion of the operating expenses consists of the maintenance costs of the ammonia production facility.

Figure 2.41 represents the cash flow data of the ammonium plant with a daily 2.5 t capacity. In cases where the price of ammonia is lower at higher capacities, the system can make a profit in a shorter time. Accordingly, the cost of ammonia was calculated

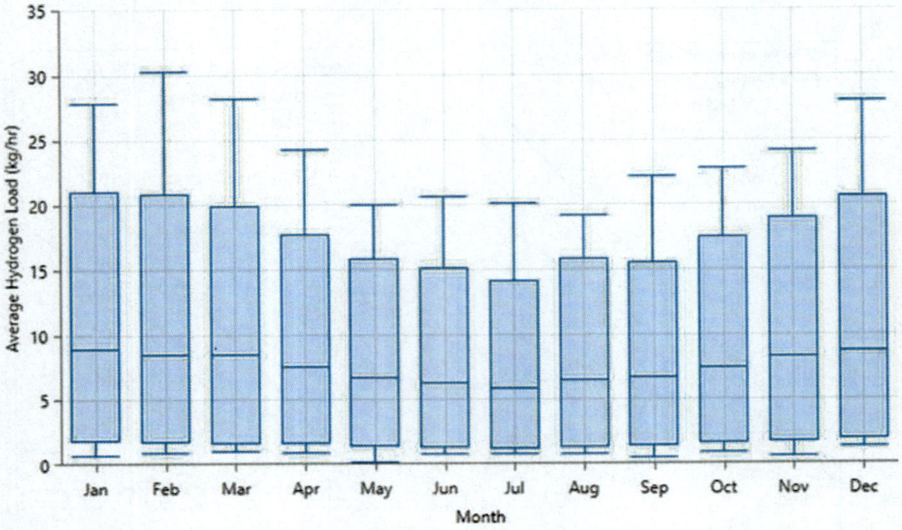

Fig. 2.39 Average hydrogen load of the 500 kg/day ammonia production facility

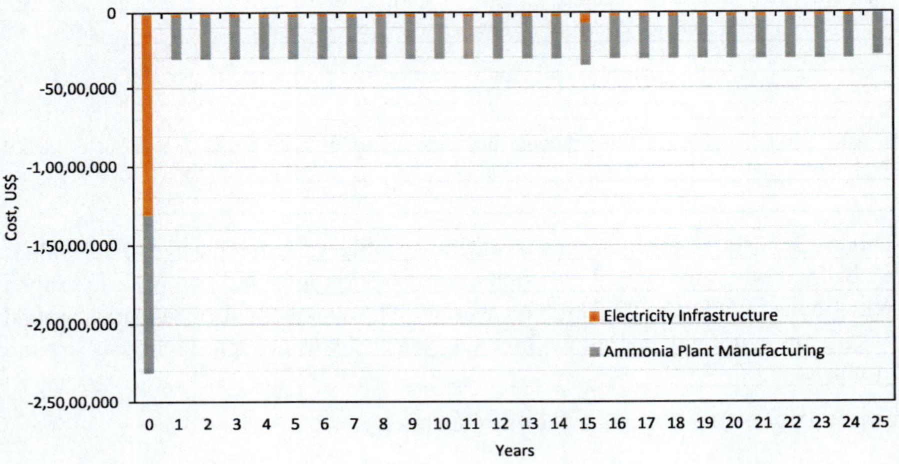

Fig. 2.40 Capital and operational costs for a 2.5 t/day ammonia production facility with electricity infrastructure

as US$3.58/kg in the case of not including the electricity infrastructure cost and US$3.89/kg when included. Although reaching a profitable point has decreased cash flow, the system still at negative in the US$1000–1500 band.

The excess electricity rate remained around 17.9%. This rate corresponds to an energy surplus of approximately 2187 MWh/year. Although the percentage of excess

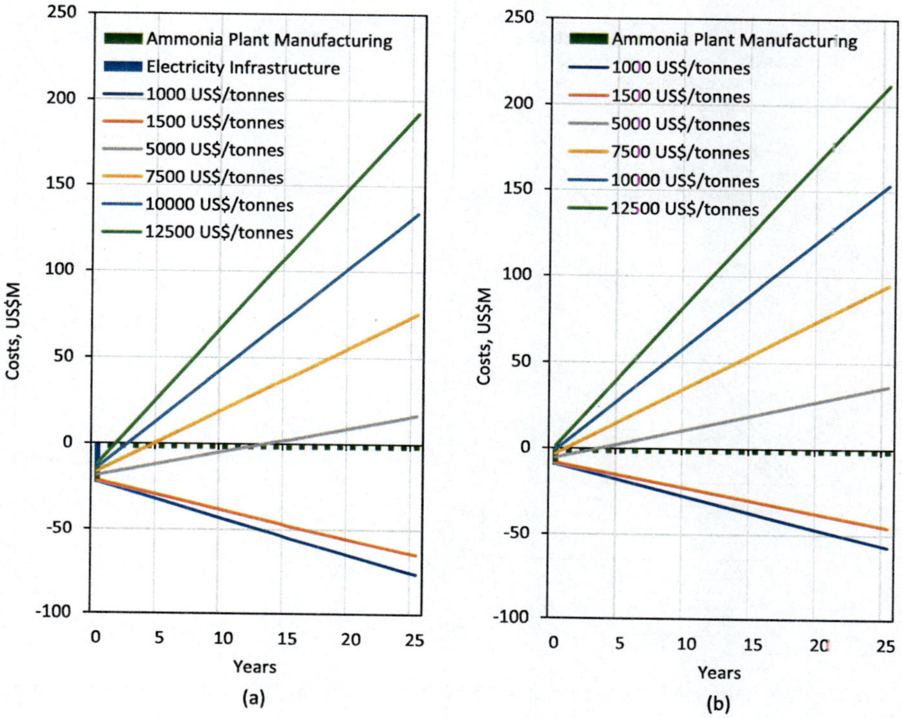

Fig. 2.41 Cash flow of 2.5 t/day ammonia production system **a** with and **b** without electrical infrastructure costs

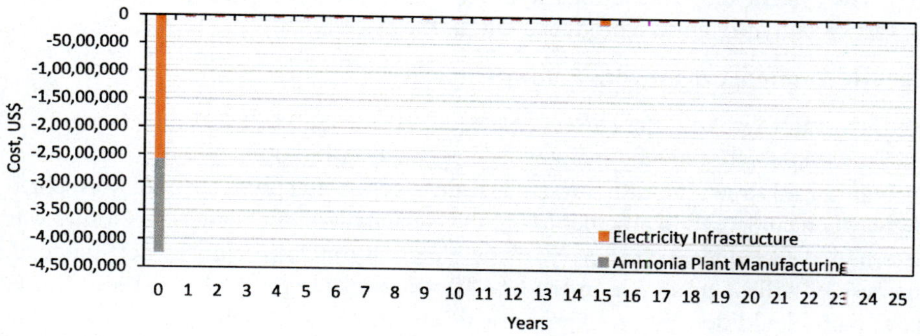

Fig. 2.42 Capital and operational costs for a 5 t/day ammonia production facility with electricity infrastructure

electricity has decreased compared to other cases, when the amount of excess electricity is considered, there is a serious increase compared to systems with 0.5 and 1 t capacity. This situation, as explained earlier, is due to the expensiveness of the

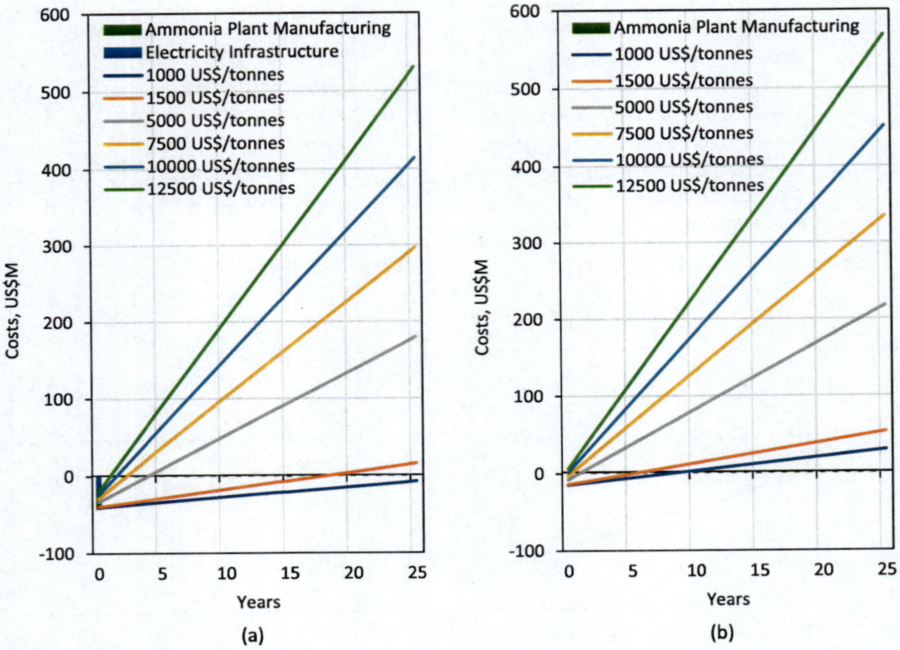

Fig. 2.43 Cash flow of 5 t/day ammonia production system **a** with and **b** without electrical infrastructure costs

infrastructure required for storage of peak productions or use in hydrogen production. The system showed a similar hydrogen production profile, and a daily amount of 440 kg of hydrogen was produced within the system.

Case 4: Ammonia Plant with 5 t/day Capacity

As shown in Fig. 2.42, the capital cost required for the electrical infrastructure is US$26,000,000. The annual maintenance cost is US$450,000.

Figure 2.43 shows the cash flow analysis in a facility that produces 5 t/day. Accordingly, when compared to all other cases, the capacity with the highest feasibility was determined like this. Accordingly, if the electricity infrastructure is not included, the system will be profitable in the first 10 years for the price of US$1000–1500/t ammonia. In addition, when the electrical infrastructure is included, it is seen that the profitability of US$1000–1500/t increases, unlike in the other cases.

2.8 Ammonia in Microgrid Applications

Smart microgrids are flexible energy distribution networks with a geographically compact design. These systems can work in integration with the main power grid, as well as independently from the grid. Each microgrid has a self-sufficient energy management system and can adapt according to the energy needs of the working region. The development of smart microgrids depends on parameters such as the climate of the region they will serve, geographical structure, available energy resources, and energy needs. For this reason, it is important to evaluate and research the case studies close to the real locations for an efficient microgrid development process [48]. Microgrids can be categorized by their capacities as presented in Table 2.20. Also, microgrids can be classified based on installation types such as military grade, campus, community, island, and remote microgrids [49] (see Table 2.20).

2.8.1 Components of a Microgrid

The microgrid design, which operates separately from the electricity grid, consists of an electrical generator, a site controller that controls the produced electricity, and batteries where the electricity is stored. Figure 2.44 depicts a microgrid system. In a microgrid, the distribution of electricity to the local grid is done through a site controller. Microgrids can be independent of the grid or, if necessary, they can be connected to the grid. Excess generation outside peak periods can be stored in batteries to prevent energy shortages during peak periods. This section gives a brief description of the components of a microgrid system.

Electrical Energy Production Systems

Electricity generation systems in microgrids can consist of renewable resources or conventional methods. Among the renewable resources, PV, small hydroelectric, and wind energy are the most frequently used methods. However, conventional methods such as diesel generators are also used in microgrids. In this study, diesel Genset is considered as an energy production system.

Table 2.20 Microgrid types by their capacities

Capacity (MW)	Type
< 2	Simple microgrid
2–5	Corporate microgrid
5–20	Feeder microgrid
> 20	Substation microgrid

Source Ray and Biswal [49]

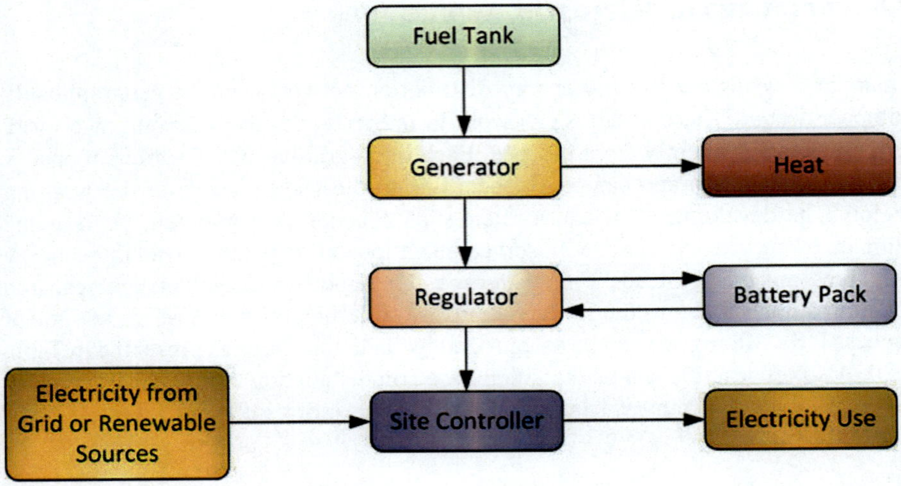

Fig. 2.44 Design of a potential microgrid

Energy Storage Systems

Electricity demand must always be met by the supply. The reliability and efficiency of a microgrid depend on successfully balancing the supply and demand. Therefore, a reliable and efficient energy storage system must be adopted to achieve an adequate power delivery system. Energy can be stored by using mechanical, chemical, electrical, and heat systems.

Table 2.21 shows the properties of various battery-based energy storage systems. Energy storage systems are important for balancing the electricity generation activity

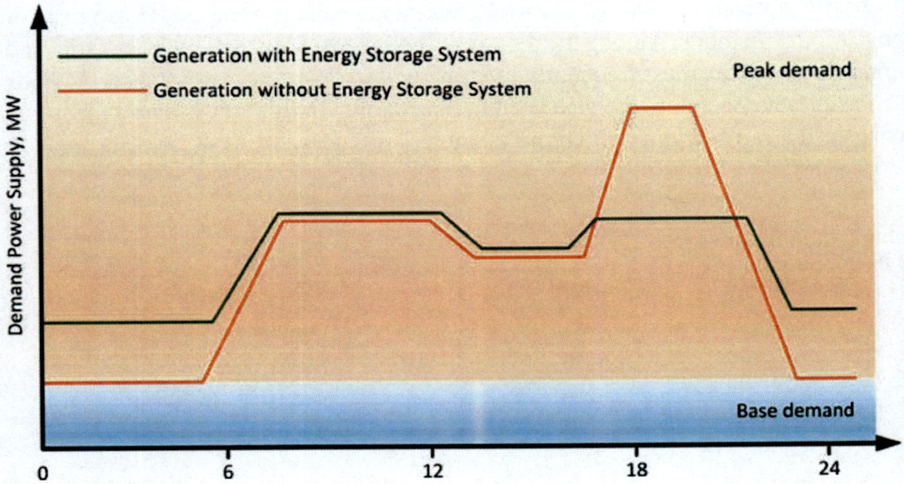

Fig. 2.45 Energy load demand profile with and without energy storage system (modified from [50])

Table 2.21 Properties of battery-based energy storage systems

Battery type	Lead-acid	Li-ion	Ni-Cd	Sodium-sulphur	Vanadium redox (VRB)
Capital cost (US$/kWh)	250–400	600–3900	800–2500	300–600	160–1100
Cycle efficiency (%)	80–90	80–97	60–85	75–85	65–85
Energy density (Wh/L)	60–90	250–500	60–180	150–350	15–40
Self-discharge (%/day)	0.1–0.2	0.1–0.3	0.2–0.7	Low	Low
Discharge time	< 4 h	< 1 h	< 5 h	< 6 h	< 8 h
Lifetime (years)	< 20	< 15	15–20	< 15	< 10
Advantages	Capital cost is low Mature technology	Quick response time Life cycle efficiency is high Energy and power density is high	Mature technology Strong and reliable	Self-discharge is low Rated capacity is high Energy density is high Highly recyclable	Faster response Efficiency is high
Disadvantages	Required high maintenance Cycling time is low Low power and energy density The performance is poor at lower temperatures	Cost is high Life cycle depends on the level of discharge Depth of discharge (DoD) has an impact on life	Impact on the environment High cost Energy density is low	Operating cost is high Temperature control devices are needed	Lower energy and power densities Construction is complex

Source Sharmeela et al. [50]

carried out in the microgrid. Lead-acid batteries are one of the most widely used battery types today. Since it is a mature technology, investment costs are very low. However, compared to other battery types, maintenance costs are high. They also do not perform well at low temperatures. Ni-Cd batteries, like lead-acid batteries, are a mature technology but their environmental impact is quite high. Nowadays, Li-ion technologies have started to be used frequently with developing technological devices. In Li-ion batteries, the frequency of charging and discharging seriously affects the battery life. Figure 2.45 shows the energy production profile when the energy storage system is used and not used. In this way, the load of the energy production system at peak hours, which occurs especially after work hours, can be reduced.

Communication Systems and Internet of Things in Microgrids

One has to note that internet of things (IoT) is a network that allows different equipment working under a specific system to communicate with each other [51]. IoT is vital for the development and optimization of microgrids. For efficient and quality energy management, some conditions must be met [50]:

- Control mechanisms should be able to operate at low bandwidth.
- Electricity needs should accurately be estimated.
- Energy production should be performed accurately according to the need.
- Energy storage methods should be used efficiently.

- It should be able to give the necessary reactions in cases such as network prices and accessibility.

Microgrid Compatible Car Design

A vehicle design that works in harmony with the Microgrid system is shown in Fig. 2.46. Microgrid compatible vehicles can work with a vehicle-to-grid approach. Accordingly, when the vehicle is parked, it can be in communication with the microgrid site controller and will supply electricity to the microgrid network when necessary or appropriate. It can provide benefits to the user, especially by discharging when the electricity price is high and charging it when it is low. On the other hand, the charging/discharging sequence might also decrease the lifetime of batteries.

Surplus Electricity Utilization in Microgrids

It is important to establish a good supply/demand relationship in renewable and nuclear energy sources. Consumer needs vary during the day. However, power networks are designed to handle the maximum load. This situation causes excess production, especially when consumption is low. In this case, the excess electricity is sold to external sources or grounded. Therefore, optimization of networks is very important to minimize the losses that caused by surplus electricity production.

Small or medium-sized grid systems are used in cases where electricity transmission with the main grid is not feasible, and the settlement is located far from the main settlements. These systems are designed to be independent or semi-independent from the main grid and are used to provide power to remote settlements or establishments such as mines or factories.

It is necessary to use different energy storage systems to effectively utilize the excess electricity produced in remote grid systems that operate independently of the main grid and use renewable energy. Today, reducing fossil fuel consumption, load shifting, storage, utilizing new loads, or power spill methods are used to manage surplus production [52].

Demiroren and Yilmaz [53] conducted a study on a combined photovoltaic, wind, and battery power system to calculate the excess energy. They reported that 64% of the produced energy is excess energy. Hafez and Bhattacharya [54] also studied a combined photovoltaic, wind, hydro, and bio-diesel generator system to found out that 25% of the produced energy is excess. They also reported that the excess energy is at 51% for the systems only using diesel generators. Daud and Ismail [55] designed an isolated power system utilizing photovoltaic, wind, battery, and diesel generators to reduce excess energy to 33%. Also, location-specific studies that combine photovoltaics, battery, and diesel generators reported varying excess energy levels between 2.3 and 35% [56–59].

There are 260 remote communities identified in Canada, consisting of settlements and commercial establishments. 17 of these communities are campuses established for commercial purposes. Figure 2.47 shows the remote communities in each Canadian province. Figure 2.48 illustrates the primary power source utilized by the remote communities in Canada. Diesel generators are the most used production method in

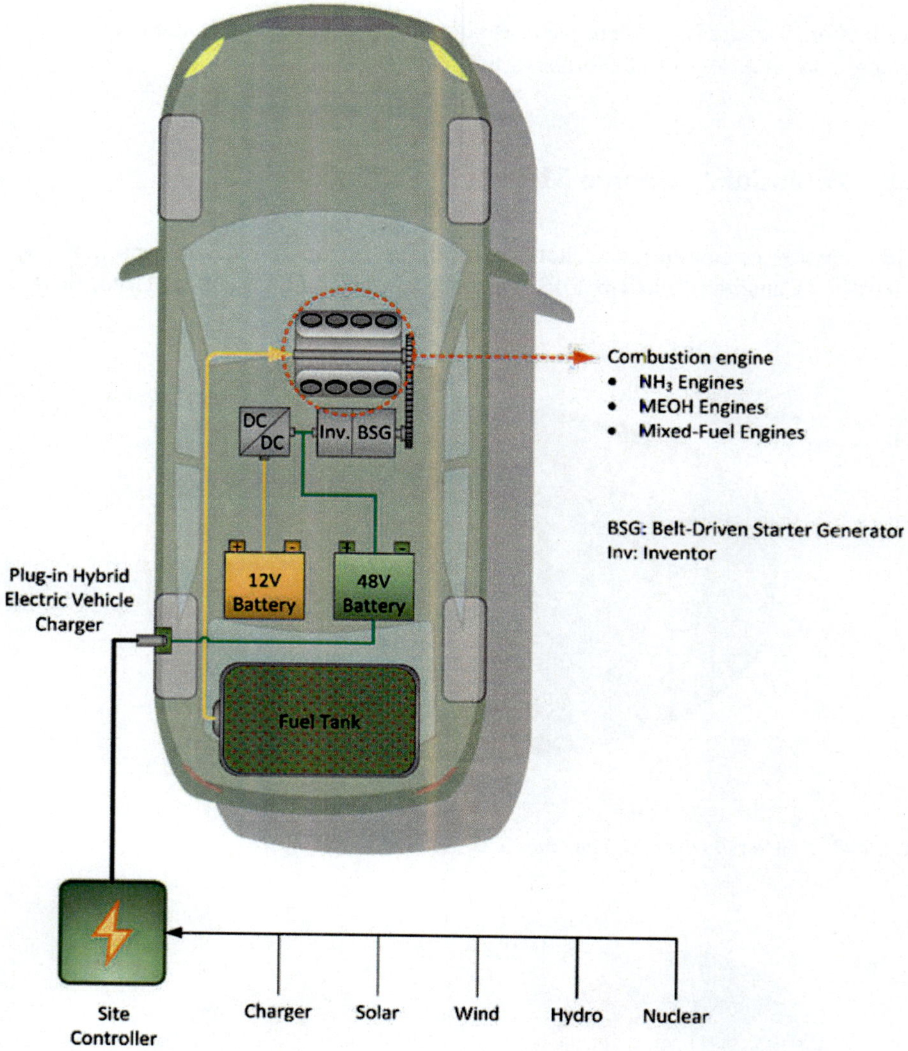

Fig. 2.46 Design of a microgrid compatible hybrid electrical vehicle

remote communities. It is followed by hydro with 14%. However, the number of communities with renewable energy generation capacity has been reported as 16 [60].

Based on the 260 remote communities identified in Canada, Fig. 2.49 shows the map of remote communities with their primary power source. As stated before, many of these communities use diesel generators for energy production. Apart from this, there are also communities fed from the mains. Table 2.22 shows the population and capacity data for these communities for communities with a population 2000 and above. Sandy Lake is identified as a suitable location to investigate local ammonia production. The area considered here is landlocked with an airport and is one of the remote communities with the highest fuel costs.

2.9 Potential Ammonia Market in Canada

The trade flow of ammonia and chemical fertilizers in Canada reached approximately 4.9 trillion Canadian dollars in 2018. The international import and export trade flow is

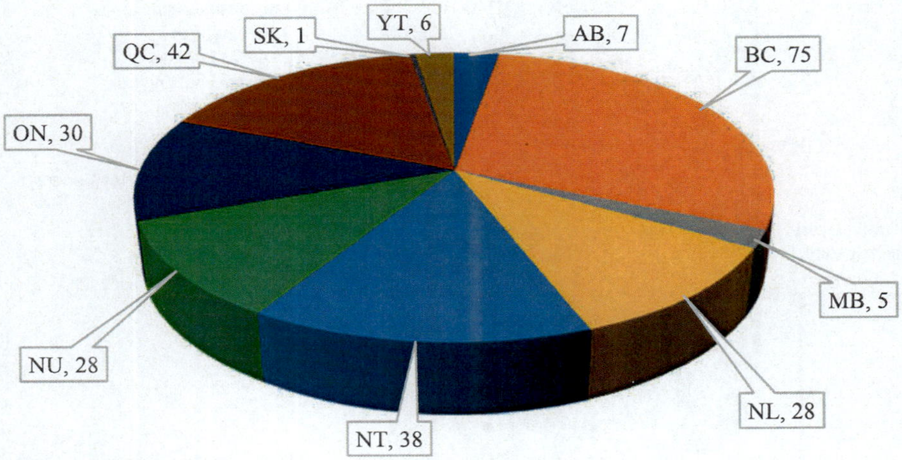

Fig. 2.47 Number of remote communities in Canada (data from [60])

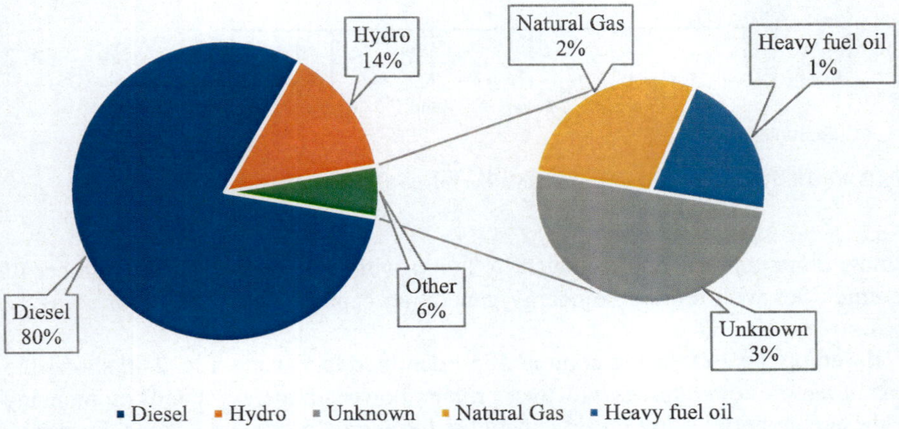

Fig. 2.48 Primary power sources that are utilized by the remote communities (data from [60])

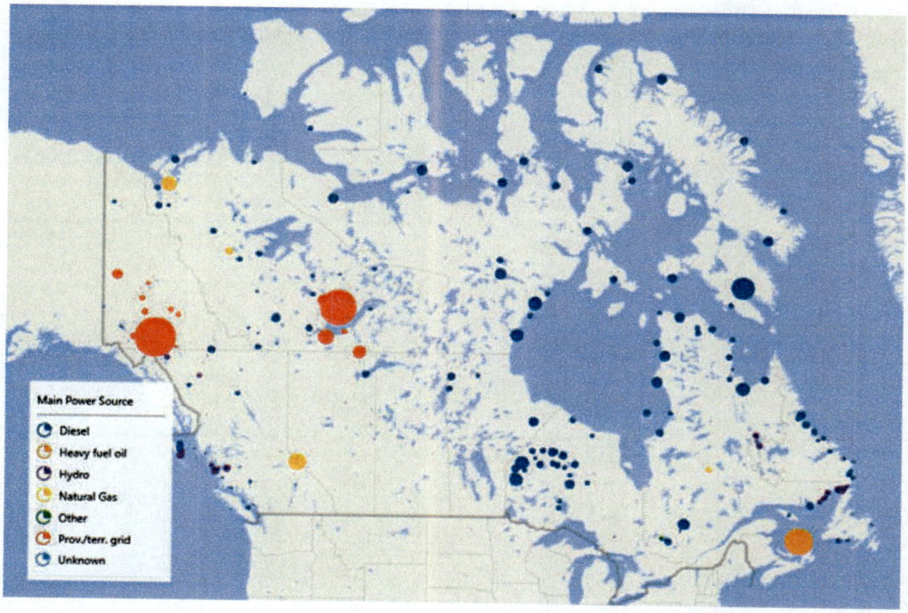

Fig. 2.49 Population distribution and main power type of remote communities of Canada (modified from [60])

at the level of 766 billion Canadian dollars and 677 billion Canadian dollars, respectively [61]. As expressed in Fig. 1.2, approximately 6% of the global ammonia trade is conducted by Canada, and Canada is one of the ten largest ammonia-producing countries.

Canada produces ammonia using conventional methods using fossil fuels. In addition, within the scope of combating climate change and sustainable environment approach, it has focused on the production of blue ammonia with reduced carbon emissions with the help of carbon capture technologies.

Figure 2.50 shows the volume of ammonia and its derivatives produced by manufacturers in Canada. Accordingly, an average of 4.5 million t of ammonia is produced annually. Although ammonia production has continued to be stagnant for the last ten years, it is seen that there is an increase in production, especially in urea. The production of ammonium nitrate and ammonium sulfate is around 1–1.5 million t on an annual basis.

The volume of ammonia exported and purchased by the local agricultural sector is presented in Fig. 2.51. Accordingly, it is seen that there is a serious increase in the trade volume of urea. The trade volume of ammonium nitrate derivatives has increased from 0.75 to 1.5 million tonnes in the last ten years.

Table 2.22 Demographics of the remote communities with 2000 population and above in Canada

Community name/Province		Community type	Population	Main power source	Fossil fuel capacity (kW)	Price of fuel at site (US$/L)	Renewable capacity (kW)
Whitehorse	YT	Settlement	25,085	Main Grid	23,000	0.330	0
Yellowknife	NT	Settlement	19,569	Main Grid	27,660	0.260	113,275
Les Îles-de-la-Madeleine	QC	Settlement	12,475	Heavy fuel oil	67,044	0.700	0
Iqaluit	NU	Settlement	7740	Diesel	25,730	1.190	0
Jasper	AB	Settlement	4590	Natural Gas	17,000	N/A	851,355
Hay River	NT	Settlement	3528	Main Grid	8340	N/A	65,011
Inuvik	NT	Settlement	3243	Natural Gas	7700	1.140	0
Rankin Inlet	NU	Settlement	2842	Diesel	6200	1.070	0
Kuujjuaq	QC	Settlement	2754	Diesel	6010	1.400	0
Pikangikum	ON	Settlement	2720	Diesel	4575	N/A	0
Arviat	NU	Settlement	2657	Diesel	3110	1.070	0
Sandy Lake	ON	Settlement	2641	Diesel	3050	1.512	0
Fort Smith	NT	Settlement	2542	Main Grid	6000	N/A	0
Baker Lake	NU	Settlement	2069	Diesel	3420	1.160	0
Obedjiwan	QC	Settlement	2019	Diesel	4900	0.900	0

Source Government of Canada [60]

2.10 Modeling Studies

In this section, ammonia production steps are modeled in the Aspen Plus software package. This section starts with the modeling of electrolytic hydrogen production. Then, the conventional hydrogen production processes are modeled. Next, the nitrogen production methods are studied. Finally, a Haber-Bosch process modeling is performed using the Aspen Plus.

2.10.1 Electrolytic Hydrogen Production Modeling in the Aspen Plus

A PEM electrolysis system was modeled in the Aspen Plus and thermodynamic analysis was performed. There is no PEM electrolysis among the ready-made models

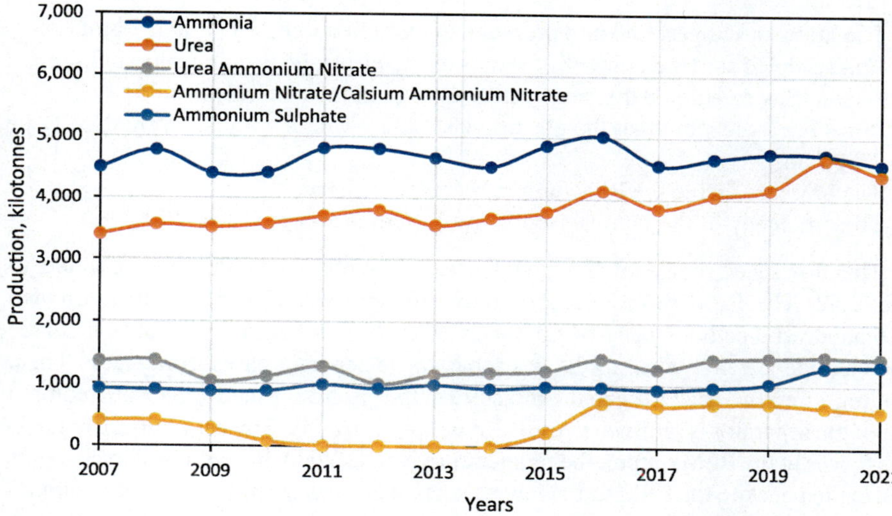

Fig. 2.50 Canadian fertilizer production by product type and years (data from [62])

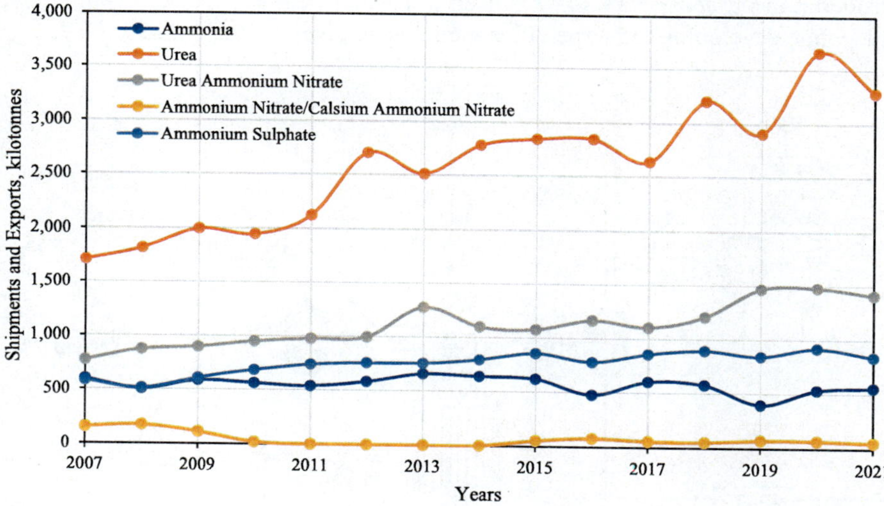

Fig. 2.51 Canadian ammonia shipments to agriculture and export markets by product type and years (data from [63])

of the Aspen Plus. For this reason, using the modeling equations presented before, a model that can work integrated with the Aspen Plus has been programmed using the Aspen Custom Modeler. The assumptions made while programming the model are same as given in the previous sections. However, some assumptions were made while modeling in the Aspen Plus:

- The ambient temperature and pressure are assumed to be 25 °C and 1 atm.
- It is accepted that the number of stacks of the electrolysis unit can be infinite.
- Aspen Plus calculated the temperature change in the electrolysis unit.
- It has been accepted that the electrolysis system works with a current density of 2.4 A/cm^2.
- The pressure inside the PEM unit is accepted as 7 bar.
- The efficiency of the pumps used in the system is accepted as 0.75.

The flowsheet image of the system created with Aspen Plus is presented in Fig. 2.52. The PEM reactor is constantly fed with water. The current is divided into anode and cathode cells within the PEM unit. Flows coming out of both anode and cathode are first taken to the gas separator to separate gases from water. Then, dehumidifiers are used to keep the moisture in the gas flow. Finally, the water coming out of the separator is returned to the PEM unit with a pump. Since the current temperature rises at the PEM output, the temperature is reduced to 30° with heat exchangers and cycled back to the PEM unit. However, a make-up water line is needed to supplement the water consumed during hydrogen production. Therefore, the NRTL method defined in Aspen Plus was used as the thermodynamic method. The NRTL method includes the activity coefficients for the liquid phase. However, the vapor phase is considered ideal, and Henry's law is used in supercritical situations. Assumptions made while developing the Aspen Plus model were given in Table 2.23.

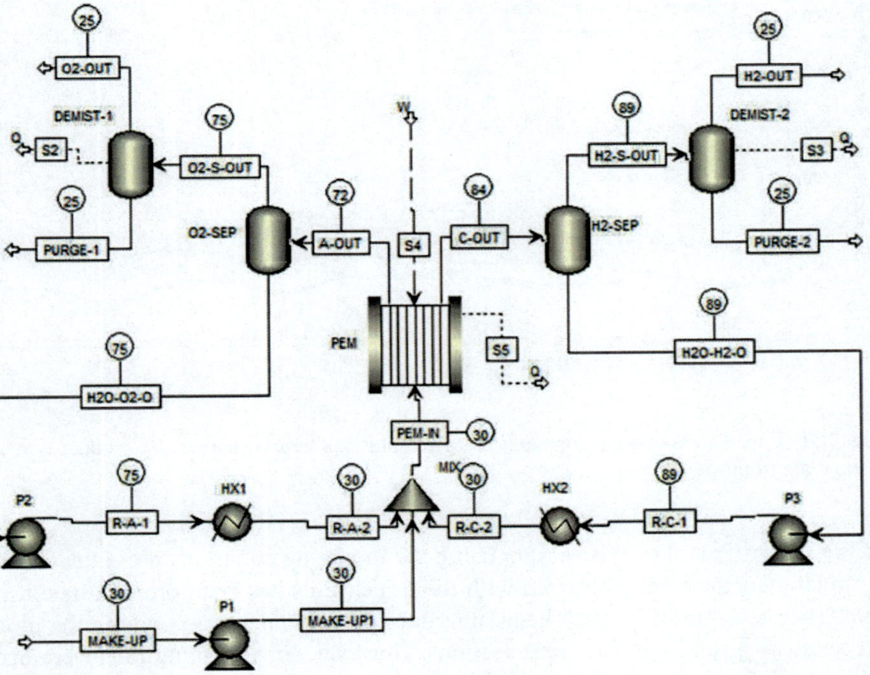

Fig. 2.52 Aspen plus flowsheet of the PEM electrolysis plant

Table 2.23 Assumptions for the Aspen Plus model

Component	Parameter	Value
PEM	Operating pressure	10 bar
	Current density	2.4 A/cm^2
	Active area	200 cm^2
Hydrogen separator	Operating pressure	9.7 bar
Oxygen separator	Operating pressure	9.7 bar
Dehumidifier-1	Operating temperature	25 °C
	Operating pressure	9.7 bar
Dehumidifier-2	Operating temperature	25 °C
	Operating pressure	9.7 bar
Pumps	Operating pressure	10 bar
	Efficiency	0.75

Analyzes were made for PEM plants with a daily production capacity of 90, 180, 440, and 890 kg. Figure 2.53 shows the relationship between current density and voltage depending on temperature. The temperature inside the reactor rises depending on the applied power. In the experiments, it was determined that the temperature went up to 80°. For this reason, it is aimed to reduce the inlet temperature by placing a heat exchanger at the reactor inlet. The increase in temperature decreases the current and voltage values. However, its effect on hydrogen production is not seen in the model. However, the decrease in current and voltage means a decrease in energy consumption. For this reason, thermal management is required, especially in high-capacity systems.

Figure 2.54 shows the hydrogen production rates with varying current density and stack numbers. As expected, the hydrogen production rate increases with increasing current density and stack number. The relationship between stack number and hydrogen yield is presented in Fig. 2.55. With the increase in the number of stacks, the energy efficiency has decreased slightly. For example, when the number of stacks is 100, the efficiency is approximately 20%, while it has decreased to 16% with the stack increase.

The formation of entropy in the system destroys exergy. Figure 2.56 shows the exergy destruction rates of the system components. The highest exergy destruction rate in the system is in the PEM cell. The exergy destruction rate was calculated as 182 kW in the system with 90 kg/day for PEM. However, it has increased to 1065 kW for the PEM electrolyzer with 890 kg/day daily capacity. Since the pumps used in the system are not pressurizing the water, they consume less energy. Also, the make-up water pump meets a flow load of less than about 1% of the main water flow. As the system capacity increased, the exergy destruction rates of peripheral equipment also increased.

The power requirements of the PEM stack are then calculated as 236, 467, 1140, and 2300 kW for the 90, 180, 440, and 890 kg/day hydrogen production capacities, respectively. In addition to these data obtained from Aspen Plus, the energy

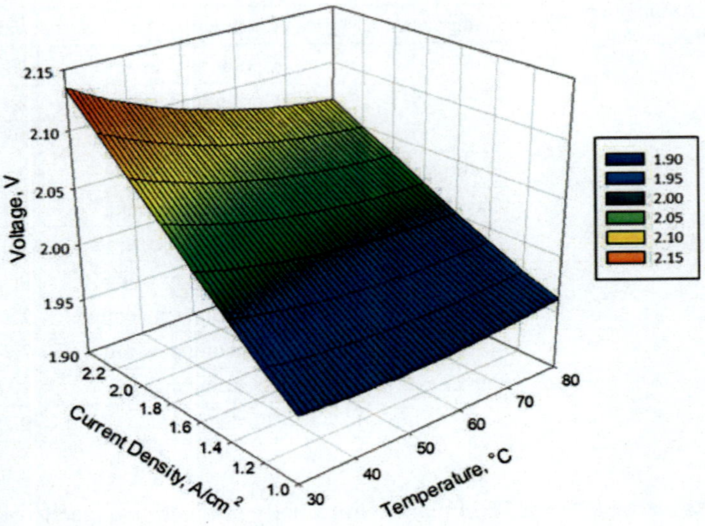

Fig. 2.53 The effect of temperature on current density and voltage

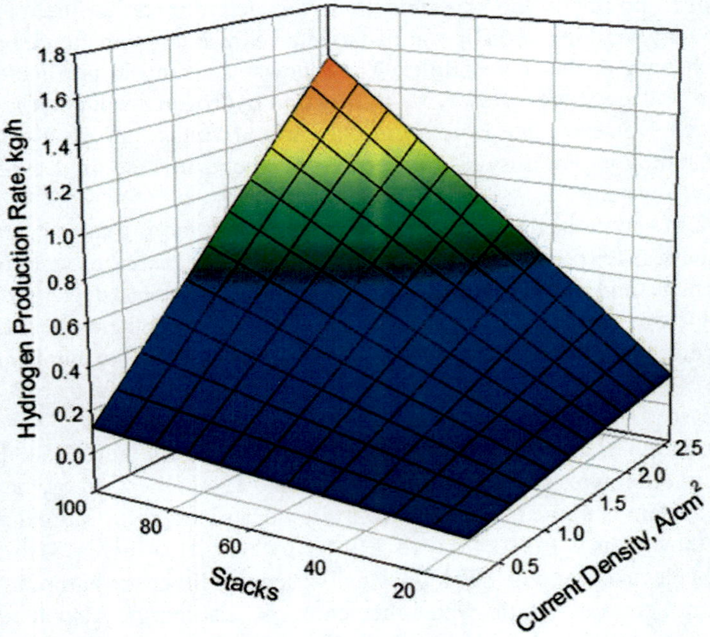

Fig. 2.54 The effect of stack number and current density on hydrogen production

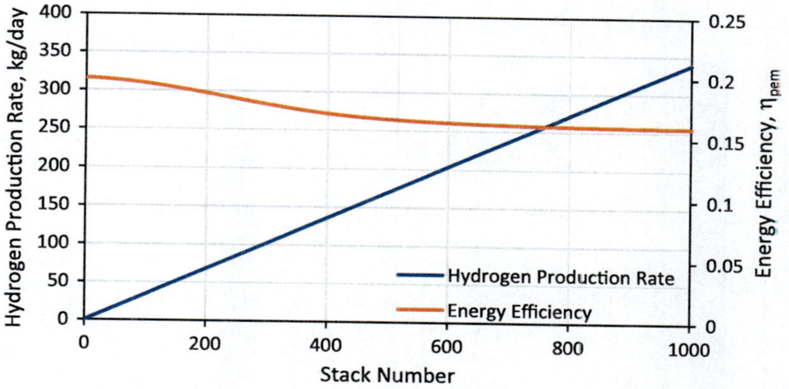

Fig. 2.55 Hydrogen production rate and energy efficiency of PEM with varying stack numbers

requirement of the balance of the plant equipment is not included in this model. The energy required for the balance of plant equipment corresponds to 10% of the electricity consumed by the electrolysis stack. When this consumption is considered, the energy consumptions of the entire system are calculated as 260, 514, 1250, and 2525 kW. Therefore, the most significant electricity consumption rate in the system belongs to the PEM unit. The energy intensity of electrolyzer units is high compared to conventional hydrogen production methods. On the other hand, the utilization of electrolysis units is necessary to curb the processes' carbon footprint. Therefore, optimizing the system and improving the membrane and electrode properties is essential to achieving economic results for electrolysis units.

Figures 2.57 and 2.58 show the energy and exergy efficiencies of the electrolysis stack in specific and the whole system in general, respectively. The current density was chosen as 2.4 A/cm^2. High current also causes the efficiency of the system to decrease. However, decreasing the current increases the amount of stack as it decreases the amount of hydrogen that the system can produce. For this reason, optimization is essential when designing electrolysis systems. In the case where the hydrogen production is 90 kg/day, the energy and exergy efficiencies of the system were found to be lower than in other cases. It is being investigated whether the model causes this situation or not. However, the exergy efficiency has increased compared to the PEM stack in the complete system analysis. This is due to heat recovery by integrating heat exchangers into the system. However, the system generally works at low temperatures compared to the Haber–Bosch production process.

The energy and exergy efficiencies of the PEM system with a capacity of 180 kg/day and higher were found in the ranges of 20–22% and 5–10%, respectively. It is known that clean hydrogen production methods have lower yields than conventional methods. The parameters used in the model were selected based on the materials used in industrial applications. When the losses in the PEM stack are ignored, the potential required for hydrogen formation is around 1.23 V within thermodynamic limits. However, voltage drops occur within the system due to electrode activation, membrane resistance, and diffusion. To provide hydrogen production, it

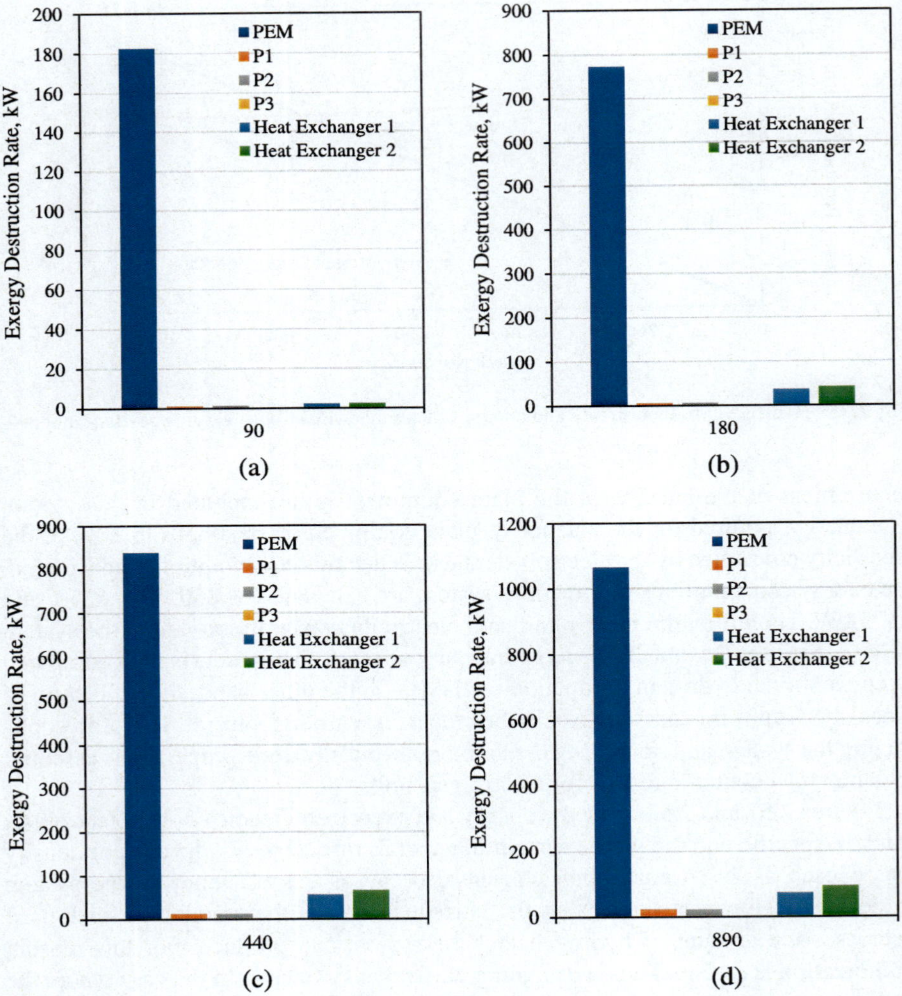

Fig. 2.56 Exergy destruction rates of the components for **a** 90 kg/day, **b** 180 kg/day, **c** 440 kg/day, and **d** 890 kg/day production capacity

is necessary to apply a higher potential to the system. This situation increases the power consumption of the system and decreases efficiency.

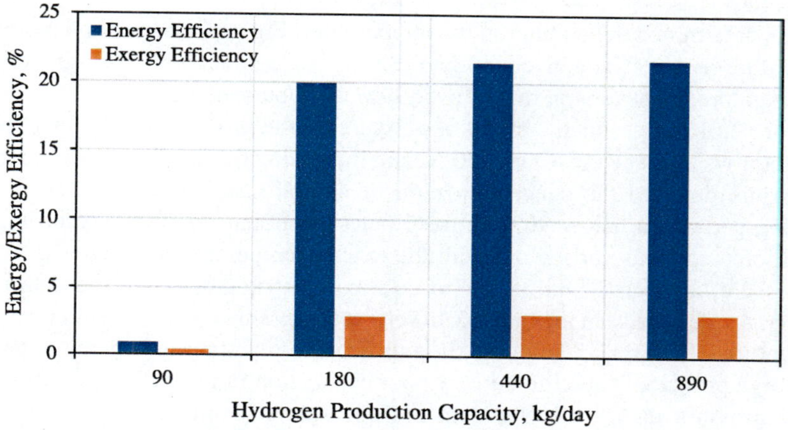

Fig. 2.57 Energy and exergy efficiencies of PEM stack

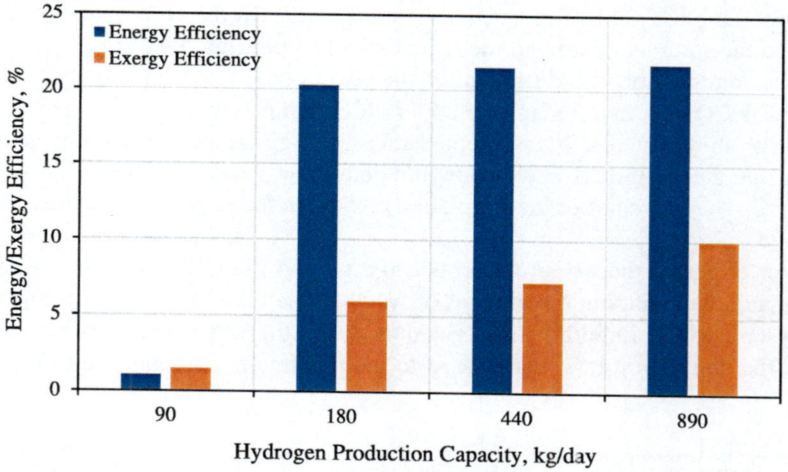

Fig. 2.58 Energy and exergy efficiencies of the system

2.10.2 *Conventional Hydrogen Production Modeling in the Aspen Plus*

Modeling of the steam methane reforming process in Aspen Plus was carried out using RGibbs reactor with Peng–Robenson property methods. Since the steam methane reforming reactions are equilibrium reactions, they can be modeled using RGibbs reactor. A flowchart of the steam methane reforming process in Aspen Plus is given in Fig. 2.59. To test the RGibbs block in Aspen Plus, the initial natural gas input was selected as 1 kmol/h, and the pressure of the reactor was set to 1 bar. The reactor temperature was changed between 500 and 900 °C and the transformation rate was determined depending on the temperature. Figure 2.60 shows the effect of temperature on the CH_4 conversion rate. Similar to the data which are given in the open literature, a conversion rate of over 99% was obtained after 750 °C.

Figure 2.61 represents the effect of pressure on the process. The temperature of the reactor was kept at a constant 750 °C and increasing the pressure of the chemical equilibrium deterred the conversion ratio of natural gas. The conversion ratio of CH_4 dropped down below 90% above 5 bar operational pressure. Accordingly, a simulation study was carried out with the reactor temperature of 750–850 °C, and the pressure between 1 and 25 bar to do a parametric study, like the data in the industry. As it can be seen in Fig. 2.62, the effect of pressure decreases with increasing temperature. However, in any case, it is difficult to maintain the chemical balance under high pressure. Therefore, a very poor conversion ratio of CH_4 was obtained.

The pressure data are lower than the operational conditions reported by the industry. The main reason for this is due to the RGibbs block used in the Aspen Plus. To understand the effect of kinetic values in equilibrium reactions, simulation can also be done with the REquil block. RGibbs block solves the reaction equilibrium by minimizing the Gibbs free energy of the species. On the other hand, REquil is based on the solution of stoichiometric chemical equilibrium kinetics.

In the continuation of the simulation, the effect of the water inlet on reducing the amount of CO was investigated by adding high-temperature and low-temperature water–gas shift reactors. It is accepted that the high-temperature water gas shift reactor operates at 350 °C, and the low-temperature water–gas shift reactor operates at 250 °C. The flowchart of the simulation created in the Aspen Plus is presented in Fig. 2.63.

Water effect on the output CO concentration and the CH_4 conversion rate was investigated by changing the amount of water in the inlet and shown in Fig. 2.64. In a system with 1 kmol/h natural gas flow, when the water inlet is 5 kmol/h and above, the conversion rate is increased to the maximum level and the CO amount

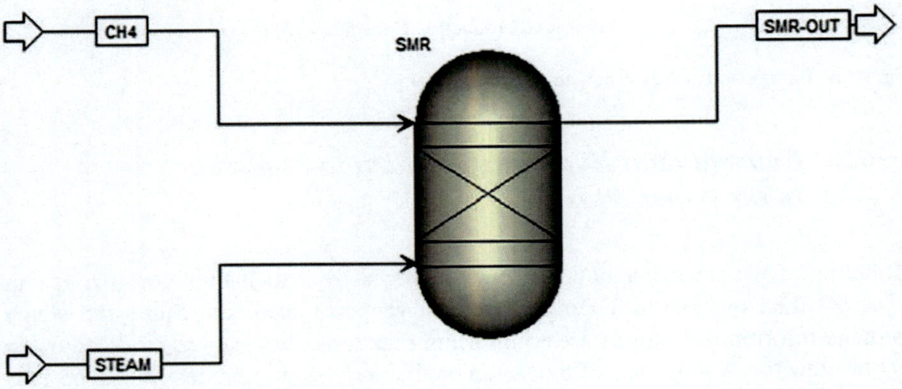

Fig. 2.59 A simple steam methane reformer model in Aspen Plus

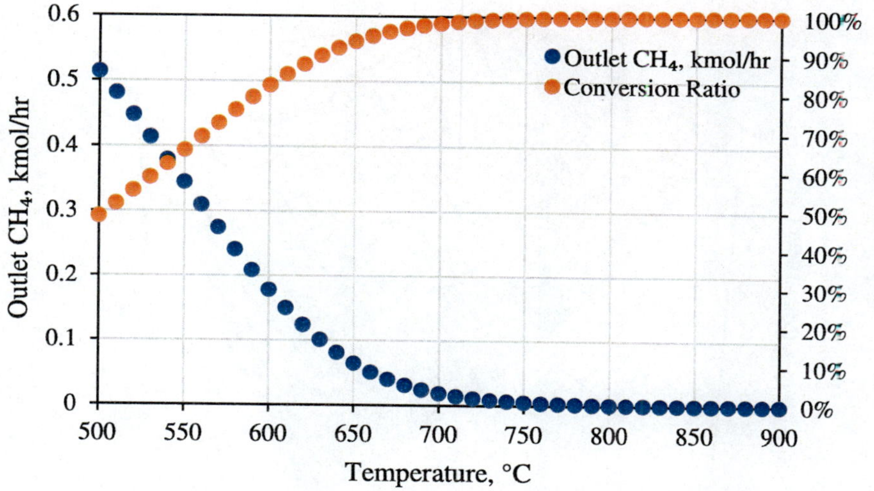

Fig. 2.60 Change of the natural gas flow rate at the steam methane reformer outlet and conversion ratio with temperature

to the minimum. However, the CO amount at the outlet is still above the industry standards. In addition, the water amount at the outlet is high.

Simulation results show that if 1 kmol of natural gas is provided per hour, the daily hydrogen production will be 193.3 kg. Accordingly, the amount of water consumed in the system is 864 kg/day. Also, the CO and CO_2 flow rates at the outlet of the water–gas shift reactor were determined as 54.35 and 970 kg/day, respectively.

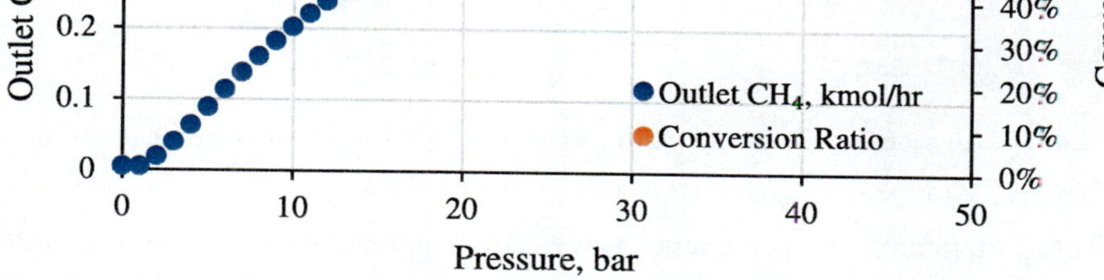

Fig. 2.61 Change of the natural gas flow rate at the steam methane reformer outlet and conversion ratio with pressure

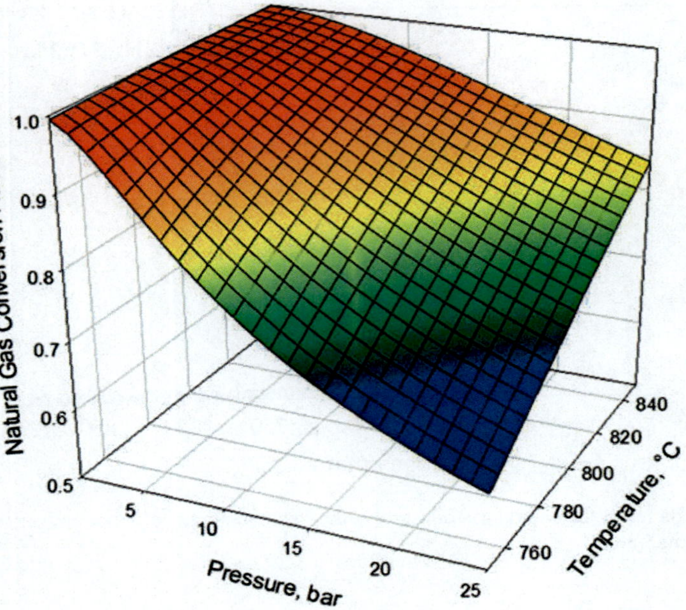

Fig. 2.62 A simple steam methane reformer model in the Aspen Plus

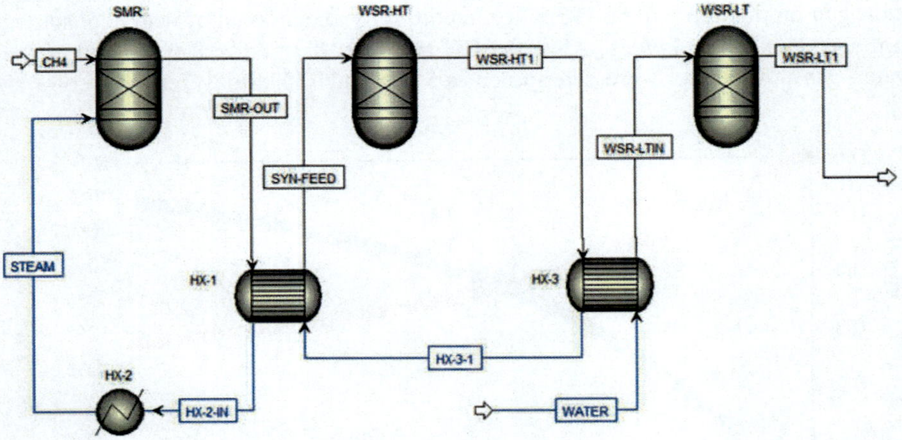

Fig. 2.63 Steam methane reformer flowchart with high and low temperature water shift reactors

In the literature, the temperature values for high-temperature water gas shift reactor (WSR-HT) and low-temperature water gas shift reactor (WSR-LT) are given as 350–500 °C, and 200–250 °C, respectively. A parametric study was carried out in Aspen Plus to coincide with these temperature ranges. The result of the parametric

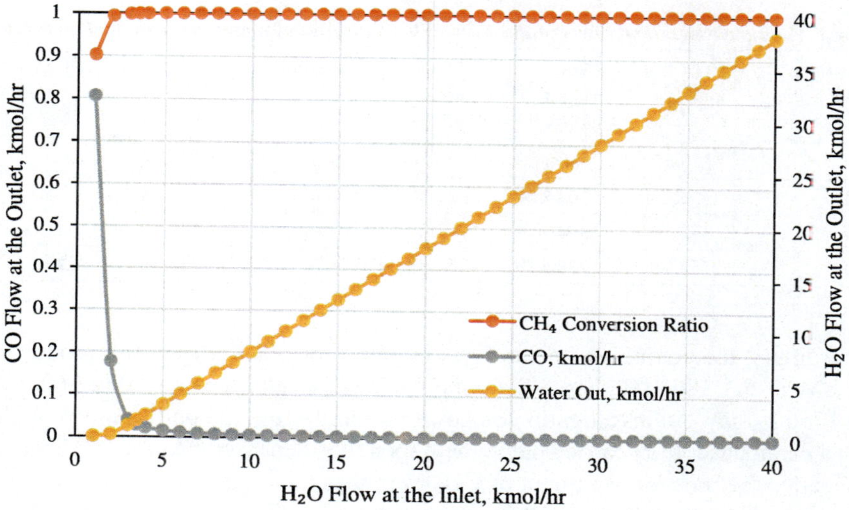

Fig. 2.64 The effect of water flow rate on natural gas conversion rate and CO concentration

study is given in Fig. 2.65. As expected, the efficiency of the conversion reaction of CO to CO_2 increases with decreasing temperature since the reaction is exothermic.

The plant feedstock data for the hydrogen production rates determined for the ammonia production plant in the previous sections are presented in Table 2.24.

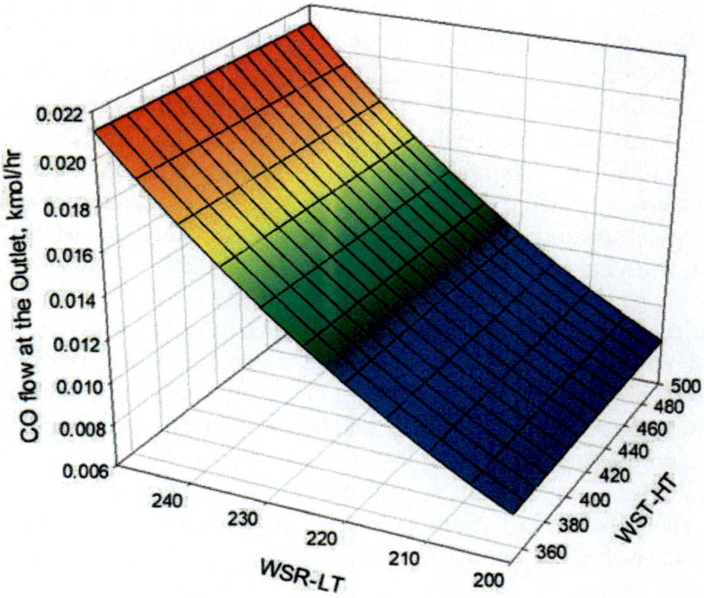

Fig. 2.65 Temperature effect on CO flow rate at the outlet

Table 2.24 Input and output flows of the Aspen Plus simulation for ammonia production cases

Hydrogen yield, kg/day	CH_4 input, kg/day	Water input, kg/day	Water output, kg/day	Consumed water, kg/day	CO_2 output, kg/day	CO output, kg/day
90	179.62	806.82	405.50	401.32	488.16	2.58
180	359.24	1613.64	811.00	802.64	976.33	5.17
440	878.14	3944.44	1982.44	1962.01	2386.58	12.64
880	1756.28	7888.89	3964.88	3924.01	4773.16	25.27

Accordingly, the amounts of natural gas required for a daily hydrogen production capacity of 90, 180, 440, and 880 kg/day have been determined as 180, 360, 878, and 1756 kg/day. An investigation on how to minimize the CO output was provided with a parametric study carried out in the Aspen Plus Sensitivity Analysis. However, to separate the remaining amount of CO_2 from the stream, it is necessary to add an extra separation process and methanation process.

2.10.3 Nitrogen Production Modeling in Aspen Plus

The ammonia production method with the Haber Bosch requires high purity nitrogen. Nitrogen production methods have been discussed in previous sections. In this section, pressure swing adsorption was chosen as the nitrogen production method. The pressure swing adsorption system works with the principle of keeping the oxygen by an adsorbent under high pressure and separating the nitrogen from the stream. The most frequently used adsorbent in industrial applications is carbon molecular sieve (CMS). Figure 2.66 shows the PSA process. The PSA systems generally consist of two adsorption towers. After the air is compressed with the help of a compressor, it is passed through the humidifier and carbon filter, and the moisture and other particles in it are separated from the line. Purified air is given under pressure to one of the adsorption towers and the oxygen is retained by the CMS. Meanwhile, the other tower, whose pressure separation process is completed, is depressurized and the oxygen trapped by the CMS is removed from the system.

The two adsorption towers are operated with cycles of compression, depressurization, and purge processes. Most Industrial CMS are produced from anthracite or hard coal. Raw materials are first oxidized in a controlled manner and then subjected to thermal treatment. The pore structure can be controlled by the adjustments made in these two processes [64].

The flowsheet created in the AspenPlus Adsorption software is presented in Fig. 2.67. The pressure loss in the pipe with the laminar flow can be calculated by the Carman-Kozeny equation [65]:

$$\frac{\partial P}{\partial z} = -1.5 \times 10^{-3} \frac{\mu (1 - \varepsilon_i)^2}{\left(2 r_p \varphi\right)^2 \varepsilon_i^3} v_g \qquad (2.31)$$

where

ϕ: Shape factor
μ: Dynamic viscosity, Pa-s
v_g: Velocity inside the vessel, m/s

The temperature changes in the two adsorption towers according to the results of dynamic modeling with the Aspen Plus Adsorption are presented in Fig. 2.68. In the operation achieved in cycles, the temperature rises above 200 °C during the pressurization process. In the depressurization phase, the temperature drops to 10 °C.

Figures 2.69 and 2.70 show the argon and water percentages at the output of an adsorption tower. While the PSA unit was operating, the argon percentage was calculated as 0.009%. However, the water content varies depending on the time, according to the retention rate of oxygen in the adsorbent. It is possible to minimize undesirable chemicals such as argon and moisture at the process output, but complete removal cannot be achieved. It has been determined that the percentage rates are found to be very low, and considered to be adequate for Haber-Bosch process..

Figure 2.71 shows the molar fraction percentages of nitrogen and oxygen gas produced. The percentage of nitrogen in the air at the entrance was accepted as 78.025%. In the output, the nitrogen percentage was determined as 78% according to the model calculations. Therefore, the separation efficiency was found to be 99.998%. This value is very close to the results in the open literature [16].

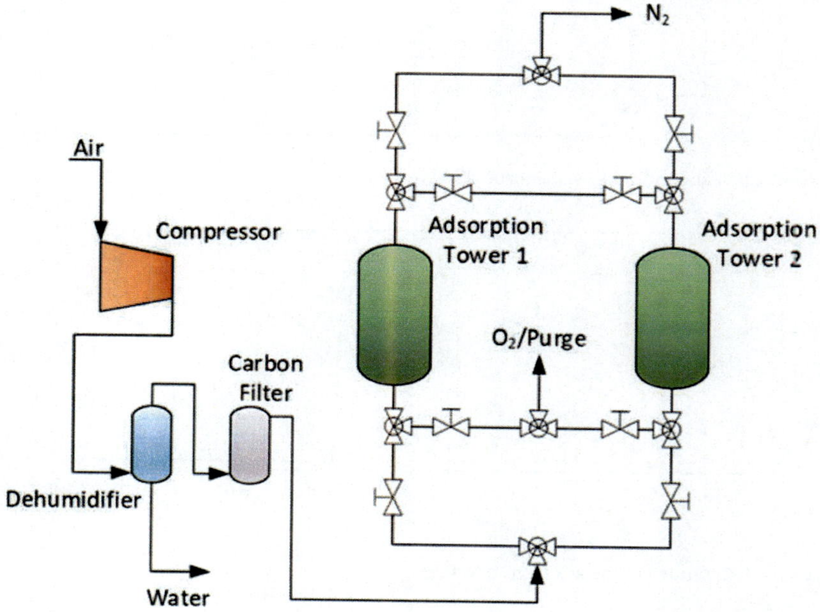

Fig. 2.66 PSA process

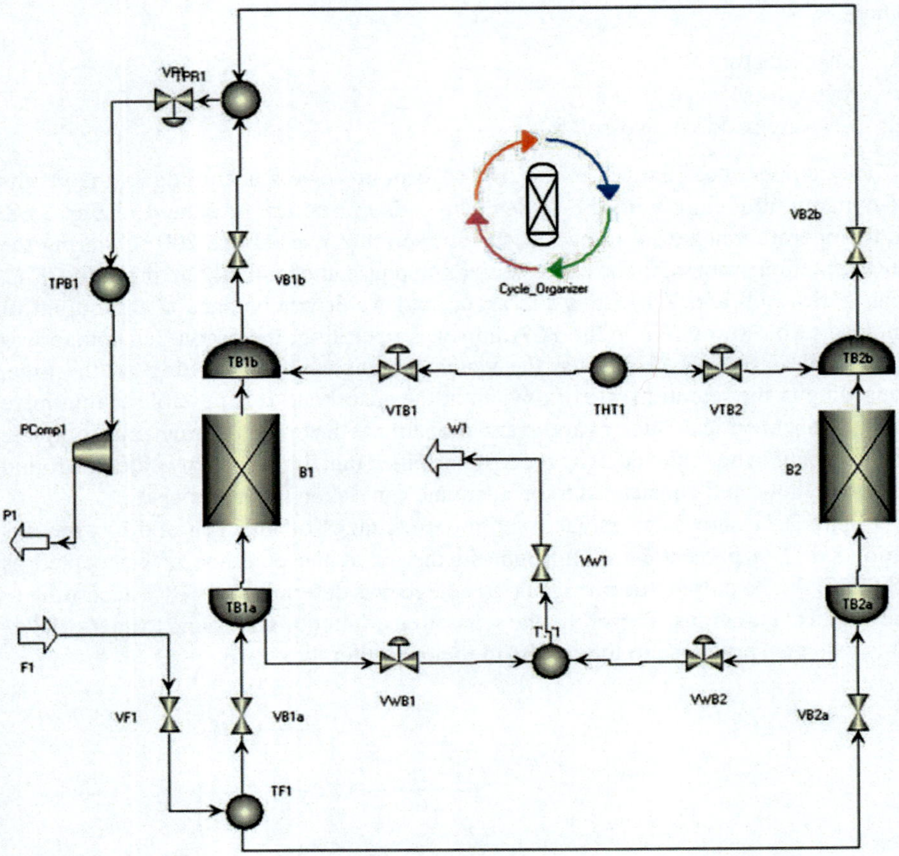

Fig. 2.67 Aspen adsorption flowsheet for PSA

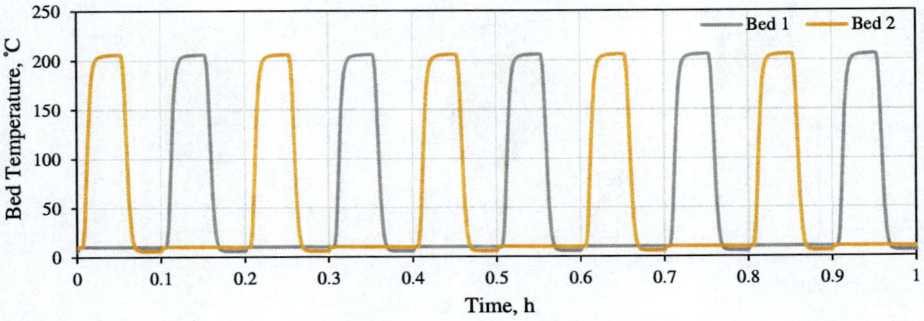

Fig. 2.68 Temperatures at the adsorption towers

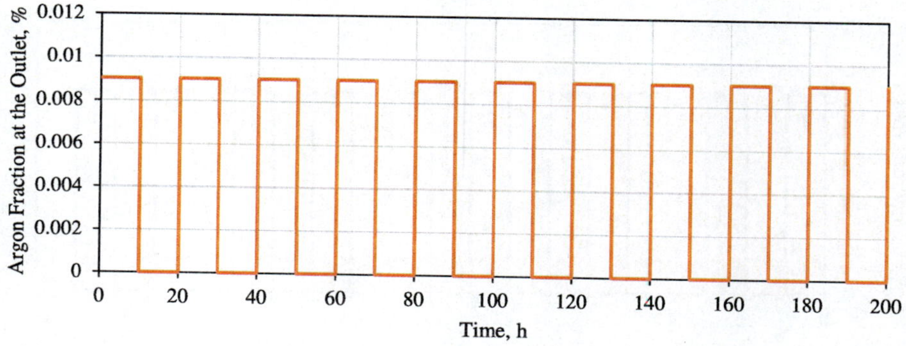

Fig. 2.69 Argon fraction at the outlet

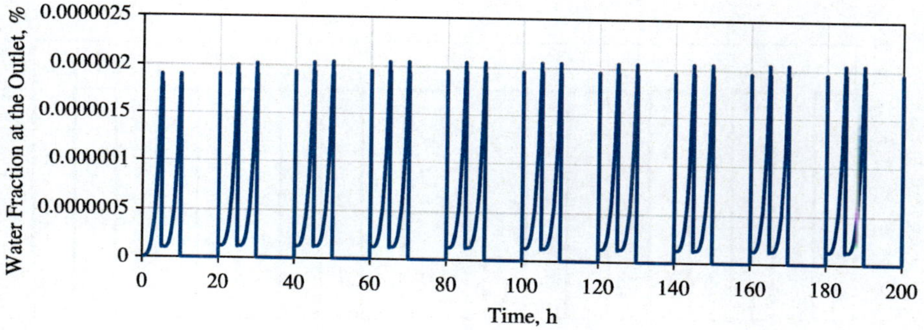

Fig. 2.70 Water fraction at the outlet

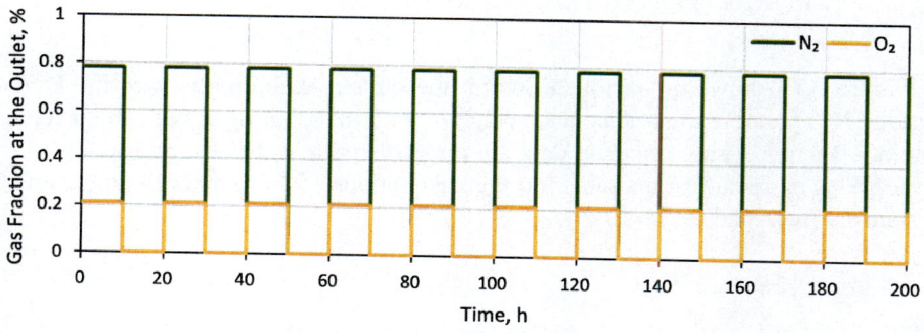

Fig. 2.71 Nitrogen and oxygen gas fractions at the outlet

The molecular flow at the PSA inlet was calculated as 90 mol/s. The purge current was found to be 14.9 mol/s. The change in the PSA input and purge current is presented in Fig. 2.72.

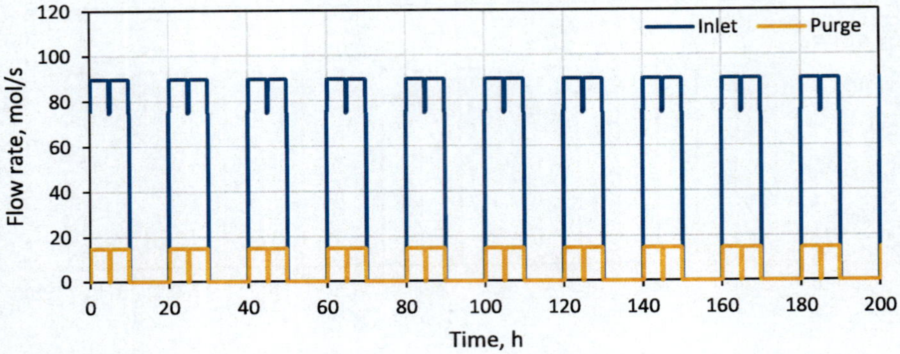

Fig. 2.72 Inlet and outlet gas flowrates of PSA

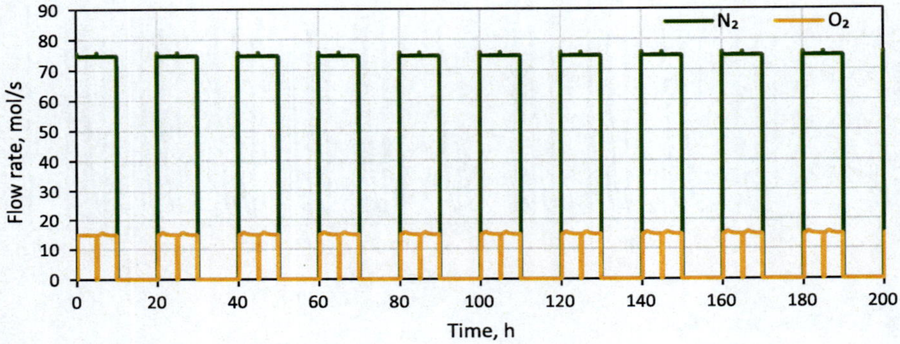

Fig. 2.73 Nitrogen and oxygen gas flow rates at the outlet

Figure 2.73 shows the nitrogen flow at the outlet. Accordingly, with the PSA system, it is possible to produce an average of 74.6 mol/s nitrogen with an air flow rate of 90 mol/s. These obtained data are thus entered into the separator blocks in Aspen Plus as specification data, and the calculations of the ammonia production system are then made accordingly.

2.10.4 *Haber–Bosch Modeling in the Aspen Plus*

To examine the ammonia production with the Haber Bosch process, a simulation was made on Aspen Plus. The flowchart of the Haber Bosch process created in Aspen Plus is presented in Fig. 2.74. The RPlug block in the Aspen Plus was used for the reactor where ammonia is produced. This study was carried out to obtain model results for the Haber-Bosch reactor in AspenPlus. In the following sections, the nitrogen and hydrogen production systems are added and the overall analysis is also performed.

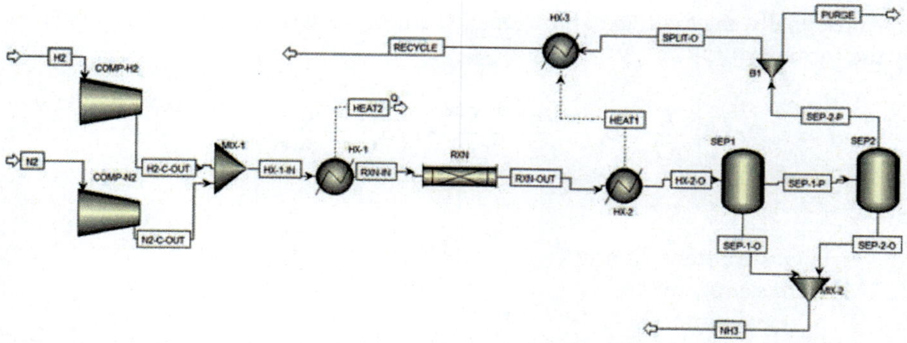

Fig. 2.74 Flow chart of Haber Bosch process simulation in the Aspen Plus

The kinetic expression of the Haber Bosch process is defined by Nielsen [66]:

$$R = A_c \left[\frac{AK\left(a_N k_{eq}^2 - \frac{a_A^2}{a_H^3}\right)}{\left(1 + K_a \frac{a_A}{a_H^W}\right)^2 \alpha} \right] \tag{2.32}$$

where

R: Kinetic expression of Haber Bosch process, kg-mole/m^{3}hr
A_c: Catalyst activity
k_{eq}: Equilibrium constant
AK: Specific rate constant
K_a: Adsorption equilibrium constant
a_N: Activity of nitrogen
a_H: Activity of hydrogen
a_A: Activity of ammonia

$$\text{Log}_{10}\text{k}_{\text{eq}} = -2.691122\,\text{Log}_{10}T - 5.519265 \cdot 10^{-5}T + 1.848863 \cdot 10^{-7}T^2 + 2001.6/T + 2.6899 \tag{2.33}$$

$$AK = 3.945 \cdot 10^{10} \exp(-5622/T) \tag{2.34}$$

$$K_a = 2.94 \cdot 10^{-4} \exp(12104/T) \tag{2.35}$$

where

T: Temperature, K

Additionally, the hydrogen, nitrogen, and ammonia activity (α) can be calculated by the formula:

$$a_i = x_i Pexp\left[\frac{P}{RT}\left(B_i - \frac{A_i}{RT} - \frac{C_i}{T^3} + \left(A_i^{0.5} - \sum X_i A_i^{0.5}\right)^2\right)\right] \quad (2.36)$$

where

R: Gas constant, L-atm/K
P: Pressure, atm
T: Temperature, K
x_i: mole fraction of the species
A, B, C: Constants given in Table 2.25

According to these values, the kinetic data required for the RPlug was calculated and the simulation was run parametrically. Ammonia production values of the reactor for single pass are given in Fig. 2.75. Accordingly, it was observed that ammonia production increased with increasing pressure. However, with the increase in temperature, a decrease is observed in ammonia production, especially after 250 °C. Accordingly, the temperature in the range of 250–300 °C is considered to be at the optimal level.

The reactions that take place with a single pass inside the reactor are presented in Fig. 2.76. The conversion rate of nitrogen and hydrogen to ammonia is very low in the reactor. Because nitrogen bonds are very difficult to break, the efficiency of ammonia production is low even at high temperature and pressure. For this reason, in industrial applications, a separation process is applied to the gas consisting of ammonia, hydrogen and nitrogen. After the ammonia is separated from the feed, the remaining hydrogen and nitrogen are converted and reacted again. Therefore, the system is operated with a recycle loop.

2.10.5 *Electrolytic Ammonia Production Modeling in the Aspen Plus*

Today, hydrogen is needed as a feedstock for ammonia production and is primarily (more than 90%) obtained from fossil fuels. Ammonia production with the Haber Bosch is a field of activity whose industrial development has continued for nearly 100 years. The facilities established to produce affordable ammonia have a daily

Table 2.25 A, B, and C values for various gases (data from [67])

	A_i	B_i	C_i
H_2	0.1975	0.02096	5.04×10^2
N_2	1.3445	0.05046	4.20×10^2
NH_3	2.393	0.03415	4.77×10^2
Ar	1.2907		
CH_4	2.2769		

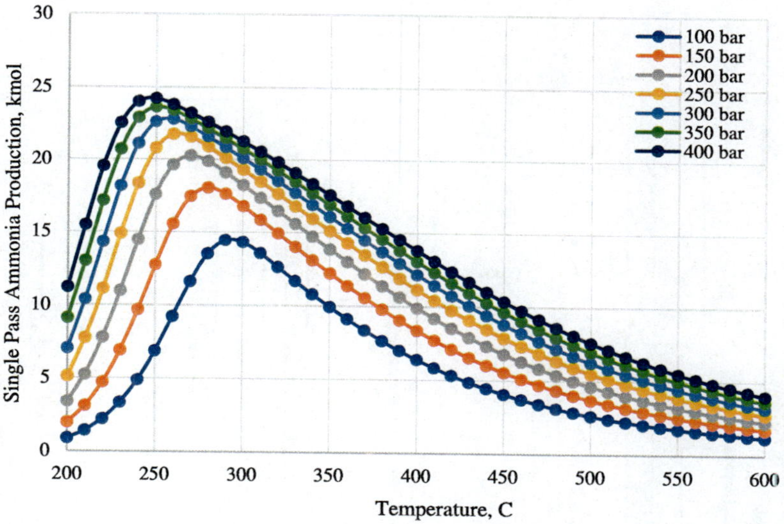

Fig. 2.75 Single pass ammonia production in the reactor for various temperatures and pressures

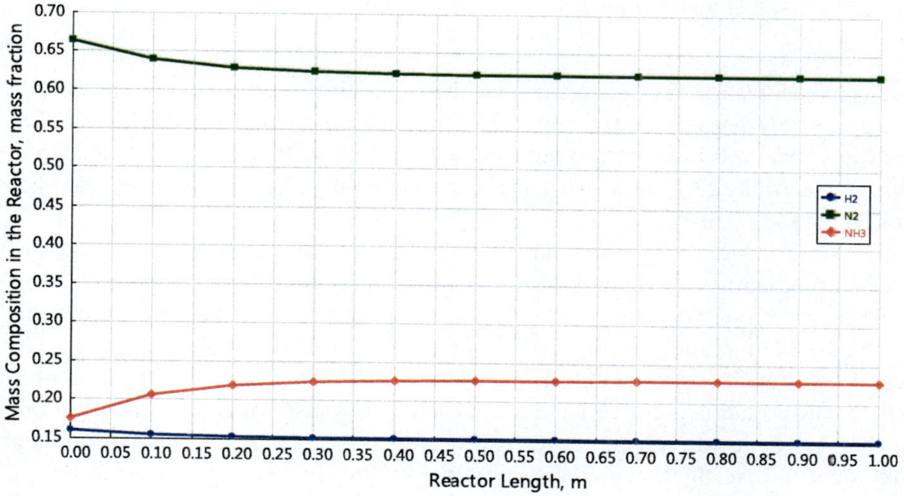

Fig. 2.76 Mass composition across the reactor length

ammonia production capacity of over 3000 t [41]. This also causes high greenhouse gas (GHG) emissions. For renewable and sustainable ammonia production, it is necessary to abandon fossil fuels and shift to cleaner alternatives in hydrogen production. Electrolysis is a commercial technology in active development. Since hydrogen is obtained only from water, it is evaluated as clean technology compared to the conventional hydrogen production methods, such as steam methane reforming

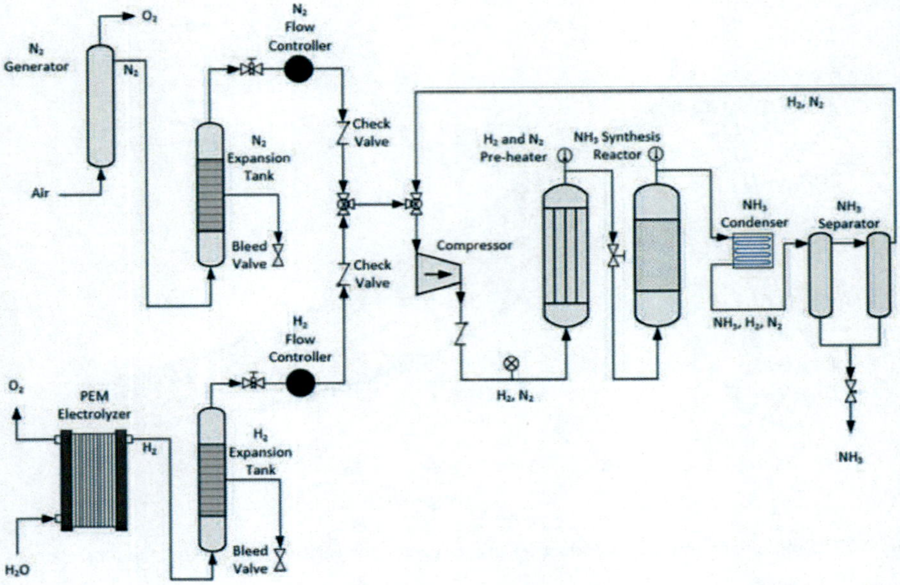

Fig. 2.77 Flow sheet of the electrolytic ammonia production system

and coal gasification [1]. However, since the electricity consumption of electrolysis is high, it increases hydrogen costs and ammonia production costs. Therefore, in the short term, it is necessary to reduce the costs with methods such as utilizing the surplus electricity generated in the grid or to increase the efficiency of the electrolysis unit in the long term.

On-site production of ammonia is becoming more attractive when the costs of transporting ammonia are high. For this reason, it is important to develop systems with low production capacity while evaluating the efficiency of the system carefully. In small-scale facilities, especially the initial investment cost is the key factor. The initial investment cost of ammonia plants to be established in remote communities where transportation is difficult and where there is no transportation in certain periods of the year seriously affects the levelized cost of ammonia. For this reason, it is important to investigate the efficiency of the facility to be established beforehand and to utilize the facility with equipment that will reduce the initial capital costs.

Therefore, the thermodynamic evaluation of the plant in the Aspen Plus is presented for the previously determined capacities. Figure 2.77 shows the flow chart of the system evaluated. Accordingly, hydrogen production is provided by PEM electrolysis, and nitrogen is obtained by pressure swing adsorption. Information on these processes has been presented in detail in previous sections. The hydrogen and nitrogen gases produced in the system are pressurized with the help of compressors in separate lines and fed to the Haber Bosch reactor at stoichiometric ratios. In kinetic modeling studies, the temperature range with the highest single pass ammonia production was found to be 250–300 °C. With the increase in pressure, the temperature required by the reaction decreases. The maximum production in the

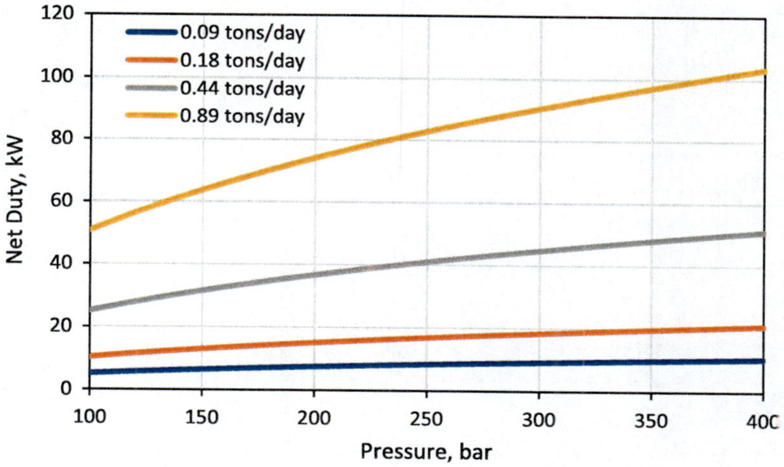

Fig. 2.78 Power consumption of hydrogen compression unit for various production rates

system operating under 100 bar is calculated as 14.4 kmol at 290 °C. However, with increasing the pressure to 400 bar, the maximum production is 24.2 kmol at 250 °C. Figure 2.78 shows the power requirement for different daily production capacities if a compressor is used. Isentropic efficiency of the compressors was assumed as 0.70. The number of compressors and the compressor energy needs increases with the daily amount of hydrogen. With the pressure increasing to 400 bar, the power requirement is calculated as 11, 21, 51, and 103 kW for the production capacity of 0.09, 0.18, 0.44, and 0.89 t/day, respectively. However, efficiency does not increase much at pressures above 200 bar in the Haber Bosch reactor. For this reason, it has been accepted that the pressure of hydrogen will be increased to 250 bar with a compressor. Accordingly, the energy requirement will be 7.4, 16.5, 41, and 82.5 kW, respectively.

Figure 2.79 shows the Aspen Plus flowsheet of the ammonia production system. The ammonia production model in Aspen Plus consists of 4 main components. These are hydrogen production, nitrogen production, pressurization of the produced nitrogen and hydrogen gas and the Haber-Bosch reactor. Information on the modeling of each component has been presented in the previous sections. The hydrogen production system was obtained by modeling the PEM electrolysis in the Aspen Custom Modeler module. The Aspen Adsorption module was used to model nitrogen production by pressure swing adsorption. The pressurization system consists of a combination of gases from the hydrogen and nitrogen production systems and recycled gas from the Haber-Bosch reactor. The model was run for predetermined ammonia production capacities of 0.5 t, 1 t, 2.5 t and 5 t per day. The results were obtained by modeling the system to produce 0.5 t of ammonia per day. Specific enthalpy, entropy and exergy values for each state point are presented in Table 2.26. The temperature of the compression process before the gas mixture reaches up to 770 °C. However,

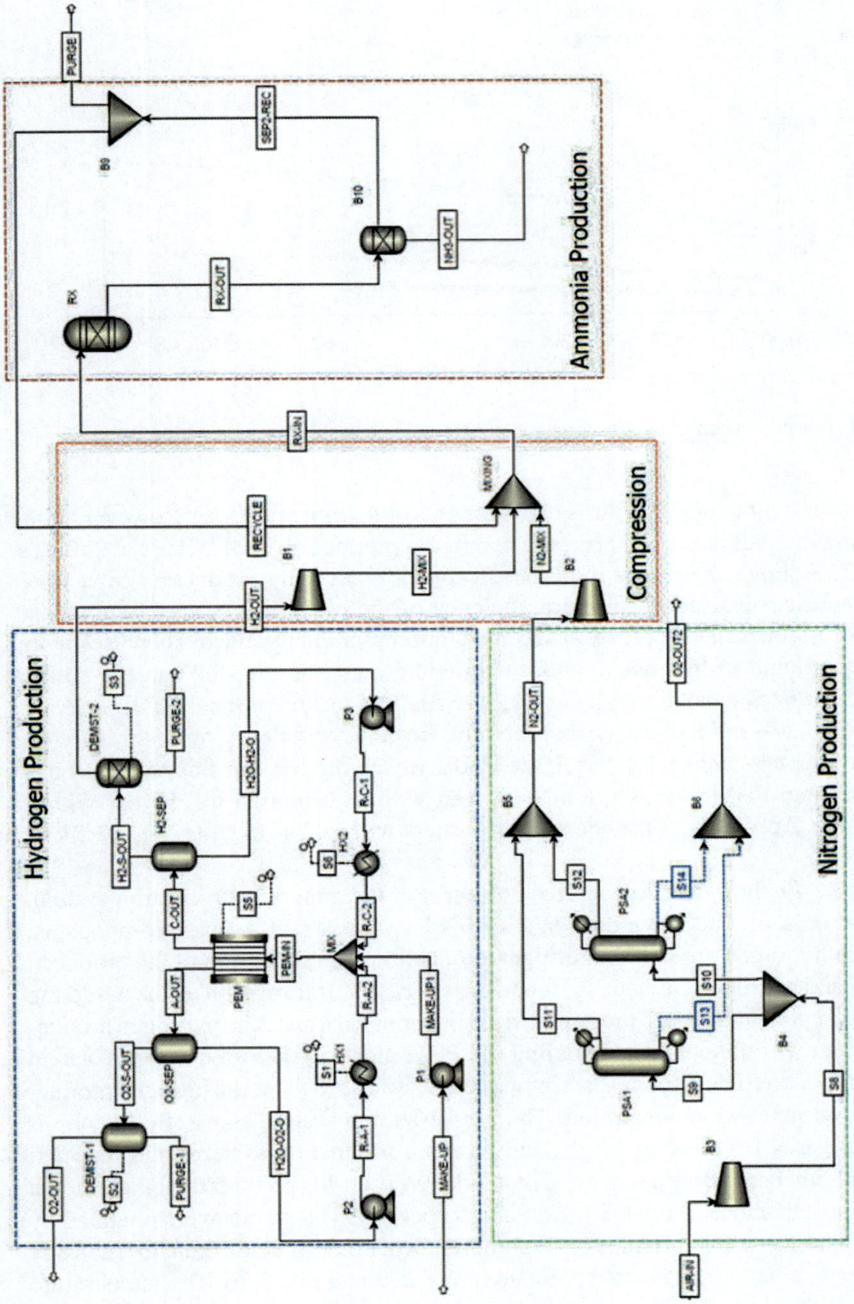

Fig. 2.79 Aspen Plus flowsheet of the ammonia production system

Table 2.26 State point results of the Aspen Plus simulation for 0.5 t/day ammonia production system

Stream name	From	To	Temperature (°C)	Pressure (bar)	Specific enthalpy (kJ/mol)	Specific entropy (kJ/kmol-K)	Mole flow rate (kmol/h)	Specific exergy (kJ/mol)
MAKE-UP		P1	30	1	−285.43	−161.91	1.98	4.61
MAKE-UP1	P1	MIX	30	10	−285.41	−161.84	1.98	5.06
PEM-IN	MIX	PEM	30	10	−285.39	−161.89	53.52	4.96
H_2O-O_2-O	O_2-SEP	P2	84	10	−281.26	−149.43	25.80	423.60
R-A-1	P2	HX1	84	10	−281.26	−149.43	25.80	423.74
H_2O-H_2-O	H_2-SEP	P3	85	10	−281.16	−149.15	25.74	441.83
R-C-1	P3	HX2	85	10	−281.16	−149.15	25.74	441.97
R-C-2	HX2	MIX	30	10	−285.39	−161.89	25.74	5.01
R-A-2	HX1	MIX	30	10	−285.39	−161.89	25.80	4.93
A-OUT	PEM	O_2-SEP	84	10	−271.59	−144.58	26.76	618.46
C-OUT	PEM	H_2-SEP	85	10	−262.45	−139.76	27.66	819.27
O_2-S-OUT	O_2-SEP	DEMIST-1	84	10	−11.95	−14.14	0.96	5773.73
O_2-OUT2	B6		383	10	10.89	6.27	0.17	9460.95
O_2-OUT	DEMIST-1		25	10	−0.79	−18.74	0.91	5599.78
PURGE-1	DEMIST-1		25	10	−285.74	−163.11	0.05	0.76
PURGE-2	DEMIST-2		85	10	−281.19	−149.17	0.11	441.42
H_2-S-OUT	H_2-SEP	DEMIST-2	85	10	−12.66	−14.19	1.93	5781.97
H_2-OUT	DEMIST-2	B1	85	10	1.74	−13.48	1.81	5754.40
AIR-IN		B3	25	1	0.00	4.69	0.77	0.00
S8	B3	B4	383	10	10.65	9.05	0.77	9355.27
S9	B4	PSA1	383	10	10.65	9.05	0.39	9355.27

(continued)

Table 2.26 (continued)

Stream name	From	To	Temperature (°C)	Pressure (bar)	Specific enthalpy (kJ/mol)	Specific entropy (kJ/kmol-K)	Mole flow rate (kmol/h)	Specific exergy (kJ/mol)
S10	B4	PSA2	383	10	10.65	9.05	0.39	9355.27
S13	PSA1	B6	383	10	10.89	6.27	0.08	9460.95
S11	PSA1	B5	383	10	10.59	4.23	0.30	9325.58
S12	PSA2	B5	383	10	10.59	4.23	0.30	9325.58
S14	PSA2	B6	383	10	10.89	6.27	0.08	9460.95
N_2-OUT	B5	B2	383	10	10.59	4.23	0.60	9325.58
N_2-MIX	B2	MIXING	767	40	22.79	7.30	0.60	20,613.48
H_2-MIX	B1	MIXING	322	40	8.67	−10.41	1.81	11,768.52
RX-IN	MIXING	RX	408	40	11.23	−2.06	2.97	13,130.97
PURGE	B9		270	250	7.15	−26.46	0.01	15,590.77
RECYCLE	B9	MIXING	270	250	7.06	−26.50	0.56	15,512.77
SEP2-REC	B10	B9	270	250	7.15	−26.46	0.56	15,590.77
RX-OUT	RX	B10	270	250	−22.35	−85.75	1.77	16,101.57
NH_3-OUT	B10		270	250	−36.22	−121.27	1.20	16,341.60

the highest specific exergy values of 16.101 and 16.341 kJ/mol were obtained at the end of the Haber-Bosch reactor and the ammonia separation process, respectively.

Table 2.27 and Fig. 2.80 show the power requirements of the equipment used in the ammonia production system. Accordingly, although it is possible to recover some energy from the system with heat exchangers, the energy consumption mainly consists of PEM electrolysis.

The energy and exergy efficiencies of the system can be calculated with the following formula [68]:

$$\eta_{NH_3} = \frac{\dot{m} \cdot LHV_{NH_3}}{\dot{E}_{in}} \tag{2.37}$$

where

η_{NH_3}: Energy efficiency
$\dot{m}$: Mass flow rate, kg/s
LHV: Lower heating value, kW/kg
$\dot{E}_{in}$: Energy input rate, kW

$$\psi = \frac{\dot{m} \cdot ex_{NH_3}}{\dot{E}x_{in}} \tag{2.38}$$

where

ψ: Exergy efficiency
$\dot{m}$: Mass flow rate, kg/s
ex_{NH_3} Chemical exergy of ammonia, kW/kg
$\dot{E}xin$ Exergy input rate, kW

Accordingly, the energy efficiency was calculated as 55% and the exergy efficiency as 47% for an electrolytic ammonia plant producing 0.5 t of ammonia per day. Accordingly, the analyzes of the facilities with a daily production capacity of 0.5, 1.0, 2.5, and 5 t, which were determined before, are presented in this section. The energy needs for all cases mainly to supply the hydrogen production in the PEM module, then in the reactor to provide the pressure and temperature. However, some of the heat of the reactor can be provided by using the high temperature that occurs especially during the pressurization processes. Table 2.28 presents the state point results for a system that produces 1 t of ammonia per day. Consequently, the power consumption of each unit was given in Fig. 2.81. Although some energy recovery is

Table 2.27 Power consumption of the units for 0.5 t/day ammonia production system

Units	Heat exchangers		Pumps			Compressors			Reactor
Block name	HX1	HX2	P1	P2	P3	B1	B2	B3	RX
Net duty (kW)	−29.62	−30.28	0.012	0.006	0.006	3.49	2.04	2.28	20.23

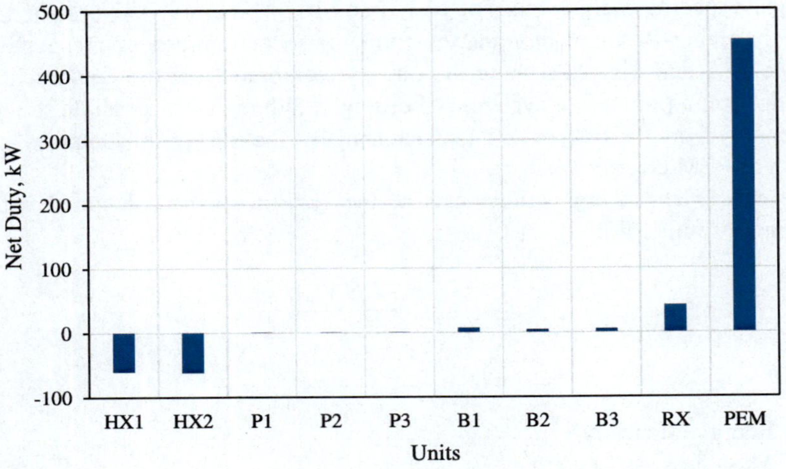

Fig. 2.80 Power consumption of the units for 0.5 t/day ammonia production system

possible with heat exchangers, the major part of the energy losses is through hydrogen production by electrolysis.

The amount of hydrogen required by the system to produce 1 t of ammonia per day was calculated as 4.35 kg/h. The system consumes water at the rate of 72.79 kg/h. The amount of air required for nitrogen production was calculated as 46.34 kg/h. The net power required to run the PEM electrolyzer to produce the required hydrogen amount was found to be 454 kW. A PEM unit with 620 cells is required to produce the required hydrogen.

Aspen Plus simulation state point values for a daily production capacity of 2.5 t are presented in Table 2.29. The power consumption of each unit is given in Fig. 2.82.

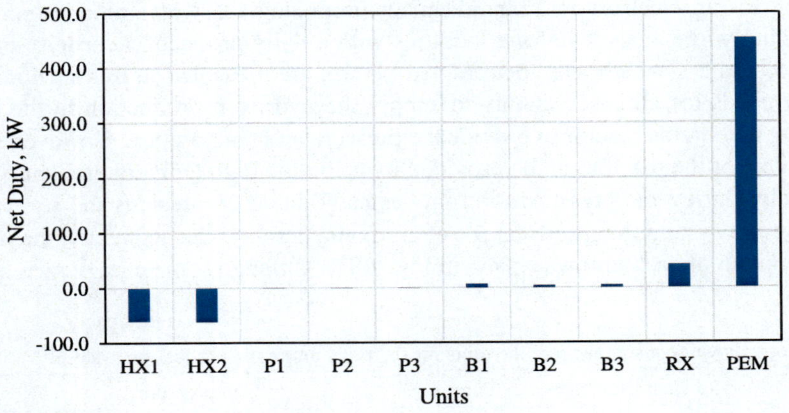

Fig. 2.81 Power consumption of the units for 1 t/day ammonia production system

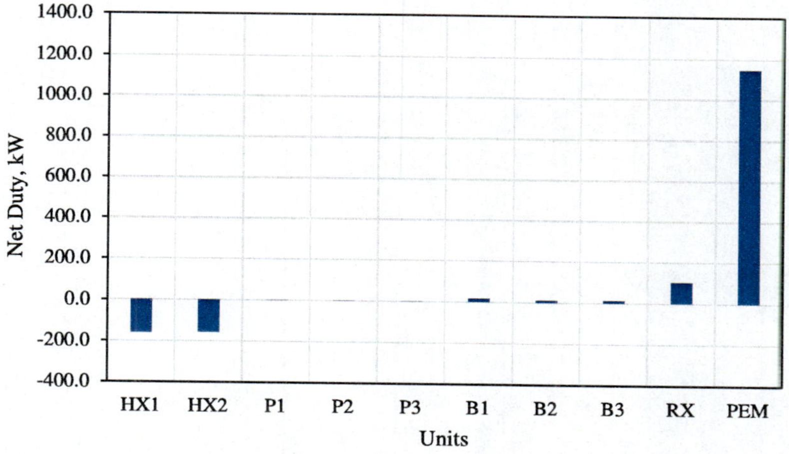

Fig. 2.82 Power consumption of the units for 2.5 t/day ammonia production system

Accordingly, to produce 2.5 t of ammonia per day, 182.5 kg/h of water is needed, and 114.40 kg/h of air is needed. The system produces 18.53 kg of hydrogen per hour and the energy requirement of the electrolysis unit is calculated as 1150 kW. The number of cells required for the PEM system was calculated as 1300.

Table 2.30 presents the data of state points in the Aspen Plus Simulation for 5 kg/day ammonia production. Also, Fig. 2.83 shows the power consumption of the components in the system. The amount of air and water required to produce 5 t of ammonia per day was found to be 227.35 and 363.8 kg/h, respectively. The system uses a PEM system that produces 37.13 kg of hydrogen per hour. The energy requirement of the PEM system is calculated as 2296 kW. The number of cells required to produce hydrogen at this capacity is calculated as 2620. It is estimated that there will be a heat loss of 147 kW in the PEM system.

Table 2.31 and Fig. 2.84 show the power requirements of the equipment used in the ammonia production system for different capacities. Accordingly, although it is possible to recover some energy from the system with heat exchangers, the energy consumption mainly consists of PEM electrolysis in all cases as expected. The water temperature at the inlet of the PEM unit is accepted as 24 °C, for all cases the water temperature rises to ~86 °C at the PEM outlet. However, it is assumed that 10% of the heat generated in the unit is lost. Due to the formation reaction of ammonia, the ammonia conversion rate in single pass reactions is around 12%. For this reason, after ammonia separation is made at the reactor outlet, the line containing nitrogen and hydrogen must be fed back to the system.

Energy efficiency for ammonia production systems is presented in Fig. 2.85. Accordingly, the energy efficiency for ammonia production was found to be around 56% for all cases. The energy requirement of the whole system is calculated as 191.5, 390, 973, and 1945 kW for 0.5, 1, 2.5, and 5 t capacity, respectively. However, optimizing by using heat exchangers in high temperature streams will increase the efficiency of the system.

Table 2.28 State point results of the Aspen Plus simulation for 1 t/day ammonia production system

Stream name	From	To	Temperature (°C)	Pressure (bar)	Specific enthalpy (kJ/mol)	Specific entropy (kJ/kmol-K)	Mole flow rate (kmol/h)	Specific exergy (kJ/mol)
MAKE-UP		P1	30.00	1.00	−285.43	−161.91	4.04	4.61
MAKE-UP1	P1	MIX	30.29	10.00	−285.41	−161.84	4.04	5.06
PEM-IN	MIX	PEM	30.01	10.00	−285.39	−161.89	107.03	4.96
H_2O-O_2-O	O_2-SEP	P2	84.55	9.70	−281.19	−149.25	51.56	435.31
R-A-1	P2	HX1	84.56	10.00	−281.19	−149.25	51.56	435.44
H_2O-H_2-O	H_2-SEP	P3	85.83	9.70	−281.09	−148.97	51.43	453.68
R-C-1	P3	HX2	85.84	10.00	−281.09	−148.96	51.43	453.82
R-C-2	HX2	MIX	30.00	10.00	−285.39	−161.89	51.43	5.01
R-A-2	HX1	MIX	30.00	10.00	−285.39	−161.89	51.56	4.92
A-OUT	PEM	O_2-SEP	84.58	10.00	−271.36	−144.32	53.52	633.54
C-OUT	PEM	H_2-SEP	85.90	10.00	−262.09	−139.43	55.36	837.49
O_2-S-OUT	O_2-SEP	DEMIST-1	84.55	9.70	−12.38	−14.11	1.96	5780.88
O_2-OUT2	B6		382.75	10.00	10.89	6.27	0.35	9460.95
O_2-OUT	DEMIST-1		25.00	9.70	−0.79	−18.74	1.85	5599.78
PURGE-1	DEMIST-1		25.00	9.70	−285.74	−163.11	0.11	0.76
PURGE-2	DEMIST-2		85.83	9.70	−281.13	−148.99	0.24	453.27
H_2-S-OUT	H_2-SEP	DEMIST-2	85.83	9.70	−13.10	−14.17	3.93	5789.44
H_2-OUT	DEMIST-2	B1	85.83	9.70	1.76	−13.41	3.68	5758.43
AIR-IN		B3	25.00	1.01	0.00	4.69	1.60	0.00

(continued)

Table 2.28 (continued)

Stream name	From	To	Temperature (°C)	Pressure (bar)	Specific enthalpy (kJ/mol)	Specific entropy (kJ/kmol-K)	Mole flow rate (kmol/h)	Specific exergy (kJ/mol)
S8	B3	B4	382.75	10.00	10.65	9.05	1.60	9355.27
S9	B4	PSA1	382.75	10.00	10.65	9.05	0.80	9355.27
S10	B4	PSA2	382.75	10.00	10.65	9.05	0.80	9355.27
S13	PSA1	B6	382.75	10.00	10.89	6.27	0.18	9460.95
S11	PSA1	B5	382.75	10.00	10.59	4.23	0.62	9325.58
S12	PSA2	B5	382.75	10.00	10.59	4.23	0.62	9325.58
S14	PSA2	B6	382.75	10.00	10.89	6.27	0.18	9460.95
N_2-OUT	B5	B2	382.75	10.00	10.59	4.23	1.25	9325.58
N_2-MIX	B2	MIXING	767.35	40.00	22.79	7.30	1.25	20,613.48
H_2-MIX	B1	MIXING	323.08	40.00	8.70	−10.34	3.68	11,788.38
RX-IN	MIXING	RX	407.57	40.00	11.24	−1.05	6.27	13,132.81
PURGE	B9		270.00	250.00	7.18	−22.84	0.01	15,598.25
RECYCLE	B9	MIXING	270.00	250.00	7.46	−22.74	1.32	15,855.75
SEP2-REC	B10	B9	270.00	250.00	7.18	−22.84	1.35	15,598.25
RX-OUT	RX	B10	270.00	250.00	−20.84	−80.99	3.81	16,078.20
NH_3-OUT	B10		270.00	250.00	−36.22	−121.27	2.46	16,341.60

Table 2.29 State point results of the Aspen Plus simulation for 2.5 t/day ammonia production system

Stream name	From	To	Temperature (°C)	Pressure (bar)	Specific enthalpy (kJ/mol)	Specific entropy (kJ/kmol-K)	Mole flow rate (kmol/h)	Specific exergy (kJ/mol)
MAKE-UP		P1	30.00	1.00	−285.43	−161.91	10.13	4.61
MAKE-UP1	P1	MIX	30.29	10.00	−285.41	−161.84	10.13	5.06
PEM-IN	MIX	PEM	30.01	10.00	−285.39	−161.89	273.72	4.95
H_2O-O_2-O	O_2-SEP	P2	85.82	9.70	−281.09	−148.97	131.96	453.51
R-A-1	P2	HX1	85.83	10.00	−281.09	−148.97	131.96	453.65
H_2O-H_2-O	H_2-SEP	P3	87.03	9.70	−281.00	−148.70	131.63	471.19
R-C-1	P3	HX2	87.04	10.00	−280.99	−148.70	131.63	471.33
R-C-2	HX2	MIX	30.00	10.00	−285.39	−161.89	131.63	5.01
R-A-2	HX1	MIX	30.00	10.00	−285.39	−161.89	131.96	4.92
A-OUT	PEM	O_2-SEP	85.86	10.00	−271.50	−144.15	136.86	647.31
C-OUT	PEM	H_2-SEP	87.10	10.00	−262.43	−139.37	141.45	846.73
O_2-S-OUT	O_2-SEP	DEMIST-1	85.82	9.70	−13.07	−14.07	4.90	5792.38
O_2-OUT2	B6		382.75	10.00	10.89	6.27	0.87	9460.95
O_2-OUT	DEMIST-1		25.00	9.70	−0.79	−18.74	4.61	5599.78
PURGE-1	DEMIST-1		25.00	9.70	−285.74	−163.11	0.29	0.76
PURGE-2	DEMIST-2		87.03	9.70	−281.03	−148.72	0.63	470.78
H_2-S-OUT	H_2-SEP	DEMIST-2	87.03	9.70	−13.78	−14.13	9.83	5800.86
H_2-OUT	DEMIST-2	B1	87.03	9.70	1.80	−13.31	9.19	5764.39
AIR-IN		B3	25.00	1.01	0.00	4.69	3.95	0.00
S8	B3	B4	382.75	10.00	10.65	9.05	3.95	9355.27
S9	B4	PSA1	382.75	10.00	10.65	9.05	1.98	9355.27

(continued)

Table 2.29 (continued)

Stream name	From	To	Temperature (°C)	Pressure (bar)	Specific enthalpy (kJ/mol)	Specific entropy (kJ/kmol-K)	Mole flow rate (kmol/h)	Specific exergy (kJ/mol)
S10	B4	PSA2	382.75	10.00	10.65	9.05	1.98	9355.27
S13	PSA1	B6	382.75	10.00	10.89	6.27	0.43	9460.95
S11	PSA1	B5	382.75	10.00	10.59	4.23	1.54	9325.58
S12	PSA2	B5	382.75	10.00	10.59	4.23	1.54	9325.58
S14	PSA2	B6	382.75	10.00	10.89	6.27	0.43	9460.95
N_2-OUT	B5	B2	382.75	10.00	10.59	4.23	3.08	9325.58
N_2-MIX	B2	MIXING	767.35	40.00	22.79	7.30	3.08	20,613.48
H_2-MIX	B1	MIXING	325.06	40.00	8.76	−10.25	9.19	11,817.40
RX-IN	MIXING	RX	403.84	40.00	11.13	−1.14	15.87	13,071.58
PURGE	B9		270.00	250.00	7.18	−23.03	0.04	15,598.69
RECYCLE	B9	MIXING	270.00	250.00	7.18	−23.03	3.60	15,601.00
SEP2-REC	B10	B9	270.00	250.00	7.18	−23.03	3.63	15,598.69
RX-OUT	RX	B10	270.00	250.00	−20.05	−79.19	9.75	16,064.91
NH_3-OUT	B10		270.00	250.00	−36.22	−121.27	6.12	16,341.60

Table 2.30 State point results of the Aspen Plus simulation for 5 t/day ammonia production system

Stream name	From	To	Temperature (°C)	Pressure (bar)	Specific enthalpy (kJ/mol)	Specific entropy (kJ/kmol-K)	Mole flow rate (kmol/h)	Specific exergy (kJ/mol)
MAKE-UP		P1	30.00	1.00	−285.43	−161.91	20.19	4.61
MAKE-UP1	P1	MIX	30.29	10.00	−285.41	−161.84	20.19	5.06
PEM-IN	MIX	PEM	30.01	10.00	−285.39	−161.89	555.86	4.96
H_2O-O_2-O	O_2-SEP	P2	85.02	9.70	−281.15	−149.15	268.15	441.96
R-A-1	P2	HX1	85.03	10.00	−281.15	−149.14	268.15	442.10
H_2O-H_2-O	H_2-SEP	P3	86.23	9.70	−281.06	−148.88	267.51	459.55
R-C-1	P3	HX2	86.24	10.00	−281.06	−148.88	267.51	459.69
R-C-2	HX2	MIX	30.00	10.00	−285.39	−161.89	267.51	5.01
R-A-2	HX1	MIX	30.00	10.00	−285.39	−161.89	268.15	4.92
A-OUT	PEM	O_2-SEP	85.05	10.00	−271.71	−144.40	277.93	632.58
C-OUT	PEM	H_2-SEP	86.30	10.00	−262.77	−139.69	287.12	829.08
O_2-S-OUT	O_2-SEP	DEMIST-1	85.02	9.70	−12.63	−14.09	9.78	5785.03
O_2-OUT2	B6		382.75	10.00	10.89	6.27	1.72	9460.95
O_2-OUT	DEMIST-1		25.00	9.70	−0.79	−18.74	9.22	5599.78
PURGE-1	DEMIST-1		25.00	9.70	−285.74	−163.11	0.55	0.76
PURGE-2	DEMIST-2		86.23	9.70	−281.09	−148.90	1.23	459.14
H_2-S-OUT	H_2-SEP	DEMIST-2	86.23	9.70	−13.33	−14.15	19.61	5793.22
H_2-OUT	DEMIST-2	B1	86.23	9.70	1.77	−13.38	18.39	5760.43
AIR-IN		B3	25.00	1.01	0.00	4.69	7.85	0.00

(continued)

Table 2.30 (continued)

Stream name	From	To	Temperature (°C)	Pressure (bar)	Specific enthalpy (kJ/mol)	Specific entropy (kJ/kmol-K)	Mole flow rate (kmol/h)	Specific exergy (kJ/mol)
S8	B3	B4	382.75	10.00	10.65	9.05	7.85	9355.27
S9	B4	PSA1	382.75	10.00	10.65	9.05	3.93	9355.27
S10	B4	PSA2	382.75	10.00	10.65	9.05	3.93	9355.27
S13	PSA1	B6	382.75	10.00	10.89	6.27	0.86	9460.95
S11	PSA1	B5	382.75	10.00	10.59	4.23	3.06	9325.58
S12	PSA2	B5	382.75	10.00	10.59	4.23	3.06	9325.58
S14	PSA2	B6	382.75	10.00	10.89	6.27	0.86	9460.95
N_2-OUT	B5	B2	382.75	10.00	10.59	4.23	6.13	9325.58
N_2-MIX	B2	MIXING	767.35	40.00	22.79	7.30	6.13	20,613.48
H_2-MIX	B1	MIXING	323.75	40.00	8.72	−10.31	18.39	11,798.17
RX-IN	MIXING	RX	412.56	40.00	11.37	−1.74	29.57	13,208.15
PURGE	B9		270.00	250.00	7.15	−25.57	0.05	15,591.32
RECYCLE	B9	MIXING	270.00	250.00	7.13	−25.44	5.05	15,528.54
SEP2-REC	B10	B9	270.00	250.00	7.15	−25.57	5.08	15,591.32
RX-OUT	RX	B10	270.00	250.00	−23.49	−88.15	17.33	16,121.41
NH_3-OUT	B10		270.00	250.00	−36.22	−121.27	12.24	16,341.60

Table 2.31 Power consumption (kW) of the units for various ammonia production systems

Units	Heat exchangers		Pumps			Compressors			Reactor	Electrolysis
Block	HX1	HX2	P1	P2	P3	B1	B2	B3	RX	PEM
0.5 t/d	−29.6	−30.3	0.012	0.006	0.006	3.5	2.0	2.3	20.2	223.3
1 t/d	−60.2	−61.5	0.024	0.011	0.011	7.1	4.2	4.7	41.6	453.7
2.5 t/d	−160.8	−157.7	0.061	0.028	0.028	17.8	10.5	11.7	103.4	1148.3
5 t/d	−315.7	−322.1	0.121	0.057	0.057	35.5	20.8	23.2	206.4	2295.7

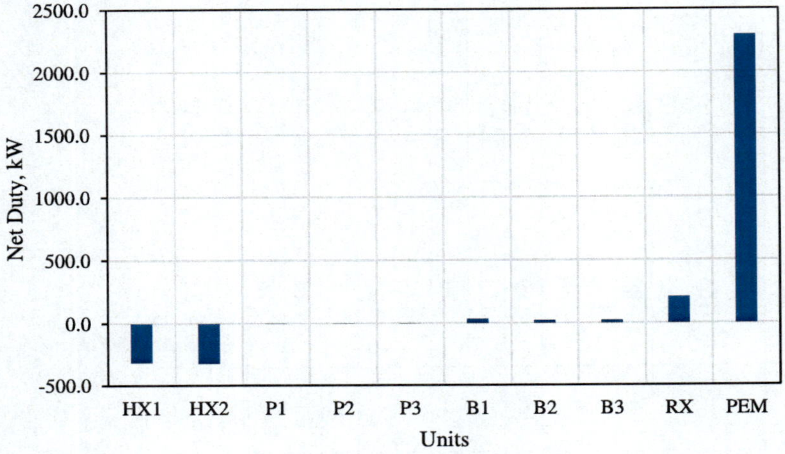

Fig. 2.83 Power consumption of the units for 5 t/day ammonia production system

The Aspen Plus modeling results of an electrolytic ammonia production system with different capacities are given in this section. According to the results obtained, the chemical and electrical needs of the system were calculated, and the energy efficiency of the system was found to be around 56%.

2.11 Case Studies

In this section, studies on the amount of surplus electricity that will be generated if renewable energy technologies are used are presented. The surplus electricity calculation is estimated by modeling with HomerPro software. Capacities and costs are estimated and presented for the case where the surplus electricity is diverted to hydrogen and ammonia production.

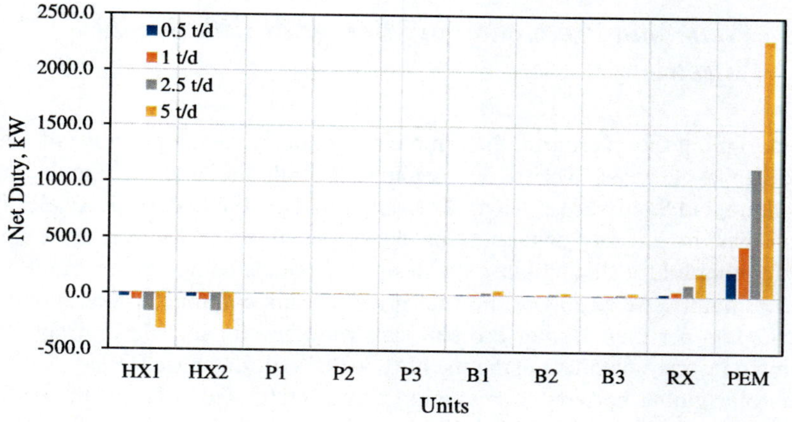

Fig. 2.84 Comparison of power consumption of the units for various ammonia production systems

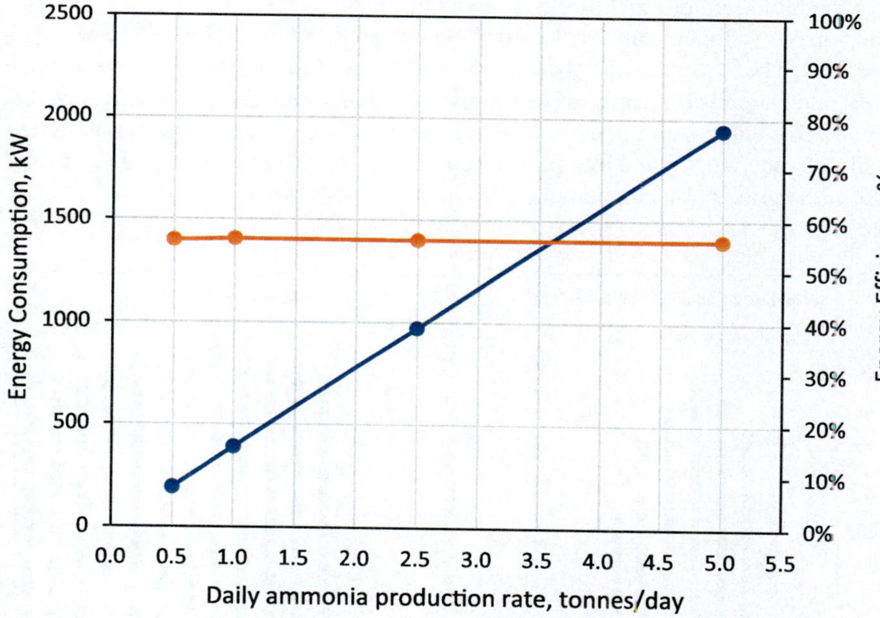

Fig. 2.85 Energy consumption and efficiencies of the ammonia production systems

2.11.1 Ammonia Production in Microgrids with Surplus Electricity

While designing the microgrid, the remote communities in the province of Ontario were taken as a basis. The two communities with the highest population are Pikangikum and Sandy Lake, where 2720 and 2641 people live, respectively. These communities are entirely dependent on diesel generators. However, since no fuel costs are reported for Pikangikum, the design was made using data from the Sandy Lake community. Accordingly, the fuel price for this community was reported as US\$1512 L^{-1} per liter. Homer Pro software was used for microgrid design. Solar and wind data were obtained using the NASA POWER database [69].

The solar global horizontal irradiation index (GHI) and wind profile of Sandy Lake are given in Fig. 2.86. Solar GHI is reported as 5.6 kWh/m^2 day in the peak summer days. However, this value decreases around 15.6 kWh/m^2 day for winter days. The wind profile of the area provides an average of 15.7 km/h, which is well above the minimum design criteria for wind turbines.

Information about the units selected while designing the microgrid is presented in Table 2.19. The system is designed to house diesel generators, wind turbines, solar panels, and batteries for storage. The electricity production is predominantly covered by renewable energy, and in cases where renewable energy is insufficient, batteries and diesel generators come into play. Accordingly, the production surplus that may occur in the grid is also calculated.

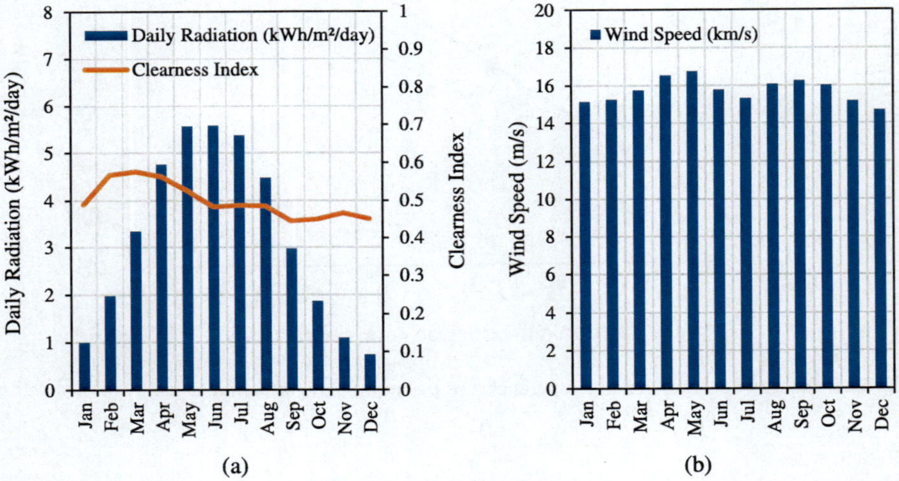

Fig. 2.86 Monthly **a** solar global horizontal irradiation index and **b** wind profile of Sandy Lakes [69]

Table 2.32 Results of Homer Pro micro grid simulation

	Power generation					Costs			
Case	PV (kW)	Wind turbines	Diesel (kW)	Battery (kW)	Converter (kW)	Net present cost (US$)	Cost of electricity (US$)	Operational costs (US$/year)	Capital costs (US$)
1	N/A	2	16	144	33.07	526,910	0.68	25,117	202,208
2	N/A	4	N/A	237	55.91	811,007	1.04	34,431	365,906
3	4.99	3	N/A	267	34.36	813,346	1.04	34,887	362,341
4	N/A	N/A	16	36	10.27	1,063,272	1.36	80,364	24,365

Since the systems currently used for power generation are diesel generators, this scenario has been accepted as the baseline scenario. According to the estimated data, combinations of solar, wind, and diesel generators were calculated, and the most cost-effective system analysis was made. The best optimal results for microgrid design are given in Table 2.32. The capital cost is US$811,007 when only wind energy is used, US$3,482,246 when PV is used, and US$1,063,272 when the diesel generator is used. The best result is obtained when the diesel generator is used together with the wind turbines. According to this scenario, the investment cost is US$202,208, and the electricity price is US$0.68.

Costing calculations of an ammonia production plant that can be established by calculating the amount of hydrogen that can be produced with surplus electricity are made. Electrolysis units used in hydrogen production can work quickly. However, a ramp up time is required in order to provide the required temperature and pressure values in the Haber Bosch process [70].

Seasonal production surpluses of microgrids designed in the previous sections were utilized in the electrolysis unit for hydrogen production estimations. Hydrogen will be stored and used when necessary. It has been accepted that the amount of nitrogen required for ammonia production is produced instantly during ammonia production. The ammonia production system assumed to be operated without shutting down, considering the ramp up time. Table 2.33 shows the summary of the cases. Surplus electricity generation is maximized when only wind case and solar and wind case. This is why the highest annual hydrogen and ammonia productions are seen in these cases. Since the diesel generator is easier to adapt to peak conditions, the system can be optimized to reduce surplus generation to almost zero when using only diesel.

Case 1: Wind Turbine and Diesel Genset

The most optimal case was found to be case 1. Case 1 includes wind turbines and diesel generators to supply electricity to the community. Also, batteries are used to store some of the surplus electricity to balance the supply/demand.

Table 2.33 Summary of hydrogen and ammonia production capacities of each case study

Case		Excess electricity production, kWh/year	Excess electricity, %	Hydrogen production, kg/year	Ammonia production, kg/year
1	Wind + Diesel	33,696	33	645	3632
2	Wind	129,324	65.5	2329	13,116
3	Solar + Wind	87,192	56.3	1420	7997
4	Diesel Genset	N/A	N/A	N/A	N/A

Figure 2.87 shows the costs when the wind turbine and diesel generator are used. In this case, the highest cost is due to wind turbines. However, compared to the base case, the costs arising from diesel fuel are reduced with the inclusion of wind turbines in the microgrid. Although the initial investment cost is high, annual operating costs are therefore reduced.

Figure 2.88 represents the monthly electrical production of wind turbines and diesel generators. In this case, wind turbines meet a load of electricity demand most of the time. Diesel generator only supplies 3.26% of the demand. Yearly 102,044 kWh electricity is produced, while 33% of this production is determined to be surplus.

Figures 2.89 and 2.91 show the heatmaps for daily and monthly surplus electricity production. Depending on the wind profile and demand, no surplus electricity is determined during peak hours. The yearly production of surplus electricity corresponds to 33,696 kWh/year.

According to the data obtained for case 1, the annual hydrogen production with surplus electricity was determined as 3632 kg. With the integration of the electrolysis system, the surplus electricity generation, which was 33%, decreased to 2.13%. Figure 2.90 shows the monthly hydrogen production amounts. Accordingly,

Fig. 2.87 Costs breakdown for case 1

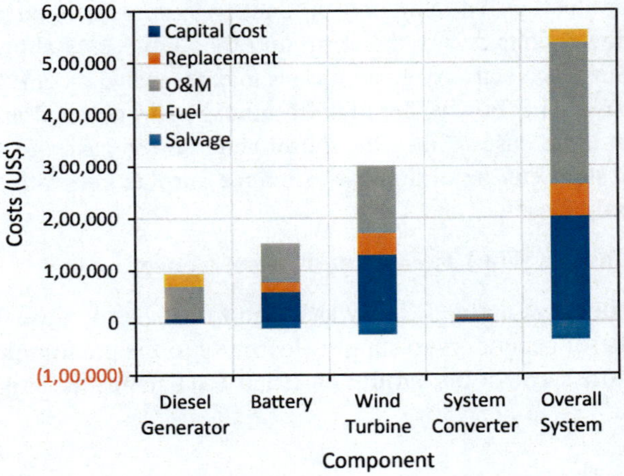

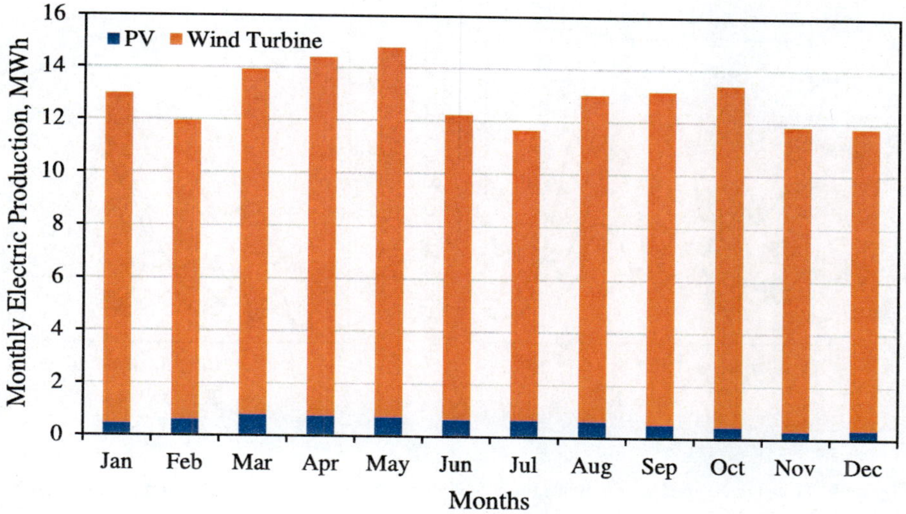

Fig. 2.88 Electricity production of the components in the microgrid for case 1

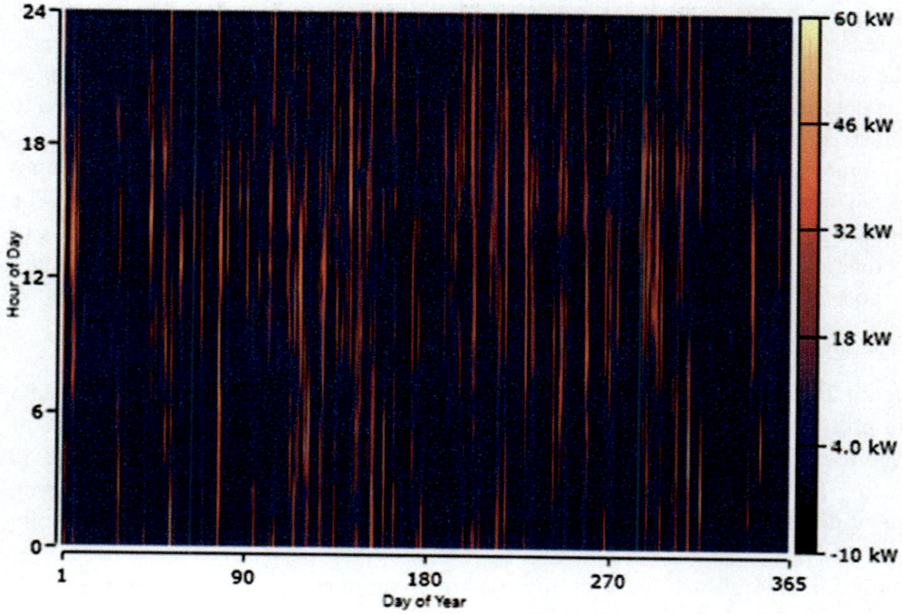

Fig. 2.89 Excess electricity production with wind and diesel generators (case 1)

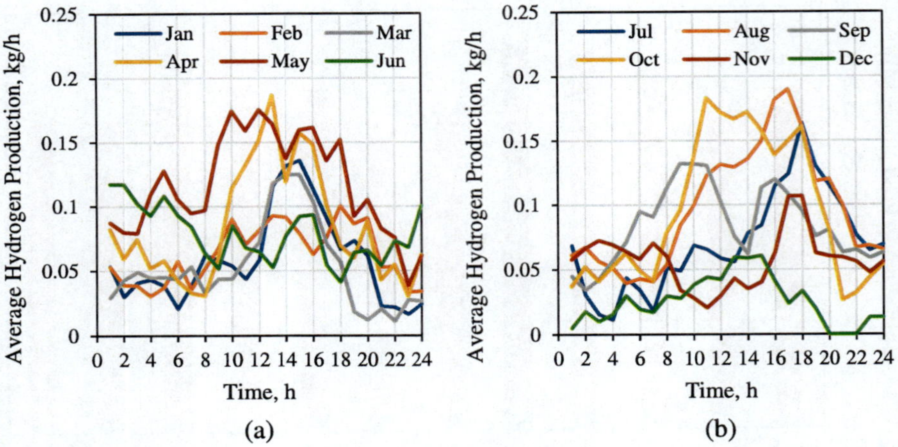

Fig. 2.90 Hourly average hydrogen production of case 1 for **a** first six months and **b** last six months

it is observed that hydrogen production is low in winter months when the heating need is high, due to the low surplus electricity generation. However, the production increases in the spring and summer months, reaching the level of 0.1–0.15 kg/h. The stoichiometric amount of ammonia that can be produced from the hydrogen was determined as 14.76 L/day. This amount of ammonia is low when compared to industrial production capacities.

Figure 2.92 represents the monthly average hydrogen and ammonia production. The monthly average ammonia production rate was found to be 300 kg. The highest production takes place between April and October. However, the amount of production decreases in the winter months due to electricity consumption for heating purposes.

Case 2: Wind Turbines

Figure 2.93 represents the costs for case 2. Only wind turbines and a battery system are preferred for this case. Compared to case one, removing the diesel generator from the microgrid doubles the wind turbine need. Therefore, the cost is increased up to US$811,007. This microgrid system requires no fuel to operate. However, operational and maintenance costs of the wind turbine increase the cost of electricity drastically.

Figure 2.94 shows the monthly electrical production of wind turbines. This case produces more electricity compared to case 1 to compensate the demand. However, this increases the surplus electricity amount without a diesel genset. In total, 197,444 kWh electricity is produced annually.

Figures 2.95 and 2.96 displays the heat map graphs of surplus electricity on daily and monthly basis, respectively. While the excess electricity rate is calculated as 33% for case 1, the surplus electricity rate is 65.5% in this case. This percentage

Days	Jan	Feb	Mar	Apr	May	Jun	Jul	Aug	Sep	Oct	Nov	Dec
1	51	0	42	0	16	0	0	0	0	24	8	0
2	27	0	0	0	47	32	0	10	0	41	7	6
3	13	0	0	44	23	22	0	35	24	6	22	17
4	35	3	0	0	48	47	0	22	42	18	31	0
5	48	0	0	7	0	47	0	28	0	0	37	0
6	49	0	0	30	0	26	2	42	25	32	13	0
7	16	4	30	32	0	0	0	38	41	18	40	0
8	35	0	0	0	0	0	0	34	46	0	22	2
9	0	0	0	41	29	0	0	0	38	0	1	0
10	0	12	0	0	34	32	6	31	30	0	40	20
11	16	19	36	0	44	44	32	44	0	2	52	27
12	3	28	18	0	36	0	9	7	0	0	0	34
13	0	50	0	17	0	0	10	0	0	0	0	46
14	0	48	0	44	33	0	23	0	0	0	0	0
15	0	0	11	30	0	35	4	0	0	0	1	0
16	0	20	0	42	37	38	37	0	0	0	41	0
17	0	0	1	37	28	0	0	0	0	25	47	45
18	7	0	17	0	42	0	39	0	20	34	0	0
19	0	24	41	8	25	0	29	0	34	46	0	11
20	4	43	50	0	17	12	37	35	44	30	1	0
21	0	40	46	0	29	0	23	45	39	44	0	4
22	0	0	0	29	4	0	0	0	7	47	13	9
23	0	39	0	35	26	0	0	39	0	27	0	9
24	0	53	42	38	42	5	31	31	9	30	1	0
25	9	12	0	34	44	36	41	21	0	41	0	2
26	6	20	0	29	17	23	41	32	0	44	0	30
27	17	0	0	46	19	23	7	7	0	25	0	0
28	22	42	0	20	43	33	37	0	0	44	0	0
29	47	0	0	47	43	13	22	10	21	0	0	0
30	0	0	39	44	41	39	0	0	39	4	0	0
31	2	0	0	0	14	0	0	0	0	0	0	0
Legend	0	5	10	15	20	25	29	34	39	44	49	54

Fig. 2.91 Heat map of monthly surplus electricity (kW) averages of case 1

corresponds to the surplus electricity of 129,324 kWh/year. Considering the monthly averages, the daily average surplus electricity generation exceeds 100 kW.

With the removal of diesel genset from the microgrid, surplus electricity generation increases. Accordingly, the surplus electricity rate increased to 65.5%. With the inclusion of the electrolysis system in the system, the surplus electricity rate has decreased to 7%. Figure 2.97 shows the hourly average hydrogen production rates

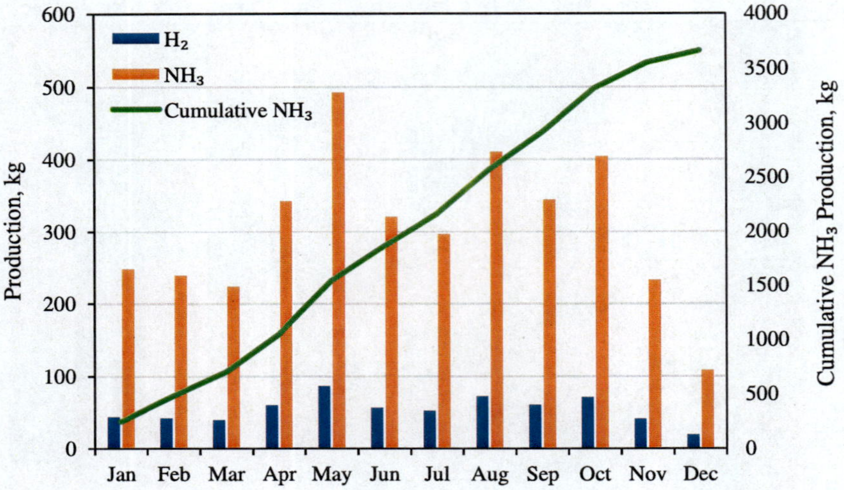

Fig. 2.92 Monthly hydrogen and ammonia production for case 1

Fig. 2.93 Costs breakdown for case 2

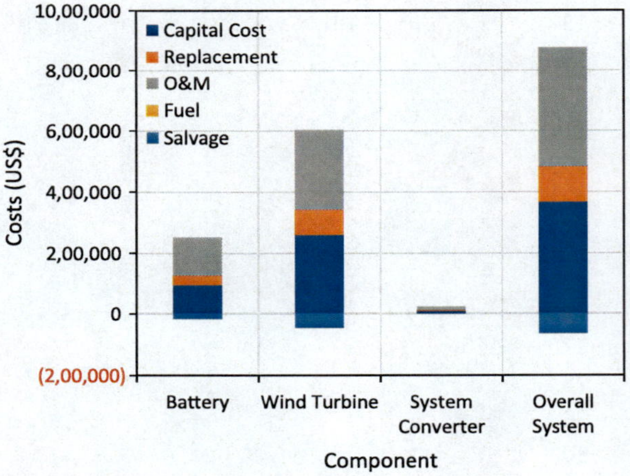

over the months. As can be observed from case 1, the amount of electricity decreases in winter months. When the averages are examined, the amount of hydrogen production is higher than in case 1. The average hydrogen production in May was determined as 0.33 kg/h. However, this value decreases to 0.20 kg/h in December.

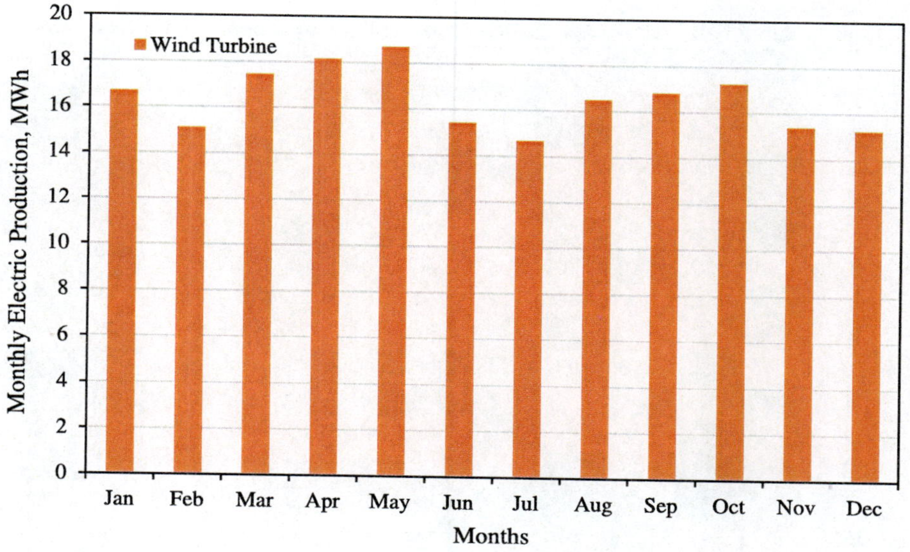

Fig. 2.94 Electricity production of the components in the microgrid for case 2

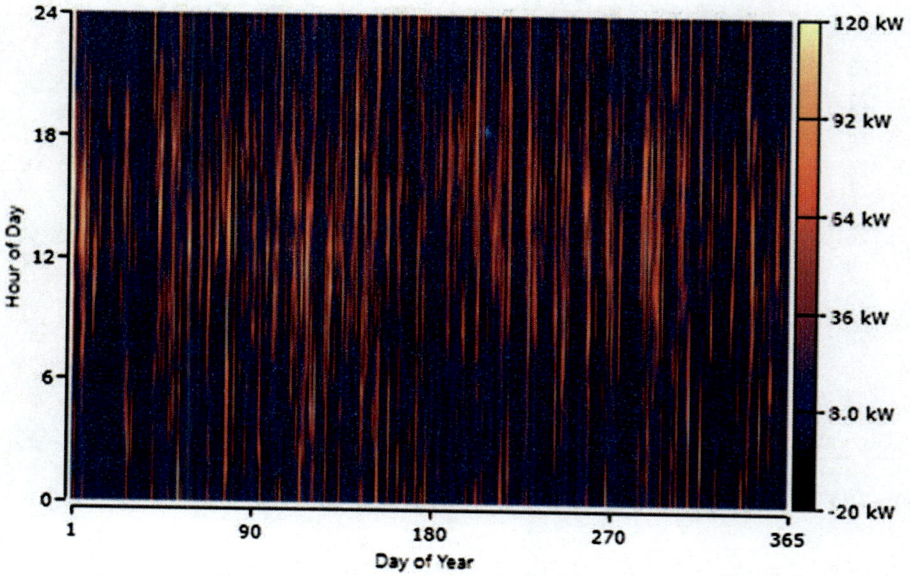

Fig. 2.95 Excess electricity production with wind turbine (case 2)

Days	Jan	Feb	Mar	Apr	May	Jun	Jul	Aug	Sep	Oct	Nov	Dec
1	109	57	94	13	61	35	16	51	0	57	94	0
2	57	0	32	67	98	75	8	47	20	90	52	85
3	51	8	30	96	72	56	49	75	90	38	54	86
4	81	28	0	0	100	96	47	72	95	53	72	47
5	105	0	42	39	0	96	0	63	0	13	78	18
6	107	0	70	72	0	57	83	86	69	78	48	48
7	38	32	84	70	0	0	40	79	88	56	94	57
8	85	0	0	68	0	0	46	78	96	0	56	72
9	0	0	0	97	82	58	0	0	89	0	44	45
10	0	92	66	64	74	94	72	79	68	0	86	103
11	104	64	86	0	95	94	75	93	0	27	107	72
12	72	79	40	34	94	23	42	17	0	39	0	80
13	0	108	0	96	0	8	48	40	0	0	0	103
14	69	105	66	97	82	30	53	0	0	29	0	42
15	9	47	98	68	32	90	64	0	42	0	40	61
16	0	56	40	96	90	86	80	15	17	46	96	61
17	0	15	84	78	63	36	27	4	52	81	100	104
18	70	50	71	0	93	9	89	38	87	89	0	0
19	0	81	91	46	70	60	64	57	80	98	18	75
20	76	98	105	3	41	58	85	86	94	69	79	0
21	29	90	101	40	67	22	48	95	86	96	24	45
22	2	42	0	82	41	50	54	0	25	99	89	80
23	0	84	15	96	60	0	51	87	39	59	0	66
24	0	111	99	86	94	79	78	69	61	70	74	0
25	62	47	0	72	96	89	91	49	10	93	51	75
26	82	46	26	84	36	54	88	73	0	97	56	78
27	74	42	63	99	57	66	36	33	41	57	25	0
28	58	101	55	43	92	70	84	42	49	94	38	50
29	98		71	100	94	55	47	57	88	0	36	43
30	17		92	97	89	82	0	14	94	32	2	27
31	74		36		44		52	0		32		4
Legend	0	10	20	30	40	50	60	71	81	91	101	111

Fig. 2.96 Heat map of monthly surplus electricity (kW) averages of case 2

Figure 2.98 shows the monthly hydrogen and ammonia production. Compared to case 1, when only wind turbines are used, the effect of consumption in the winter months is less noticeable due to the increase in surplus electricity production. The average monthly ammonia production was found to be 1100 kg.

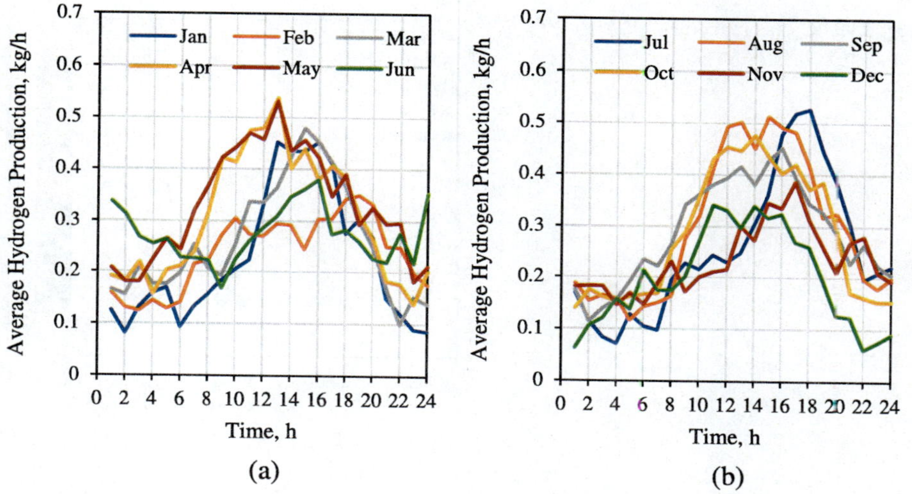

Fig. 2.97 Hourly average hydrogen production of case 2 for **a** first six months and **b** last six months

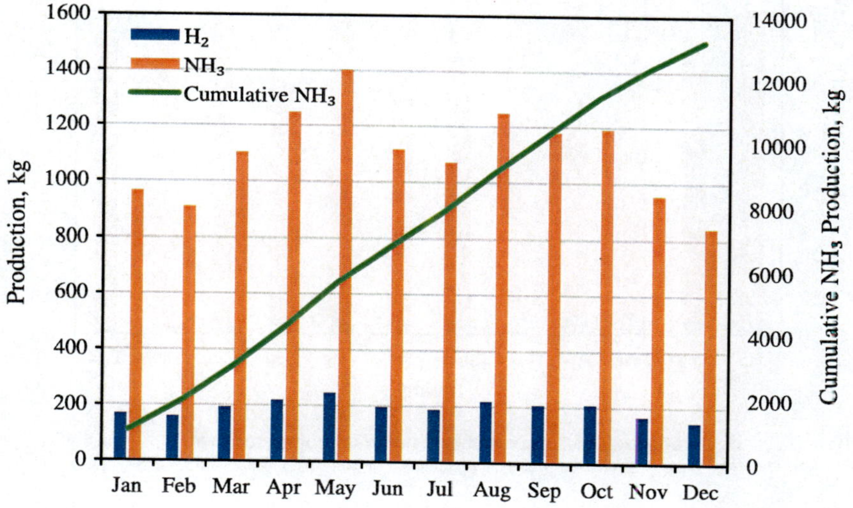

Fig. 2.98 Monthly hydrogen and ammonia production for case 2

Case 3: Wind and PV

The cost breakdown for case 3 was presented in Fig. 2.99. Case 3 utilizes PV panels instead of diesel generators. Since Canada's remote communities are generally in the north, solar irradiance and air clearing index in these regions may not yield efficient solar power generation in every period of the year. This creates high costs as it increases the number of panels to be installed.

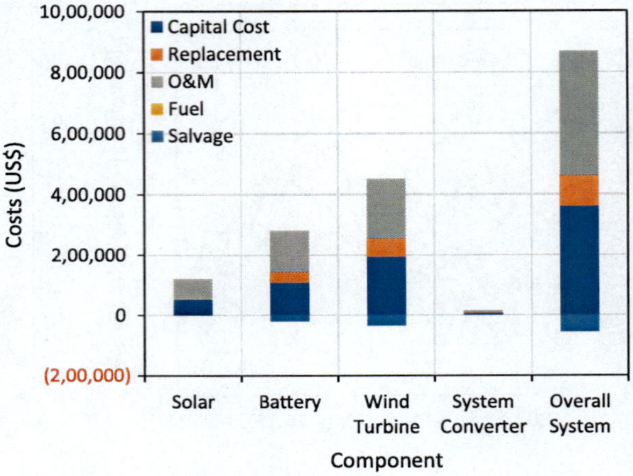

Fig. 2.99 Costs breakdown for case 3

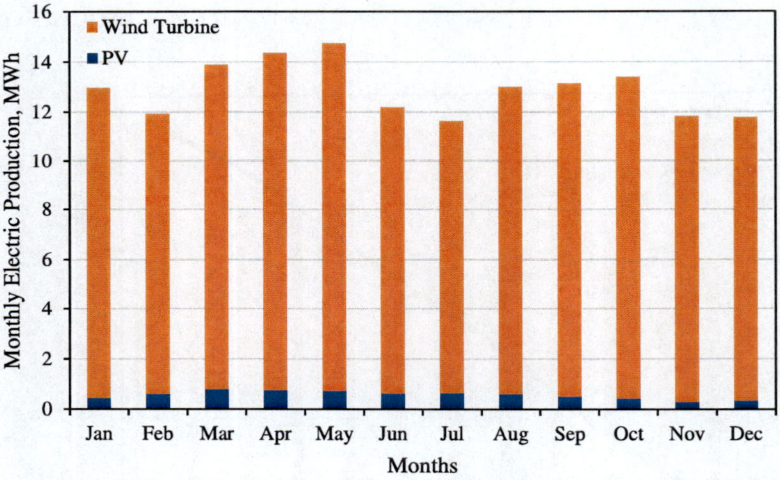

Fig. 2.100 Electricity production of the components in the microgrid for case 3

Figure 2.100 shows the monthly electric production of wind turbines and PV panels. Introducing PV to the microgrid decreases the wind turbine requirement, thus decreasing the total production. In total, yearly 155,000 kWh of electricity is produced in the case 3. The production contribution of PV panels is calculated as 4.43%. Which is a little higher than the diesel genset contribution in case 1.

Heat maps for daily and monthly surplus electricity amounts are given in Fig. 2.101 and Fig. 2.103. Using PV panels in the microgrid decreases the surplus electricity production compared to case 2. However, the surplus electricity production is still

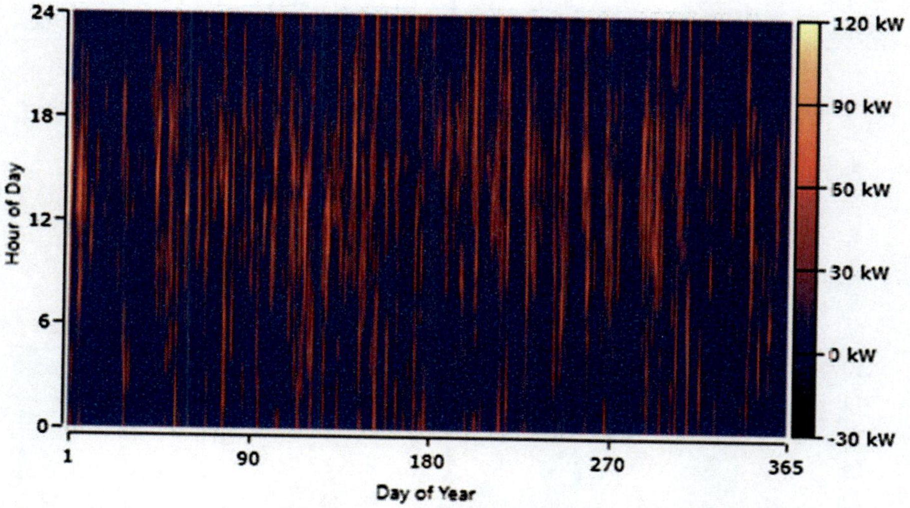

Fig. 2.101 Excess electricity production with wind turbine and PV (case 3)

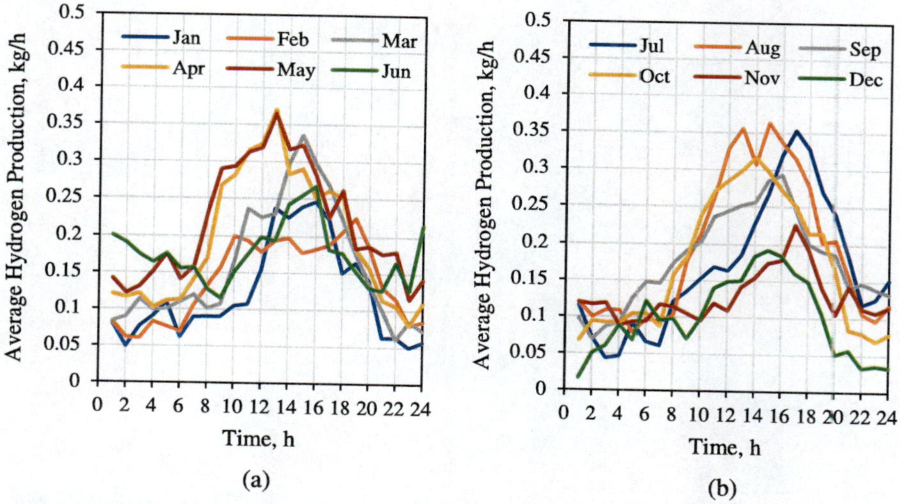

Fig. 2.102 Hourly average hydrogen production of case 3 for **a** first six months and **b** last six months

higher than the case 1 with 87,192 kWh/year. The ratio of surplus electricity calculated as 56.3%. Since the solar irradiation is low on the northern parts, the contribution of PV has no effect on the price of electricity.

The average hydrogen production rates for case 3 are presented in Fig. 2.102. The production profile is like case 1 and case 2. However, the annual hydrogen production

Days	Jan	Feb	Mar	Apr	May	Jun	Jul	Aug	Sep	Oct	Nov	Dec
1	80	44	71	0	45	25	10	35	0	41	65	0
2	42	0	8	41	73	57	0	33	0	68	35	29
3	37	13	17	73	52	39	35	56	64	29	40	60
4	58	26	0	0	74	72	31	54	71	36	51	24
5	77	0	18	31	0	72	2	48	0	7	58	4
6	82	0	53	52	0	43	61	64	50	57	37	28
7	31	28	66	52	0	0	28	58	65	42	68	9
8	64	0	0	50	0	0	31	57	74	0	39	31
9	0	0	6	74	57	42	0	0	63	0	31	21
10	0	37	50	49	57	71	55	56	53	0	63	74
11	73	16	66	0	71	70	56	72	0	25	80	53
12	49	55	29	24	66	7	32	12	0	9	0	62
13	0	80	0	74	0	5	34	23	0	0	0	75
14	45	76	30	73	61	23	42	0	0	7	0	27
15	0	30	75	52	26	65	46	0	0	0	25	48
16	0	44	18	70	65	64	61	8	9	1	65	42
17	0	0	60	58	46	25	19	0	37	55	72	77
18	32	33	55	0	68	0	66	24	63	64	0	0
19	0	58	67	32	52	44	48	42	61	74	0	38
20	27	71	80	1	32	43	63	61	72	50	36	0
21	4	66	79	28	51	8	36	73	63	73	0	26
22	0	28	0	56	29	37	33	0	21	73	62	58
23	0	62	16	73	45	0	37	66	24	46	0	43
24	0	82	74	65	71	57	56	50	45	52	52	0
25	33	33	0	53	73	64	67	35	0	67	36	48
26	27	34	0	64	27	40	66	56	0	75	39	60
27	52	23	37	76	42	48	26	27	27	41	0	0
28	43	74	42	32	68	54	61	33	34	70	5	34
29	72	0	49	74	71	39	35	41	64	0	0	28
30	0	0	70	75	69	61	0	7	69	27	0	9
31	54	0	19	0	32	0	38	0	0	0	0	9
Legend	0	7	15	22	30	37	45	52	60	67	75	82

Fig. 2.103 Heat map of monthly surplus electricity (kW) averages of case 3

rate is around 1420 kg. The rate of excess electricity, which was 56.3%, decreased to 10%.

The hydrogen production with surplus electricity is rather low compared to dedicated hydrogen production systems. Even though the surplus electricity ratio is up to 65.5% in some cases, hydrogen production was found to be 2.3 tonnes/year. However, stoichiometric ammonia production did not exceed 13 tonnes/year. The monthly

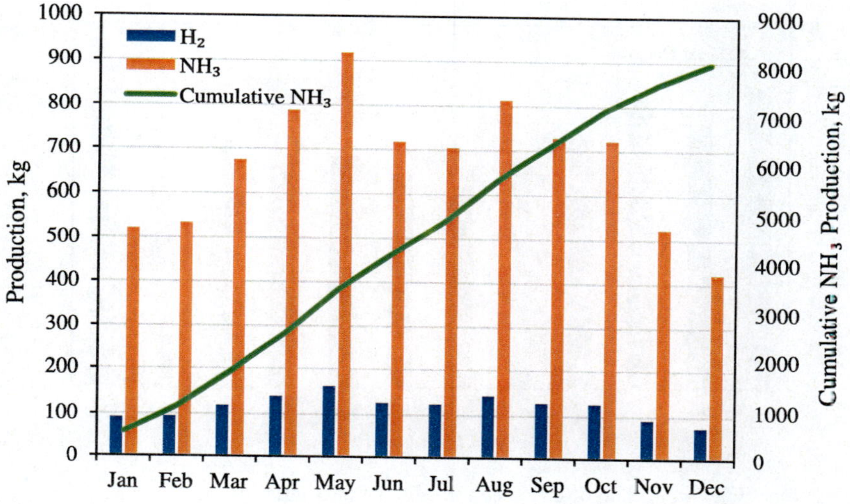

Fig. 2.104 Monthly hydrogen and ammonia production for case 3

average hydrogen and ammonia production for case 3 is shown in Fig. 2.104. The average ammonia production was found to be as much as 650 kg NH_3/month.

Case 4: Diesel Genset

Case 4 is calculated for comparison with the current situation where only diesel generators are used. Results for the costing study was given in Fig. 2.105. The costs breakdown shows that the operational and maintenance costs and fuel costs are high. Although the initial investment cost is low, the expenses in the long run rise to a very high level compared to other cases.

Fig. 2.105 Costs breakdown for case 4

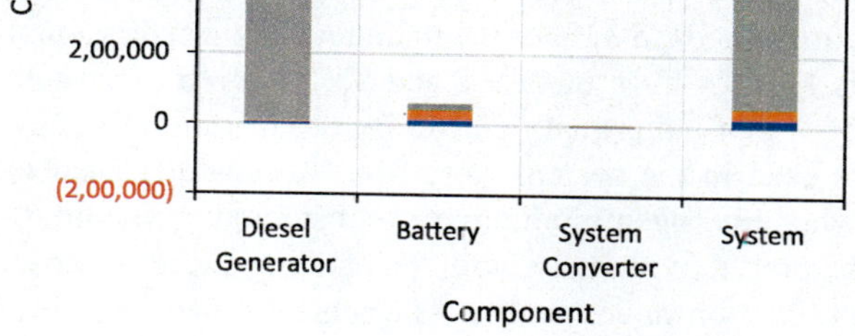

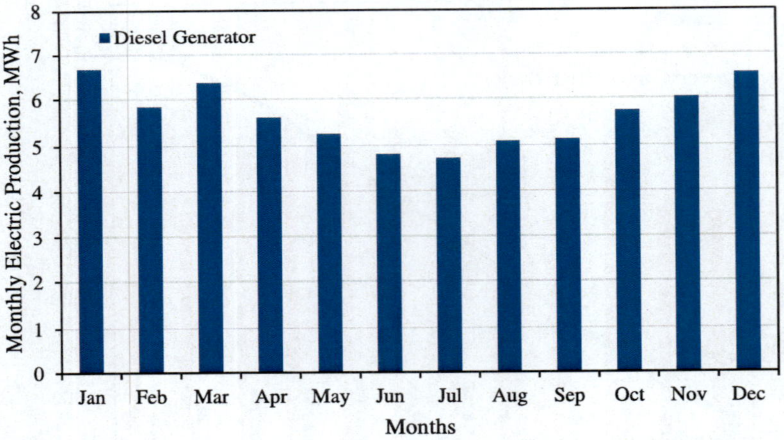

Fig. 2.106 Electricity production of the components in the microgrid for case 4

The electricity production of the diesel generator is presented in Fig. 2.106. Because the diesel generator can be run on-demand, using it with a battery system minimizes the electrical production compared to other cases. However, the need for diesel fuel increases the price of the electricity. The price of the diesel fuel in Sandy Lake is high compared to most of the other remote communities. This increases the price of the electricity drastically.

Excess electricity is an issue that needs to be emphasized when examining renewable energy sources. Since diesel generators can operate on-demand, excess electricity generation is considered zero. However, solar panels and wind turbines continue to work regardless of the load. For this reason, especially when the need is low, excess production is generated. The use of this excess electricity in the production of ammonia is being evaluated.

For this reason, studies on excess electricity production have been carried out. The use of diesel generators reduces excess electricity production. Fluctuations caused by the irregularities of renewable energy generation, together with the on-demand operating profile of the diesel generator, gave more efficient results than other cases. However, the instantaneous surplus electricity generation can be up to 60 kW. In case 2 and case 3, the instantaneous surplus electricity production can be up to 120 kW. For case 1, excess electricity generation is calculated as 33,696 kWh/year. This value corresponds to 33% of the main load. However, annual excess electricity production is 129,324 kWh in case 2 and 87,192 kWh in case 3. These values cover 65.5 and 56.3% of the primary electricity production, respectively. Although battery storage is used in the system, very high surplus electrical values are obtained. However, when hydrogen production is considered, a maximum of 1700 kg of hydrogen can be produced annually with the existing excess electricity. Since the target hydrogen production values could not be achieved, new sources should be added to the existing renewable energy sources to achieve the targeted ammonia production. In short, it is not possible to obtain high amounts of ammonia alone with the surplus electricity obtained from a system that will be installed only for the community's needs.

2.11.2 Economics of Ammonia Production in Microgrids with Surplus Electricity

Determining electrolysis units for corresponding capacities were carried out through modeling. Active area of the electrolysis selected as 400 cm^2. Specific work of the electrolyzer units were found to be 55.5 and 50 kWh/kg for PEM and AEL, respectively. Table 2.34 shows the modeling results. For cases 1, 2, and 3, the stack sizes are determined as 4, 15, and 9, respectively. The corresponding daily hydrogen productions are 1.77, 6.38 and 3.89 kg, respectively.

The economic analysis results were primarily obtained for both PEM and AEL units. Figure 2.107 summarises the results of capital costs for PEM and AEL units. Since the production capacity is low, the capital costs were also found to be low compared to previous studies.

Table 2.34 Modeling results for electrolysis units

Case	Stack size	P_{AEL}, kW	P_{PEM}, kW	Capacity (kg/day)
1	4	3.6	4.008	1.77
2	15	13.5	15.03	6.38
3	9	8.1	9.02	3.89

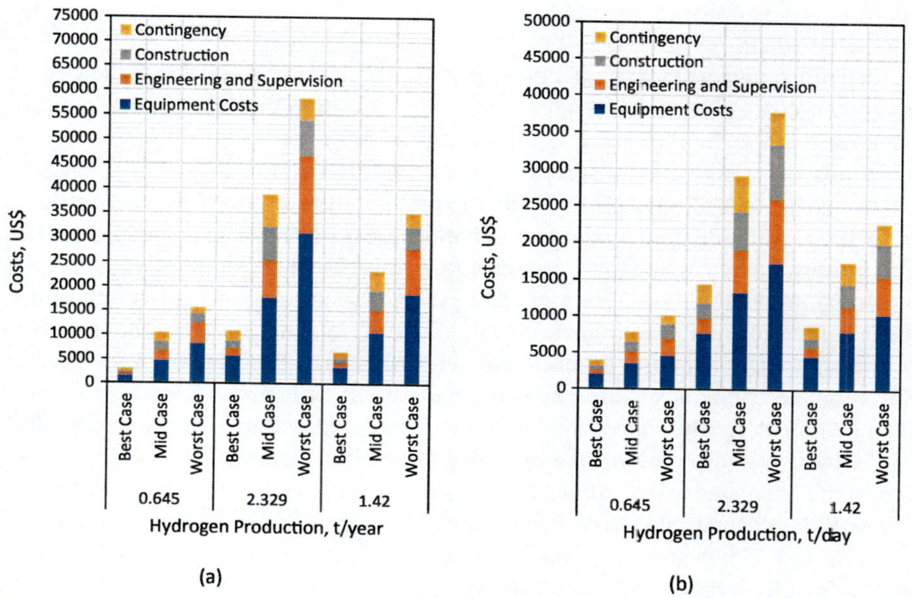

Fig. 2.107 Capital costs for **a** PEM and **b** AEL electrolyser

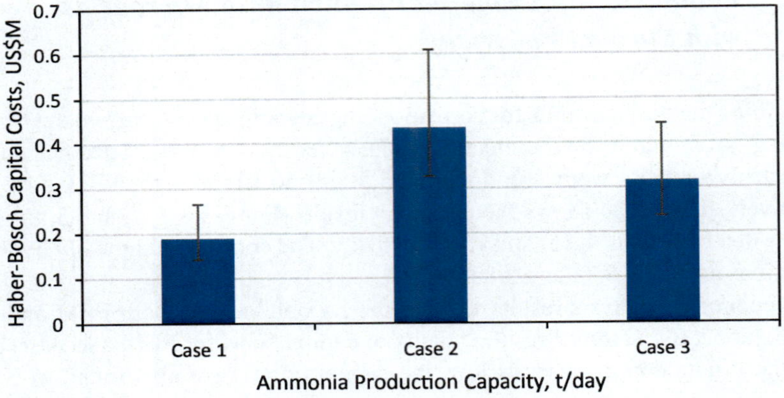

Fig. 2.108 Capital costs for Haber–Bosch unit for the cases 1–3

Figure 2.108 shows the calculated capital costs for the Haber Bosch unit for each case. Average capital cost for the Haber Bosch unit with a capacity of 9.95, 35.93, 21.91 kg/day were found as US$189,000, US$434,000, and US$315,000, respectively. Unfortunately, scaling of the capital costs for the Haber Bosch unit gives an exponential curve. Therefore, although the capital cost decreases with the decrease in capacity, this decrease does occur exponentially rather than proportionally. In addition, sizing down to a portable solution increases the uncertainty in the capital cost calculations.

One of the biggest factors affecting the cost of ammonia production is the cost of hydrogen. When hydrogen is produced by coal gasification and fossil fuel reforming, the costs are less than US$1/kg H_2. This value is between US$3 and 5.7/kg H_2 for electrolysis [1]. High energy consumption of electrolysis units drastically increases the unit price of hydrogen. On the other hand, the utilization of surplus electricity for hydrogen or ammonia production in remote areas that have no connection to the grid is a sustainable option to recover energy. In this case, it is possible to make clean ammonia production cheaper by reducing the electricity cost, which is one of the biggest items in ammonia production costs. Tables 2.35 and 2.36 shows the summary of annual operational costs for each case with AEL and PEM options, respectively. Since the electricity is supplied by surplus electricity, it is not included in the final calculations. Also, the size of the facility is considered portable, hence making the operation easy and decreasing the operating labor costs. Although the system size is very small compared to industrial examples, it has been accepted that the depreciation rate will not change. Similarly, maintenance and repair costs are assumed to be at a similar rate to industrial systems. However, fixed costs of the units in the system make the biggest impact on annual costs.

Ammonia costs calculated for cases 1–3 are presented in Figs. 2.109 and 2.110. Electricity cost is not included in the calculations as it is covered by excess electricity. The capital costs of the electrolysis units do not have a major impact on the cost when the energy expenditures are not included. Without electricity costs, the initial investment cost of the Haber–Bosch unit has the greatest impact on the cost of ammonia. When it is assumed that the facility will operate for 25 years, the annual effect of capital cost on operational costs varies between US$11,850 and US$52,500. Among the costs, it is one of the parameters that most affect ammonia cost. Maintenance and repairs have the second most impact on costs. Considering most of the equipment operates at high temperatures, this is an expected result. Also, the operational and maintenance costs for the Haber Bosch unit are high due to high capital costs. In addition, the supply of bulk equipment in large-capacity facilities also reduces costs.

However, ammonia production can become reasonable with these methods, especially for remote communities where transportation is difficult and logistics costs are high. Fuel prices can go from US$0.25/L up to US$1.17/L depending on the distance

Table 2.35 Annual operational costs for an ammonia production facility with alkaline electrolyzer

	Case I		Case II		Case III	
	Best	Worst	Best	Worst	Best	Worst
Fixed costs						
Air separation unit	US$9044	US$11,951	US$13,308	US$16,421	US$11,819	US$14,800
Haber–Bosch unit	US$142,148	US$274,819	US$326,018	US$630,301	US$236,768	US$457,751
Alkaline electrolyzer	US$3837	US$10,069	US$14,389	US$37,760	US$8634	US$22,656
Sub-total	US$155,028	US$296,839	US$353,714	US$684,481	US$257,220	US$495,207
O&M costs, annual						
Direct costs						
Operating labor	US$2000	US$2000	US$2000	US$2000	US$2000	US$2000
Supervision	US$350	US$350	US$350	US$350	US$350	US$350
Maintenance and repairs	US$3101	US$5937	US$7074	US$13,690	US$5144	US$9904
Stack replacement	US$195	US$195	US$732	US$732	US$439	US$439
Operating supplies	US$465	US$891	US$1061	US$2053	US$772	US$1486
Laboratory charges	US$300	US$300	US$300	US$300	US$300	US$300
Fixed costs						

(continued)

Table 2.35 (continued)

	Case I		Case II		Case III	
	Best	Worst	Best	Worst	Best	Worst
Depreciation	US$7751	US$14,842	US$17,686	US$34,224	US$12,861	US$24,760
Taxes and insurance	US$4961	US$9499	US$11,319	US$21,903	US$8231	US$15,847
Plant overhead	US$3270	US$4972	US$5655	US$9624	US$4497	US$7352
Total	US$22,394	US$38,985	US$46,177	US$84,877	US$34,594	US$62,439

Table 2.36 Annual operational costs for ammonia production facility with PEM electrolyzer

	Case I		Case II		Case III	
	Best	Worst	Best	Worst	Best	Worst
Fixed costs						
Air separation unit	US$9044	US$11,951	US$13,308	US$16,421	US$11,819	US$14,800
Haber–Bosch unit	US$142,148	US$274,819	US$326,018	US$630,301	US$236,768	US$457,751
PEM electrolyzer	US$2875	US$15,533	US$10,780	US$58,249	US$6468	US$34,949
Sub-total	US$154,066	US$302,302	US$350,105	US$704,970	US$255,054	US$507,500
O&M costs, annual						
Direct costs						
Operating labor	US$2000	US$2000	US$2000	US$2000	US$2000	US$2000
Supervision	US$350	US$350	US$350	US$350	US$350	US$350
Maintenance and repairs	US$3081	US$6046	US$7002	US$14,099	US$5101	US$10,150
Stack replacement	US$332	US$332	US$1246	US$1246	US$748	US$748
Operating supplies	US$462	US$907	US$1050	US$2115	US$765	US$1522
Laboratory charges	US$300	US$300	US$300	US$300	US$300	US$300
Fixed costs						
Depreciation	US$7703	US$15,115	US$17,505	US$35,248	US$12,753	US$25,375
Taxes and insurance	US$4930	US$9674	US$11,203	US$22,559	US$8162	US$16,240
Plant overhead	US$3259	US$5038	US$5611	US$9870	US$4471	US$7500
Total	US$22,394	US$38,985	US$46,177	US$84,877	US$34,594	US$62,439

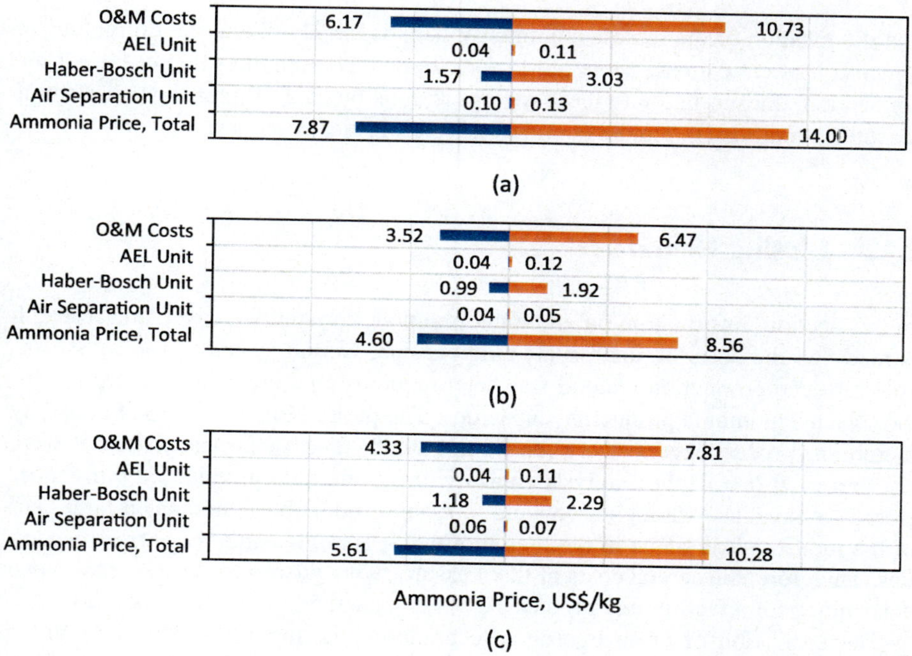

Fig. 2.109 Ammonia price breakdown for an ammonia plant with AEL. **a** Case 1, **b** case 2, **c** case 3

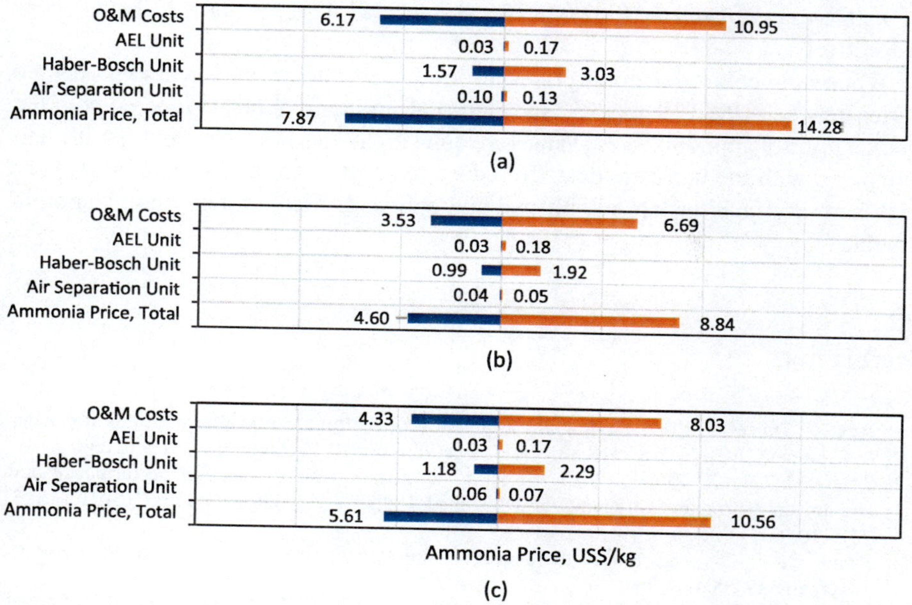

Fig. 2.110 Ammonia price breakdown for an ammonia plant with PEM. **a** Case 1, **b** case 2, **c** case 3

and the location of the remote community. This is largely due to the difficulties of logistics in remote areas. Since most of the remote communities do not have a sea connection, transportation to these regions is done by land. Therefore, the effect of seasonal conditions in remote communities on logistics costs is significant.

2.12 Closing Remarks

In this section, information on ammonia production systems is given and research is made on the costs of small-scale ammonia production systems. The most suitable nitrogen production method was determined as pressure swing adsorption for the selected ammonia production capacities. The purity of nitrogen gas obtained by membrane processes varies between 94 and 99.5%. The membrane processes were not preferred due to the catalysts' sensitivity in ammonia production with Haber Bosch to oxygen. Although cryogenic nitrogen production dominates a large part of the industry, it is stated in the literature that it is not feasible for current capacities. Therefore, the capital costs of the pressure swing adsorption system have been determined for a facility that produces 0.5, 1, 2.5, and 5 t of ammonia per day.

The utilization of clean hydrogen technologies is important to achieve green ammonia production. Therefore, AEL and PEM units were modeled, and their cost assessment was done. Accordingly, AEL performed better in mid and worst-case scenarios. The cost of an electrolysis plant with a daily hydrogen production capacity of 0.89 t was calculated as US$2.04–5.35 M and US$1.47–8.02 M for AEL and PEM, respectively.

When ammonia production prices are evaluated, the price of ammonia per kg varies between US$2.19 and 5.92 in small-scale systems. When compared with the market price of ammonia, these values are quite high. This is an expected result when compared with the literature data. To reduce the cost of ammonia production, very high-capacity facilities are established and operated. Also, system optimization is crucial.

References

1. Acar C, Dincer I (2019) Review and evaluation of hydrogen production options for better environment. J Clean Prod 218:835–849. https://doi.org/10.1016/j.jclepro.2019.02.046
2. Zhang H, Guo X, Zhu Z (2017) Effect of temperature on gasification performance and sodium transformation of Zhundong coal. Fuel 189:301–311. https://doi.org/10.1016/j.fuel.2016.10.097
3. Penner SS (2006) Steps toward the hydrogen economy. Energy 31:33–43. https://doi.org/10.1016/j.energy.2004.04.060
4. Minutillo M, Perna A, Di Trolio P, Di Micco S, Jannelli E (2021) Techno-economics of novel refueling stations based on ammonia-to-hydrogen route and SOFC technology. Int J Hydrogen Energy 46:10059–10071. https://doi.org/10.1016/j.ijhydene.2020.03.113
5. IRENA (2019) Hydrogen: a renewable energy perspective. Tokyo

6. Dincer I, Acar C (2015) A review on clean energy solutions for better sustainability. Int J Energy Res 39:585–606. https://doi.org/10.1002/er.3329
7. International Renewable Energy Agency (IRENA) (2020) Green hydrogen cost reduction scaling up electrolysers to meet the 1.5 °C climate goal. Abu Dhabi
8. U.S. Department of Energy (DOE) (n.d.) DOE technical targets for hydrogen production from electrolysis. https://www.energy.gov/eere/fuelcells/doe-technical-targets-hydrogen-production-electrolysis#:~:text. Accessed 20 Feb 2021
9. Razi F, Dincer I, Gabriel K (2020) A specific exergy costing assessment of the integrated copper-chlorine cycle for hydrogen production. Int J Hydrogen Energy. https://doi.org/10.1016/j.ijhydene.2020.08.269
10. Ozcan H, Dincer I (2014) Energy and exergy analyses of a solar driven Mg–Cl hybrid thermochemical cycle for co-production of power and hydrogen. Int J Hydrogen Energy 39:15330–15341. https://doi.org/10.1016/j.ijhydene.2014.07.089
11. Tkemaladze GS, Makhashvili KA (2016) Climate changes and photosynthesis. Ann Agrar Sci 14:119–26. https://doi.org/10.1016/j.aasci.2016.05.012
12. Kerry FG (2007) Industrial gas handbook: gas separation and purification. CRC Press
13. Baker RW (2004) Gas separation. Membr Technol Appl 301–353. https://doi.org/10.1002/0470020393.ch8
14. Morgan ER (2013) Techno-economic feasibility study of ammonia plants powered by offshore wind. University of Massachusetts Amherst
15. Smith AR, Klosek J (2001) A review of air separation technologies and their integration with energy conversion processes. Fuel Process Technol 70:115–134. https://doi.org/10.1016/S0378-3820(01)00131-X
16. Ruthven DM, Farooq S (1990) Air separation by pressure swing adsorption. Gas Sep Purif 4:141–148. https://doi.org/10.1016/0950-4214(90)80016-E
17. Bernardo P, Drioli E, Golemme G (2009) Membrane gas separation: a review/state of the art. Ind Eng Chem Res 48:4638–4663. https://doi.org/10.1021/ie8019032
18. Wiley-VCH (ed) (2007) Ullmann's agrochemicals, 2 vols. Wiley, New York
19. Morgan ER, Manwell JF, McGowan JG (2017) Sustainable ammonia production from US offshore wind farms: a techno-economic review. ACS Sustain Chem Eng 5:9554–9567
20. Christiansen LJ (1995) Thermodynamic properties in ammonia synthesis BT—ammonia: catalysis and manufacture. In: Nielsen A (ed) Ammonia. Springer, Berlin, pp 1–15. https://doi.org/10.1007/978-3-642-79197-0_1
21. Rossetti I (2020) Reactor design, modeling and process intensification for ammonia synthesis BT. In: Inamuddin, Boddula R, Asiri AM (eds) Sustainable ammonia production. Springer International Publishing, Cham, pp 17–48. https://doi.org/10.1007/978-3-030-35106-9_2
22. Hooper CW (1991) Ammonia synthesis: commercial practice BT. In: Jennings JR (ed) Catalytic ammonia synthesis: fundamentals and practice. Springer US, Boston, pp 253–283. https://doi.org/10.1007/978-1-4757-9592-9_7
23. Nazemi M, El-Sayed MA (2021) Managing the nitrogen cycle via plasmonic (photo) electrocatalysis: toward circular economy. Acc Chem Res 54:4294–4304. https://doi.org/10.1021/acs.accounts.1c00446
24. Nazemi M, Ou P, Alabbady A, Soule L, Liu A, Song J et al (2020) Electrosynthesis of ammonia using porous bimetallic Pd–Ag nanocatalysts in liquid-and gas-phase systems. ACS Catal 10:10197–10206
25. Zhou F, Azofra LM, Ali M, Kar M, Simonov AN, McDonnell-Worth C et al (2017) Electrosynthesis of ammonia from nitrogen at ambient temperature and pressure in ionic liquids. Energy Environ Sci 10:2516–2520
26. O'Connor D (2017) Inductively heated microchannel reactor
27. Appl M (1999) Storage and shipping. Ammonia 213–220. https://doi.org/10.1002/9783527613885.ch09
28. Fertilizers Europe (2014) Guidance for inspection of atmospheric, refrigerated ammonia storage tanks

29. Turton R, Bailie RC, Whiting WB, Shaeiwitz JA (2008) Analysis, synthesis and design of chemical processes. Pearson Education
30. Green DW et al (1997) Perry's chemical engineers' handbook. McGraw-Hill
31. Olivier P, Bourasseau C, Bouamama PB (2017) Low-temperature electrolysis system modeling: a review. Renew Sustain Energy Rev 78:280–300. https://doi.org/10.1016/j.rser.2017.03.099
32. Colbertaldo P, Gómez Aláez SL, Campanari S (2017) Zero-dimensional dynamic modeling of PEM electrolyzers. Energy Procedia 142:1468–1473. https://doi.org/10.1016/j.egypro.2017.12.594
33. Tugirumubano A, Shin H-J, Go S-H, Lee M-S, Kwac LK, Kim H-G (2016) Electrochemical performance analysis of a PEM water electrolysis with cathode feed mode based on flow passage shape of titanium plates. Int J Precis Eng Manuf 17:1073–1078. https://doi.org/10.1007/s12541-016-0130-9
34. Marangio F, Santarelli M, Calì M (2009) Theoretical model and experimental analysis of a high pressure PEM water electrolyser for hydrogen production. Int J Hydrogen Energy 34:1143–1158. https://doi.org/10.1016/j.ijhydene.2008.11.083
35. See DM, White RE (1997) Temperature and concentration dependence of the specific conductivity of concentrated solutions of potassium hydroxide. J Chem Eng Data 42:1266–1268. https://doi.org/10.1021/je970140x
36. Christensen A (2020) Assessment of hydrogen production costs from electrolysis. USE
37. Ivy J (2004) Summary of electrolytic hydrogen production. Small 27
38. Peters MS, Timmerhaus KD, West RE (2003) Plant design and economics for chemical engineers. McGraw-Hill Education
39. Dincer I, Acar C (2014) Review and evaluation of hydrogen production methods for better sustainability. Int J Hydrogen Energy 40:11094–11111. https://doi.org/10.1016/j.ijhydene.2014.12.035
40. Bartels JR (2007) A feasibility study of implementing an ammonia economy. Iowa State University
41. Chehade G, Dincer I (2021) Advanced kinetic modeling and simulation of a new small modular ammonia production unit. Chem Eng Sci 236:116512. https://doi.org/10.1016/j.ces.2021.116512
42. Dybkjaer I (1995) Ammonia production processes BT. In: Nielsen A (ed) Ammonia: catalysis and manufacture. Springer, Berlin, pp 199–327. https://doi.org/10.1007/978-3-642-79197-0_6
43. Market and Market (2018) Catalyst fertilizers market
44. Echemi (n.d.) Ammonia price market analysis
45. Bičáková O, Straka P (2012) Production of hydrogen from renewable resources and its effectiveness. Int J Hydrogen Energy 37:11563–11578. https://doi.org/10.1016/j.ijhydene.2012.05.047
46. Acar C, Dincer I (2015) Impact assessment and efficiency evaluation of hydrogen production methods. Int J Energy Res. https://doi.org/10.1002/er.3302
47. Al-Breiki M, Bicer Y (2021) Comparative life cycle assessment of sustainable energy carriers including production, storage, overseas transport and utilization. J Clean Prod 279:123481. https://doi.org/10.1016/j.jclepro.2020.123481
48. Farhangi H, Joos G (2019) Microgrid planning and design: a concise guide. Wiley
49. Ray P, Biswal M (2020) Microgrid: operation, control, monitoring and protection. Springer, Singapore
50. Sharmeela C, Sivaraman P, Sanjeevikumar P, Holm-Nielsen JB (2021) Microgrid technologies. Wiley
51. Nord JH, Koohang A, Paliszkiewicz J (2019) The internet of things: review and theoretical framework. Expert Syst Appl 133:97–108. https://doi.org/10.1016/j.eswa.2019.05.014
52. Pensini A, Rasmussen CN, Kempton W (2014) Economic analysis of using excess renewable electricity to displace heating fuels. Appl Energy 131:530–543. https://doi.org/10.1016/j.apenergy.2014.04.111
53. Demiroren A, Yilmaz U (2010) Analysis of change in electric energy cost with using renewable energy sources in Gökceada, Turkey: an island example. Renew Sustain Energy Rev 14:323–333

54. Hafez O, Bhattacharya K (2012) Optimal planning and design of a renewable energy based supply system for microgrids. Renew Energy 45:7–15
55. Daud A-K, Ismail MS (2012) Design of isolated hybrid systems minimizing costs and pollutant emissions. Renew Energy 44:215–224
56. Shaahid SM, Al-Hadhrami LM, Rahman MK (2014) Review of economic assessment of hybrid photovoltaic-diesel-battery power systems for residential loads for different provinces of Saudi Arabia. Renew Sustain Energy Rev 31:174–181
57. Rehman S, Al-Hadhrami LM (2010) Study of a solar PV–diesel–battery hybrid power system for a remotely located population near Rafha, Saudi Arabia. Energy 35:4986–4995
58. Khatib T, Mohamed A, Sopian K, Mahmoud M (2011) Optimal sizing of building integrated hybrid PV/diesel generator system for zero load rejection for Malaysia. Energy Build 43:3430–3435
59. Ismail MS, Moghavvemi M, Mahlia TMI (2013) Techno-economic analysis of an optimized photovoltaic and diesel generator hybrid power system for remote houses in a tropical climate. Energy Convers Manag 69:163–173
60. Government of Canada (n.d.) The atlas of Canada—remote communities energy database. https://atlas.gc.ca/rced-bdece/en/index.html. Accessed 10 Nov 2021
61. Statistics Canada (2022) Interprovincial and international trade flows, basic prices, detail level (x1000)
62. Statistics Canada (2022) Canadian fertilizer production, by product type and fertilizer year, cumulative data (x1000). https://www150.statcan.gc.ca/t1/tbl1/en/tv.action?pid=3210003701&pickMembers%5B0%5D=1.1&cubeTimeFrame.startYear=2006+%2F+2007&cubeTimeFrame.endYear=2021+%2F+2022&referencePeriods=20060101%2C20210101
63. Statistics Canada (2022) Fertilizer shipments to Canadian agriculture and export markets, by product type and fertilizer year, cumulative data (x1000). https://www150.statcan.gc.ca/t1/tbl1/en/tv.action?pid=3210003801&pickMembers%5B0%5D=1.1&cubeTimeFrame.startYear=2006+%2F+2007&cubeTimeFrame.endYear=2021+%2F+2022&referencePeriods=20060101%2C20210101
64. Ruthven DM (1984) Principles of adsorption and adsorption processes. Wiley
65. Ishaq H, Dincer I (2020) Design and simulation of a new cascaded ammonia synthesis system driven by renewables. Sustain Energy Technol Assess 40:100725. https://doi.org/10.1016/j.seta.2020.100725
66. Nielsen A (1950) Investigation on promoted iron catalysts for the synthesis of ammonia
67. Mok L (1982) Sensitivity study of energy consumption in ammonia plant operation
68. Qureshy AMMI, Dincer I (2020) Energy and exergy analyses of an integrated renewable energy system for hydrogen production. Energy 204:117945. https://doi.org/10.1016/j.energy.2020.117945
69. NASA (2020) NASA Power: prediction of worldwide energy resources. https://power.larc.nasa.gov/data-access-viewer/. Accessed 25 Nov 2021
70. Verleysen K, Parente A, Contino F (2021) How sensitive is a dynamic ammonia synthesis process? Global sensitivity analysis of a dynamic Haber-Bosch process (for flexible seasonal energy storage). Energy 232:121016. https://doi.org/10.1016/j.energy.2021.121016

Chapter 3
Ammonia and Alternative Fuels

Nomenclature

Acronyms

1,4-DB	1,4-dichlorobenzene
ADF	Abiotic depletion factor
ADP	Abiotic resource depletion potential
AECL	Atomic energy of Canada limited
AEL	Alkaline electrolysis
AP	Acidification potential
BEV	Battery electric vehicle
BG	Biomass gasification
BTMS	Battery thermal management systems
CACs	Criteria air contaminants
CAD	Canadian dollar
CCS	Carbon capture and storage
CEPCI	Chemical engineering plant cost index
CERL	Clean energy research lab
CFC-11	Trichlorofluoromethane
CFC	Chlorofluorocarbon
CG	Coal gasification
CI	Compression ignition
CML	Institute of Environmental Sciences, Leiden University
CNG	Compressed natural gas
CO	Carbon monoxide
CO_2	Carbon dioxide
DALY	Disability-adjusted life year
DME	Dimethyl ether
EIA	Energy information administration

© The Author(s), under exclusive license to Springer Nature Switzerland AG 2022

I. Dincer et al., *Ammonia Energy Technologies*, Lecture Notes in Energy 91,
https://doi.org/10.1007/978-3-031-13532-3_3

EP	Eutrophication potential
EPD	Environmental product declaration
EPFL	French Federal Polytechnic School of Lausanne
EVs	Electric vehicles
GHGs	Greenhouse gases
GWP	Global warming potential
HBEFA	Handbook Emission Factors for Road Transport
HEVs	Hybrid electric vehicles
HHV	Higher heating value
HTP	Human toxicity potentials
ICE	Internal combustion engine
ICEV	Internal combustion engine vehicle
IEA	International Energy Agency
IPCC	Intergovernmental panel on climate change
LCA	Life cycle assessment
LCI	Life cycle inventory
LCIA	Life cycle impact assessment
LNG	Liquefied natural gas
LPG	Liquefied petroleum gas
MJ	Megajoules
NG	Natural gas
NO_x	Nitrous oxides
NP	Nitrification potential
NREL	National Renewable Energy Laboratory
ODP	Ozone depletion potential
PEM	Proton exchange membrane electrolysis
PHEVs	Plug-in hybrid electric vehicles
PM	Particulate matter
PO_4	Phosphate
PSA	Pressure swing adsorption
PV	Photovoltaic
RAD	Radioactive radiation
RAINS	Regional air pollution information and simulation
SATIM	Assumptions and Methodologies in the South African Times
SMR	Steam methane reforming
SO_2	Sulfur dioxide
SOEC	Solid oxide electrolysis cell
SOFC	Solid oxide fuel cell
TES	Thermal energy storage
UCG	Underground coal gasification
UCTE	Union for the co-ordination of transmission of electricity
VOCs	Volatile organic compounds
WMO	World Meteorological Organization

In this chapter, environmental impacts of ammonia and alternative fuels are discussed. The first section provides some general information regarding life cycle assessment (LCA). In the next section, a cradle-to-grave LCA study of ammonia is carried out and then various case studies are presented. The research addresses all aspects and domains of ammonia, including diverse ammonia production processes and their applications. Furthermore, various case studies on ammonia usage on a sectoral basis are conducted.

Obvious problems with fossil fuels and their widespread use in many industries have had harmful effects on human health, the ecosystem and the environment. There is therefore an urgent need for some affordable, effective and environmentally benign possible replacements to be developed and employed successfully. Ammonia maybe the solution to wrapping up the growing search for an alternative fuel. Ammonia has been widely utilized in a variety of sectors and for a variety of purposes. It can be used as an energy source, a major chemical in the manufacturing of feedstock and chemicals, a primary ingredient in cleaning materials, an engine fuel, and a coolant for cooling systems. Because of its versatility and wide variety of applications, ammonia is an important component.

3.1 Introduction

An intimate relationship exists between energy, the environment and sustainable development. Ideally, a community pursuing sustainable development would only use energy products that do not have any environmental impacts. However, because all energy resources have some impact on the environment, it is fair to say that some of the concerns about the constraints placed by environmental emissions on sustainable development and their harmful impacts can be partially resolved by increasing energy efficiency. Clean, boundless and highly competitive energy sources are renewable energy sources. They vary primarily in their variety, availability and capacity for use anywhere on the planet from fossil fuels, but primarily in that, they generate neither greenhouse gases nor polluting emissions that cause climate change. Their expenditures are also declining and at a sustainable rate, despite their current uncertainty, the general cost trend for fossil fuels is in the opposite direction [1–3].

3.2 Environmental Impacts of Ammonia and Other Fuels

In this section, some general information regarding LCA is provided. LCA is one of the most comprehensive approaches for assessing a product's environmental impact over its entire life cycle. Different applications have different goals for conducting

LCA studies. The main goal of using the LCA methodology is to reduce the environmental impact of evaluated items in order to find more sustainable alternatives during the decision making process.

3.2.1 Background and Literature Review

An LCA is an approach to cradle-to-grave assessment that helps to make productive and reasonable environmental choices. An LCA is often used, which starts from raw material production and progressing to final disposal, to determine the environmental effect of a product or process over its entire life. To measure the overall environmental effects, an LCA considers all life phases of the product or process. An LCA is often performed to establish the impacts of alternative items, but it may also identify areas of major concern and assist optimize processes. There are good number of studies on hydrogen production methods were conducted in the open literature, however; LCA of ammonia production methods and utilization on different applications was not widely researched. Prof. Dincer's group [4–13] has numerous publications on LCA of ammonia, ammonia production and its utilization and applications. They have investigated different aspects of ammonia and its possible environmental impacts on ecosystem, human health and resources.

Bicer et al. [6] conducted an LCA study on impact assessment and environmental evaluation of various ammonia production processes. The research compares fossil fuels-based ammonia generation routes to a detailed life cycle evaluation. The chosen ammonia production options range from steam methane reforming to heavy oil partial oxidation. The most popular commercially available Haber–Bosch method is the selected ammonia synthesis process. Different conventional resources, such as coal, nuclear, natural gas and heavy oil, are used to provide the necessary energy input for the methods. The environmental impacts of selected approaches are defined and quantified using cradle to gate approach. The LCA results of conventional resource-based ammonia production routes indicate that the method of producing ammonia based on nuclear electrolysis produces the lowest impacts of global warming and climate change, while the options for coal-based electrolysis lead to higher environmental issues. For nuclear-based electrolysis, the measured greenhouse gas emission is 0.48 kg CO_2 equivalent, whereas for coal-based electrolysis, it is 13.6 kg CO_2 per kg of ammonia.

Bicer et al. [12] performed a comparative LCA on various ammonia production methods. Four distinct methods of ammonia processing are comparatively examined in the study. The proposed ammonia production systems are made up of an electrolyzer for the production of hydrogen, and a Haber–Bosch ammonia synthesis facility. The energy required for the processes are applied from different sources such as hydropower, nuclear, biomass and municipal waste. The methodology of LCA is used to classify and measure environmental impacts across the life cycle of the processes in the categories of global warming potential, human toxicity and abiotic depletion of each technique. The suggested non-conventional alternatives for the production of ammonia are comparatively evaluated as relevant parameters for realistic applications for the production of ammonia in terms of environmental impacts and both energy and energy efficiencies. The LCA findings for the ammonia production processes suggest that ammonia production methods focused on urban waste incineration plants and hydropower have lower emissions than other selected methods. The estimated greenhouse gas (GHG) emissions from the processing methods of ammonia based on hydropower, nuclear, biomass and urban waste was measured as 0.38, 0.84, 0.85 and 0.34 kg CO_2-eq for each kg of ammonia.

Bicer and Dincer [5] carried out a comparative evaluation study on LCA of nuclear-based hydrogen and ammonia production options. Through LCA tool, nuclear energy-based hydrogen and ammonia production methods ranging from thermochemical cycles to high-temperature electrolysis are comparatively assessed in the study. The results of the LCA for the chosen ammonia production methods indicate that the ammonia production method based on nuclear electrolysis produces lower global warming and climate change impacts, while the thermochemical options generate higher values of abiotic degradation and acidification.

Bicer and Khalid [8] presented a study on life cycle environmental impact comparison of solid oxide fuel cells fueled by natural gas, hydrogen, ammonia and methanol for combined heat and power generation. The goal of this study is to provide a thorough assessment of the environmental life cycle of heat and power generation through solid oxide fuel cells (SOFCs) fueled by different chemical feeds, namely natural gas, hydrogen, ammonia and methanol. The LCA involves the complete phases from the extraction of raw materials or the synthesis of chemical fuel to the usage as a cradle-to-grave method in the electrochemical reaction. They used GaBi LCA software for analysis. The results of the study indicate that feeding carbon-free SOFC systems reduces greenhouse gases during operation, but more steps needed for hydrogen, ammonia and methanol conversions of natural gas make the entire process more environmentally troublesome. However, if hydrogen and ammonia are derived from renewable sources such as wind-based energy, the environmental effect is greatly reduced, resulting in approximately 0.05 and 0.16 kg of CO_2 eq per kWh of SOFC electricity production, respectively.

Singh et al. [11] published a book section on LCA of ammonia production methods. The section provides a unique and detailed LCA of ammonia production methods to determine their relative performances. The section considers and contrasts five strategies and seven routes of ammonia. The methods they considered are based on fossil fuels, gasification of biomass and water electrolysis using wind and solar PV power. SimaPro 7, which is commonly used software for LCA studies, was used to conduct the impact assessment. The life cycle inventory (LCI) was analyzed using two methods: CML 2001 and Eco-indicator 99. The findings of the analysis are primarily analogous to emissions of carbon dioxide, acidification, eco-toxicity, human toxicity, carcinogens, and abiotic depletion.

Several LCA studies have recently been conducted on the production and usage of ammonia [12–30]. LCAs are primarily carried out based on input and output energy and material flows. Setting a boundary for the system in an LCA is important. The LCA is recognized as a four-step process. These steps include goal and scope definition, inventory analysis, impact assessment and interpretation where potential improvements are identified. The interpretation step relates to each of the four components. This indicates that the LCA is a feedback mechanism and that it is feasible to utilize the outcomes of subsequent phases to strengthen the prior step.

Chisalita et al. [14] conducted a study on environmental evaluation of European ammonia production considering various hydrogen supply chains. In this study, a cradle-to-gate environmental impact study of conventional and environmentally friendly hydrogen production pathways integrated with ammonia production is performed. The production of hydrogen by conventional SMR coupled with the absorption of gas–liquid carbon dioxide by amines and chilled ammonia, electrolysis and chemical looping based on iron was analyzed and compared. They used mass and energy balances from process flow modeling as inputs for the environmental assessment using LCA methodology. The system boundaries considered in this study are including: (a) main processes: production of ammonia, production of hydrogen, separation of carbon dioxide; (b) upstream processes: nitrogen, natural gas, ilmenite, methyl-di-ethanol-amine, chilled ammonia and supply chain catalysts; (c) downstream processes: compression, transport and storage of carbon dioxide, solvent degradation/disposal/oxygen carrier. The findings suggest that the synthesis of natural gas-based ammonia integrated with the production of chemical looping hydrogen offers the greatest reduction in global warming potential, while six out of the nine effect indicators examined suffer a rise of 30–60%.

Gomez et al. [31] presented a study on techno-economic analysis and life cycle assessment for electrochemical ammonia production using proton conducting membrane. To gain insight regarding potential energy savings for metric tonne of ammonia, diverse approaches using alternative approaches for nitrogen generation and hydrogen production were examined. A theoretically viable route for green ammonia was identified to be electrochemical synthesis utilizing electrolysis for hydrogen and cryogenic nitrogen. For this route, the profitability metrics such discounted cash flow rate of return, present value and discounted payback period were calculated to be 8%, \$40 MM, and 4–6 years respectively.

Makhlouf et al. [22] their finding show that total energy demand is of 51.945 $\times 10^3$ MJ/t of ammonia, which is higher than the global average. Global Warming Potential (GWP) is of 1.44 t CO_2 eq/t of ammonia; this value is lower than the global average. Tropospheric ozone layer depletion and acidification are also studied, and their values are 549.3×10^{-6} t NMVOC eq and 259.3×10^{-6} t SO_2 eq respectively.

Young et al. [25] conducted a study on comparative environmental life cycle assessment of carbon capture for petroleum refining, ammonia production, and thermoelectric power generation in the United States. Their study showed that the potential for the composition of particulate matter, eutrophication potential and water usage increased in all sectors due to the installation and activity of carbon capture and storage (CCS) technologies per kg of CO_2e decreased, while the impact on the propensity for acidification and the potential for the formation of particulate matter was mixed. The discrepancies between systems in trade-offs are mainly influenced by three factors: the combustion emissions from the fuel used to power the capture unit, the upstream supply chain to procure that fuel, and the relative effect of the carbon capture unit on the base emissions of the flue gas.

Cox and Treyer [19] carried out a study on environmental and economic assessment of a cracked ammonia fueled alkaline fuel-cell for off-grid power applications. Their results indicate that ammonia production leads the results of the LCA, and that routes of sources of renewable ammonia significantly improve environmental efficiency. Sensitivity analyses show that cell power density and lifetime and system performance are the fuel cell parameters that most influence system cost and environmental burdens. They observed that recycling of anode catalyst and electrode substrate materials has a significant effect on environmental performance, but without significant cost incentives. The alkaline fuel cell is estimated to generate electricity with 1.08 kg kWh^{-1} CO_2 emissions, which is 23% lower than that of a diesel generator.

Kim et al. [18] worked on a study regarding LCA of CO_2 assessment of an ammonia-based carbon capture and storage system. The specific goal of their LCA study is determine the net CO_2 reduction efficiency of a simulated ammonia-based CCS system and to identify high CO_2 emitting processes to be improved with priority. Just 64% of the CO_2 emitted from a power plant is trapped by the CCS system, which is slightly lower than the CO_2 capture output (95%), since CO_2 equivalent to the gap is subsequently emitted due to the use of energy and materials for CCS operations.

Sadeek et al. [32] performed a study on the influence of raw material availability and utility power consumption on the sustainability of the ammonia process. The findings showed that, due to higher steam demand for CO_2 removal, the real energy demand rises as the gas charge decreases. The results also showed that there was indeed a direct loss of profit of US$ 10.3 million per year below a threshold of 98% gas charge, with a rise in real energy demand of up to 10%.

Cai et al. [33] conducted a study on mitigating NO_x emissions from an ammonia-fueled micro-power system with a perforated plate implemented. Results indicate that NO emissions are decreased by up to 73.3% with those in the lack of perforated plates when an optimized perforated plate is introduced. The reduction is mainly due to the development of a low flame-temperature recirculation region. It is shown that increasing width plays a positive role in lowering the generation of NO, whereas location contributes to a reverse trend resulting from the recirculation zone's variation in scale. The plate material, by comparison, has a marginal effect on NO_x emissions.

3.2.2 *Life Cycle Assessment (LCA)*

In this section, a detailed discussion of an LCA methodology is provided. The life cycle steps involved are detailed in accordance with ISO standards. In addition, there is also a full explanation of the different environmental impact categories and their significance.

Description of Life Cycle Assessment

LCA is one of the most comprehensive approaches for assessing a product's environmental impact over its entire life cycle. Various applications have different goals for conducting LCA studies. LCA is used in many applications for various goals. Using LCA technique to reduce the environmental impact of certain items under consideration for more sustainable solutions through decision-making process is of paramount importance [34, 35].

The most important stage of an LCA study is life cycle impact assessment (LCIA), in which the LCI results of individual compounds linked with a given system study are transformed into understandable impact categories that represent the environmental impact. Several LCIA methods have been created to deal with the emissions impacts modeling on the environment using various approaches.

Why LCA?

The need to measure product environmental emissions in order to classify probable area for improvement through lower resource consumption and fewer environmental impacts throughout the life cycle prompted the creation of LCA studies. LCA can be used for a variety of purposes, and each of these goals may require a different set of data collection details. The data is likely to be very simplified or detailed, depending on the application and purpose for which the LCA tool is used [36, 37]. Some significant LCA goals are explained in the following sections.

Product Development

LCA-based product development is referred to as sustainable process design. During the design phase of any given product, there are usually several possibilities for selecting resources and materials. Using LCA during product development is critical since any decision about resources and materials has an impact on following life cycle phases [38]. The more effectively the previous LCA is used in the design phase, the lower the environmental impact will be. In this case, the LCA methodology may involve extensive data collection, which might be time-consuming. Making LCA simpler in this circumstance can be beneficial. Since the focus, research, and study will be on the resources and materials that may have a significant environmental impact, and then try to discover replacement preliminary design solutions before moving further with the development process [39].

Product Improvement

In terms of data collecting, it is possible to improve an existing process or product. When using LCA as a method for product improvement, it is critical to focus primarily on the materials and resources that have a substantial impact on the product. As a result, different items can be compared from an environmental standpoint, where the environmental impact of each specific product can be estimated and compared to another case in the same categorization. Substitute solutions for resources or commodities that have a higher environmental impact during the life cycle stages (production, operation, and disposal) are then merged and assembled for re-evaluation of the entire solution [40].

Marketing

Marketing is the process of aligning product features with customer demands and needs in order to meet specified quality standards. With the rise in environmental consciousness, consumers are paying more attention to certain environmental qualities of services and goods. Environmental labelling (Eco-labelling) is widely known as the most significant sort of LCA technique for environmental marketing [41]. Environmental labelling is used as proof to determine whether a product is environmentally friendly. When a product meets Eco-labelling criteria, it receives an Eco-label, which is why it is appealing for marketing purposes and ecologically friendly items are more appealing to consumers. LCA is a crucial criterion for the advancement of Eco-label criteria, according to EU Eco-label regulation [42]. Another marketing purpose type of Eco label scheme is environmental product declaration (EPD). EPDs are comprised of a variety of information relating to the environmental features and components of a specific product, as defined by the ISO 14025 standard [43, 44]. The overall impression of an EPD is to use a bar diagram to give a graphic representation to a product of a present environmental impact number. The graphical presentation is easy to understand for sustainable products and services and professional buyers, but it may not be for all consumers. Though it appears that EPD is superior to Eco-labeling in terms of allowing conversant decisions rather than passing judgments, the level of detail regarding the establishment of EPD is higher than that required for the Eco-labeling scheme, which poses a challenge in terms of its implementation and operation [45].

General LCA Framework

This section discusses some general LCA-related terms and meanings. The ISO 14040 standards give a defined structure for using an LCA approach. The subject and intended application of the work determine the scope, considered boundaries, and detail level of an LCA research. Depending on the purpose of a particular LCA study, the complexity and scope of the study can vary greatly. To give accurate results for direct applications, all LCA studies should adhere to ISO standards [46, 47].

Figure 3.1 displays LCA framework according to ISO 14040 standard. The arrows in Fig. 3.1 point out the iterative and cooperative nature of LCA. For example, during the impact assessment, one may find that specific information is ambiguous or missing, indicating also that inventory analysis needs to be improved. In another case, the results interpreted during the interpretation stage may be confusing or insufficient to meet the application requirements, indicating that the scope and goal definitions may need to be improved or modified. The ISO standard [46, 47] outlines these four processes for conducting an LCA evaluation efficiently, and they are discussed in the following subsections.

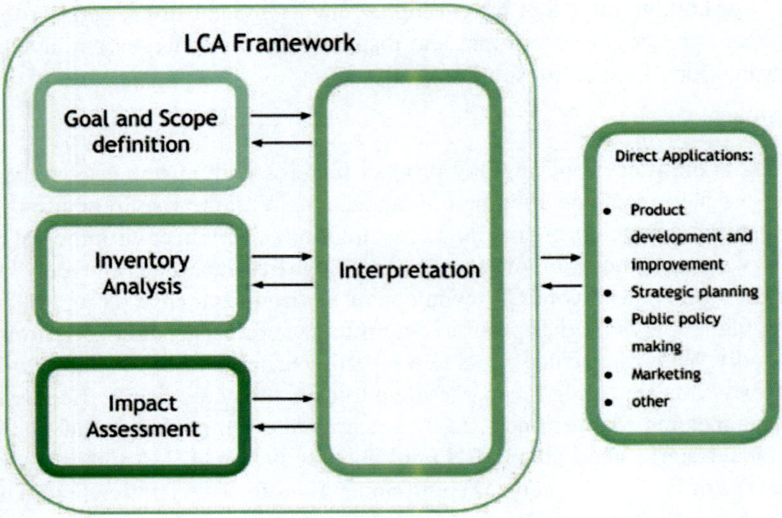

Fig. 3.1 LCA framework according to ISO 14040 standard (modified from [48])

Goal and Scope Definition

The goal and scope definition step of an LCA analysis is where the product is defined and the assessment context is established. This is a critical stage in the LCA process. This phase informs the impact assessment stage by outlining a variety of criteria such as the study's aim, the time and resources required the intended application, the assessment methodology, system boundaries, and a wide range of restrictions and assumptions. As a result, aim and scope definition will guide the entire LCA process to ensure the best possible outcomes [47].

Inventory Analysis

The inventory analysis results are presented in the form of a list, which includes the quantities of materials and energy utilised during each stage of a product's life cycle. The accompanying materials and energy flow based on the individual product are thus specified during the inventory analysis stage to symbolise the product and the overall inputs and outputs to and from the appropriate natural environment [49]. The LCI analysis is heavily dependent on the amounts and types of multiple natural resources such as water, electricity, and others, as well as transportation methods, manufacturing materials, product usage over its lifetime, and final disposal. Depending on the locality, the deliberation and special impacts of the described factors may differ. For various scenarios, such as a region with insufficient resources for certain product production, or another region with various technologies for certain materials production, or it may be more reliant on fossil fuels or renewable energy, these variances may cause the LCA assumptions and limitations to be disrupted.

Impact Assessment

In this step of the impact assessment, a compiled LCI list containing linked product studies based on spent energy quantities and matching materials is analysed and turned into comprehensible impact indicators. The metrics listed below show the intensity of the impact categories' contribution to the environmental burden. Some steps within the listed steps are mandatory, while the remaining steps are optional, according to ISO standard 14,044. Impact categories formulation and categorization, as well as characterization, are required processes, but normalisation and weighting are optional [46].

Interpretation of the Results

The final step in the LCA analysis is life cycle interpretation. The LCI and LCIA findings are categorized, quantified, reviewed, and measured in a systematic approach. Relevant assumptions and suggestions will result in interpretations. Interpretation should classify the important problem on the basis of the LCI and the LCIA performance. A study should be assessed by interpretation, taking into account the study's completeness, accuracy and sensitivity. Most significantly, interpretation of the life cycle should draw a conclusion, research limitations, and recommendations.

3.2.3 LCIA Methods

The LCIA stage is considered the most important step in an LCA study since it deals with a large amount of data that is displayed in the inventory analysis results. After passing through compound environmental modeling focused on the normalization and characterization of environmental and natural science while acknowledging social, political, and ethical issues. The results of the inventory analysis are transformed into easily understandable impact indicators coinciding within the impact categories. Because of this intricacy in the LCIA step, methods to simplify and improve the LCIA process have been developed. LCIA approaches are defined as the methods used to relate LCI findings to the associated environmental consequences, with the LCI findings being classified into impact classifications with a category indication. The creation of LCIA methodology speeds up the trade-offs and differences among many alternative goods, simplifies the process for LCA practitioners, and facilitates benchmarking.

Many LCIA methods have been developed over time to identify and describe the system's environmental impacts, such as Eco-indicator 99, EDIP 2003, CML 2001, IMPACT 2002+, ReCiPe Endpoint, CML 2 baseline 2000, BEES, TRACI 2, EDIP 2.

CML 2001 Assessment Method

In 2001, a group of scientists coordinated by the Institute of Environmental Sciences at Leiden University introduced CML, a midpoint oriented method encompassing an effect categories group and impact assessment stage characterization methodologies. Standardization is provided, but no weighting or addition is included. The CML approach's legitimacy in provincial effect categories is universal, with the exception of acidification and photo-oxidant formation. Because it is a problem-focused approach, the CML 2001 method is the most extensively employed in LCA investigations.

Acidification Potential (AP)

The acid indication of acidifying contaminants on groundwater, soil, surface water, ecosystems, biological organisms, and substances is known as acidification potential (AP). SO_2, NO_x, and NH_x are the main acidifying pollutants. The natural habitat, human health, the human-made environment, and natural resources are all protected zones. Groundwater, soil, surface water, ecosystems, species, and components are all affected by acidifying compounds. The regional air pollution information and simulation 10 (RAINS) model is used to determine the AP for air discharges.

Eutrophication Potential (EP)

In both terrestrial and aquatic environments, nutrient upgrading can prevent unwanted changes in species composition and increase biomass production. Eutrophication encompasses all possible effects of extremely high macronutrient concentrations in the environment, namely nitrogen and phosphorus. All emissions with the same effects are grouped together in the "eutrophication" impact category. The eutrophication potential (EP) is mostly based on the Heijungs stoichiometric technique, which is expressed as a ratio of kg PO_4 equivalents to kg emission, with the geographical scale varying between continental and local.

Abiotic Depletion Potential (ADP)

Non-living natural commodities like crude oil and copper ore, as well as energy sources like wind energy, are examples of abiotic resources. One of the most often discussed effect categories is abiotic depletion potential (ADP), and there are a variety of approaches for describing contributions in this category. The extraction of non-renewable raw materials is primarily considered. The unit indicator for this impact category is the kg antimony (Sb) equivalent. The ecology and human health are the two most important focuses of the mentioned category, both of which are impacted by mineral and fossil fuel extraction as system inputs.

Global Warming Potential (GWP)

The influence of emissions on various radiative forcing of the atmosphere, such as heat radiation absorption, is represented by the global warming potential (GWP). Global warming has an impact on the environment and human health, resulting in climate change. The majority of such releases increase radiative forcing, which results in an increase in temperature at the earth's surface, known as the "greenhouse effect". GHGs released into the atmosphere have an impact on climate change. Climate change has the potential to have negative consequences for the ecology, human health, and substantial welfare. The intergovernmental panel on climate change (IPCC) developed the classification model, and it was carefully chosen for the advancement of characterisation criteria. The GWP is measured in kilograms of carbon dioxide emitted over a 500-year time period (GWP500).

Ozone Depletion Potential (ODP)

Because of released emissions, the stratospheric ozone layer is becoming thinner, and this phenomenon is known as stratospheric ozone depletion. Because of the thinning, more solar UV-B radiation reaches the earth's surface, endangering ecology, human health, marine and terrestrial eco systems, animal health, materials, and biochemical cycles. On a global scale, this classification is output-related. The world meteorological organization (WMO) designed the description model, and it expresses distinct gases by their ozone depletion potential (kg CFC-11 equivalent/kg emission).

Human Toxicity Potential (HTP)

The focus of this category is toxic elements in the ecology that impair human health and the environment. The dangers to human health in the operational environment are not included in the discussed category. LCA evaluates human toxicity potentials (HTP) and categorization criteria, as well as exposure, describing destiny, and toxic substance impacts, for an indefinite period. Each harmful substance is expressed in 1,4-dichlorobenzene equivalents/kg emissions.

Ecotoxicity Potential

Toxicity in the environment is quantified using three different effect categories: freshwater, marine, and terrestrial. Some compounds, such as heavy metals, might have a negative impact on the ecosystem. Maximum tolerated concentrations in water for ecosystems have been used to assess toxicity. The USES-LCA, which is based on EUSES, the EU's toxicity model, is used to calculate ecotoxicity potentials. This gives a way for describing the fate, exposure, and environmental impacts of dangerous chemicals. The reference unit is kg 1,4-dichlorobenzene equivalent (1,4-DB), and characterisation parameters are determined individually for harmful substances' effects on:

- Fresh-water aquatic ecosystems
- Marine ecosystems
- Terrestrial ecosystems.

Photochemical Oxidation Potential

Stratospheric ozone layer is protecting with in stratosphere, but it is hazardous to people at ground level at large concentrations. The interaction of volatile organic molecules with nitrogen oxides in the presence of heat and sunshine produces photochemical ozone, often known as "ground level ozone". The impact category is determined by the amount of carbon monoxide (CO), sulphur dioxide (SO_2), nitrogen oxide (NO), ammonium, and NMVOC (non-methane volatile organic compounds). Summer smog is a photochemical ozone production potential for substance emission into the air.

Eco-Indicator 99 Method

During the Dutch NOH programme in 1995, PRe consultants developed the Eco-indicator approach, which is an endpoint-oriented method. The Eco-indicator approach identifies the environmental impact through scores and statistics. It simplifies LCA clarification by containing the weighting technique. It encourages assigning a single score to each process or product after weighing, which is then compared to the environmental impact. The score is given on a scale of one to ten, with one point representing a regular citizen's annual environmental burden.

Human Health

This category includes the duration and number of diseases, as well as the amount of life years lost as a result of permanent mortality caused by environmental deterioration. Climate change, carcinogenic effects, ozone layer depletion, ionisation, and respiratory effects all play a role in the described effects on a worldwide scale.

Ecosystem Quality

At the regional scale, this category includes species diversity, ecotoxicity, and acidification, land-use, and eutrophication impacts.

Resources

This category includes the depletion of energy supplies and basic commodities. It is estimated in terms of the additional future energy needed to extract lower-quality minerals and energy on a global scale. The land use category includes depletion due to agricultural resource depletion.

IMPACT 2002+ Method

IMPACT 2002+ is an approach that focuses on the middle and end of the spectrum. EPFL, the French Federal Polytechnic School of Lausanne, and the Swiss Federal Institute of Technology developed this method. IMPACT 2002+ combines four methods: IMPACT 2002, CML, Eco-indicator 99, and IPCC. The methodology takes into account human toxicity, ionising radiation, respiratory effects, photochemical oxidant formation, ozone depletion, terrestrial ecotoxicity, global warming, aquatic ecotoxicity, mineral extraction, terrestrial and aquatic eutrophication and acidification, land occupation, and non-renewable energy. Endpoint via midway categories, which are also environmental protection zones, are related with four damage categories based on inventory results: human health, climate change, ecosystem quality, and resources. Four different techniques alter categorization factors: Impact 2002, CML, Eco-indicator 99 and IPCC.

3.3 Life Cycle Assessment of Ammonia and Other Fuels

In this section, various cradle-to-grave LCA studies of ammonia and other fuels are carried out. The studies consider all aspects and domains of ammonia by considering various methods for ammonia production and their use in different applications. In addition, numerous case studies is carried out for ammonia utilization in sectoral base. The results of the study are illustrated in following impact categories; acidification potential, global warming potential, eutrophication potential, abiotic depletion potential, ozone layer depletion potential, human toxicity potential, ionising radiation, and photochemical oxidation potential.

Figure 3.2 compares environmental impacts of various fuels. Ammonia has lowest global warming potential impact when produced from clean hydrogen sources. The other alternative fuels show lower global warming impact as well when they come from renewables or from clean hydrogen. On the other hand, conventional fuels such as gasoline and diesel emit very high CO_2 emissions in comparison with ammonia. Global warming effect of ammonia mainly comes from chemical factory and electricity production plants.

When ammonia is produced from clean hydrogen, its global warming potential impact value comes out as 0.37 kg CO_2 eq/kg fuel, but when produced from fossil

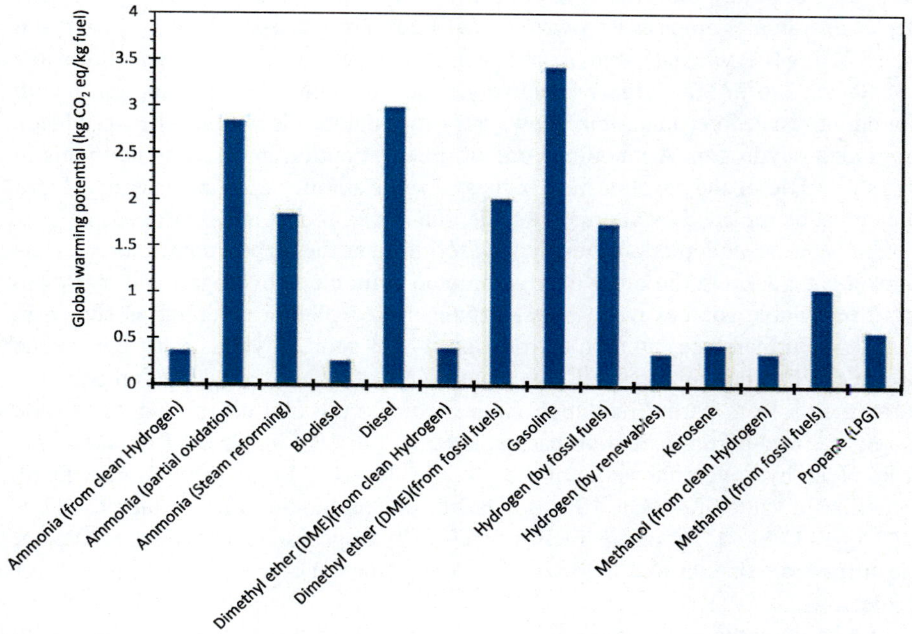

Fig. 3.2 LCA results of global warming potential of various fuels

fuels methods like steam reforming the value comes out as 1.85 kg CO_2 eq/kg fuel. On the other hand, global warming value of gasoline is calculated as 3.43 kg CO_2 eq/kg fuel. Thus, the LCA results of various fuels points that switching from fossil fuels to green ammonia fuel will provide great benefits to the environment in terms of global warming and climate change.

Green ammonia, in other words ammonia from green hydrogen, is recognized as the most environmentally benign option in all ammonia production methods that examined in this study. Several companies are working on generating green ammonia, a method of producing ammonia in which hydrogen obtained from water electrolysis powered by renewable energy replaces hydrocarbon-based hydrogen, effectively eliminating carbon dioxide emissions from the process.

The primary issue of abiotic depletion is the impact of mineral and fossil extraction as system inputs on human and ecosystem health. The abiotic depletion factor (ADF) is calculated for each mineral and fossil fuel extraction. This parameter is connected to concentration reserves and rate of de-accumulation on a global scale. The key concern of this category is the human and ecosystem health that is affected by the extraction of minerals and fossil as inputs to the system. For each extraction of minerals and fossil fuels, the abiotic depletion factor is defined. This indicator has globe scale where it is related to concentration reserves and rate of de-accumulation.

As shown in Fig. 3.3 production of ammonia-fuel yields lower abiotic depletion compared to petrol, diesel, dimethyl ether (DME) and propane production. Abiotic depletion value of ammonia shows 6.63 MJ/kg fuel when from clean hydrogen and 41.17 MJ/kg fuel when it comes from fossil fuels. However, diesel and gasoline show 332.28 and 296.77 MJ/kg fuel respectively, which are high values in comparison with ammonia. Therefore, ammonia shows very low abiotic depletion when produced from clean hydrogen. Ammonia is one of greenest option in all examined fuels in this study. Diesel and gasoline have biggest share in abiotic depletion category. Thus, they must be replaced by alternative fuels, such as ammonia and hydrogen.

For ozone layer depletion, diesel and gasoline make the highest contribution in this impact category. On the other hand, ammonia from clean hydrogen and hydrogen from renewable sources have lowest ozone layer depletion potential as shown in Fig. 3.4. Furthermore, ammonia from fossil fuel sources yields very low ozone layer depletion potential as well. The main source of ozone layer depletion potential in ammonia from partial oxidation comes from heavy fuel oil production. While ammonia from partial oxidation has value of 5.11×10^{-7} kg CFC-11 eq, ammonia from clean hydrogen has value of 4.85×10^{-8} kg CFC-11 eq. On the other hand, diesel has a value of 4.28×10^{-6} kg CFC-11 eq and biodiesel has value of 3.87×10^{-8} kg CFC-11 eq. Therefore, fuels from clean hydrogen or renewables are potential clean future fuels that will help tackling environmental concerns over ozone layer depletion.

While ammonia has lowest environmental impact in most categories, its human toxicity LCA results comes out a bit high, but still performs better than gasoline. Most importantly human toxicity of ammonia can be mitigated when clean hydrogen is

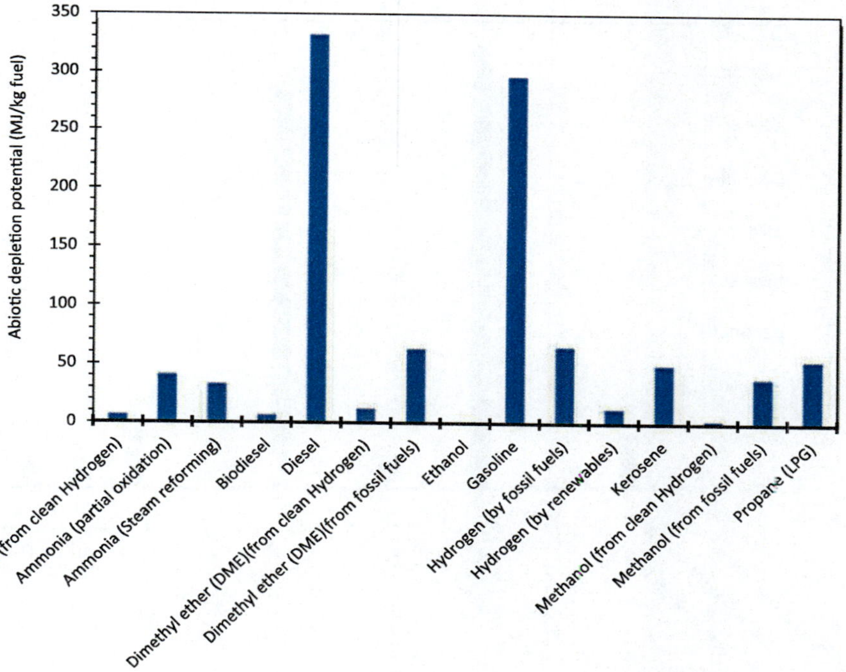

Fig. 3.3 Abiotic depletion potential LCA results of various fuels

used for ammonia production. Human toxicity values are calculated as 0.28 kg 1,4 DB eq/kg fuel for ammonia from hydrogen, and 2.26 kg 1,4 DB eq/kg fuel for ammonia from partial oxidation. For gasoline 2.57 kg 1,4 DB eq/kg fuel, and 1.96 kg 1,4 DB eq/kg fuel for diesel were calculated. Figure 3.5 shows that ammonia from clean hydrogen drastically minimizes the human toxicity emissions and makes ammonia a promising future fuel. Furthermore, when there is more than 5 ppm of ammonia in the air, it has a very heavy, unpleasant odor that can easily be detected/felt. As a result, one may probably smell ammonia until they are exposed to an excessive level of concentration. Ammonia levels in the air that may cause serious effects over people are much higher than those one may encounter at home or at work.

When fuels are produced from clean energy sources, they have huge benefits for the environment. As seen in Fig. 3.6, the ammonia derived from clean hydrogen has one of the lowest acidification impacts compared to other fuels. Ammonia, which is produced with partial oxidation methods releases higher acidification emissions, but still shows much lower results than diesel and gasoline fuels. Fuels from clean sources such as hydrogen, biodiesel, DME, ethanol, and methanol yield much lower environmental impacts in comparison with conventional fuels like, gasoline, diesel, and propane. One aspect of global climate change is ocean acidification. Anything we

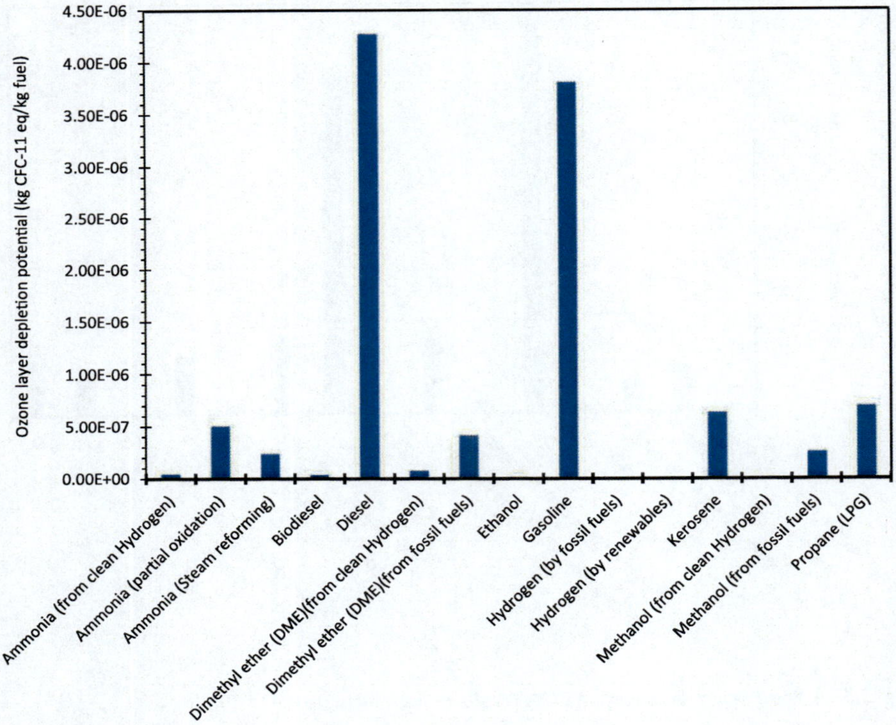

Fig. 3.4 Ozone layer depletion potential LCA results of various fuels

do now to combat climate change will help the ocean's future as well. The analysis of the possible impacts of ocean acidification has received a lot of attention in the ocean science community over the last decade. It is currently difficult to foresee how ocean acidification would affect the overall structure of marine ecosystems and cascade across the marine food web. Scientists, resource managers, and politicians understand the critical need to reinforce research as a foundation for sound decision-making and action as the rate of ocean acidification accelerates.

Similar to other impact categories fossil fuel-based fuels have highest eutrophication impact then renewable based fuels. Ethanol and hydrogen from renewables have most promising LCA results as displayed in Fig. 3.7. The eutrophication results for ammonia from clean hydrogen is 0.00017, 0.0015 kg PO_4-eq/kg fuel is calculated for ammonia from partial oxidation. For diesel and gasoline, the values of 0.0047, 0.0045 kg PO_4-eq/kg fuel are calculated respectively. As fossil fuels are burned, nitrogen oxides (NO_x) are released into the atmosphere. Smog and acid rain are caused by NO_x emissions. Rain and snow re-deposit NO_x on land and in bodies of water. The main sources of NO_x are coal-fired power plants and car, bus, and truck exhaust.

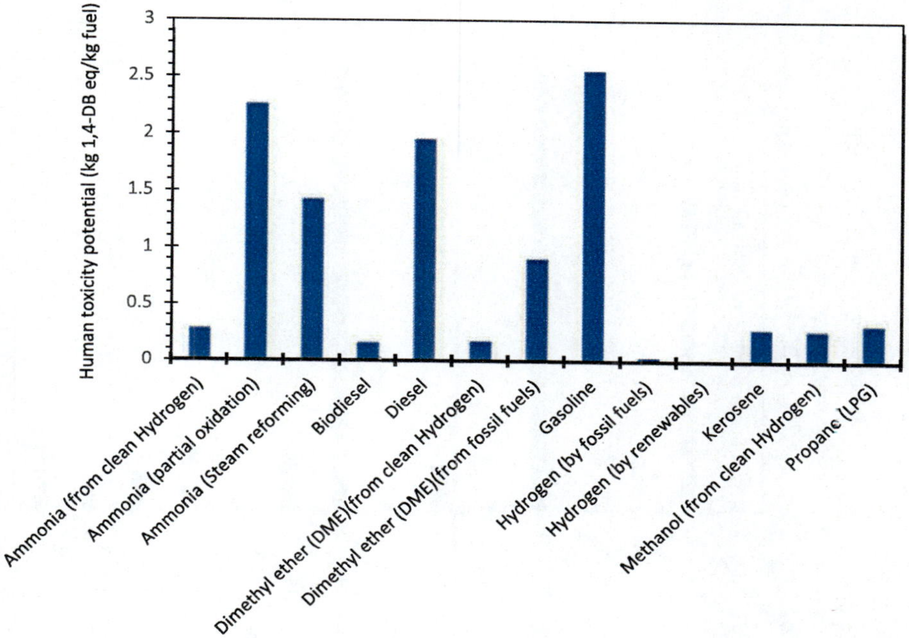

Fig. 3.5 Human toxicity potential LCA results of various fuels

3.3.1 LCA Analysis of Ammonia Production Methods

In the current study, for comparative assessment purposes, different ammonia production methods are selected from both conventional and renewable resources. Haber–Bosch will be used for ammonia synthesis. For hydrogen production, following methods will be used; electrolysis, thermolysis, thermo-chemical, photo-electrochemical, fossil fuel reforming, and photo-catalysis. In addition, SMR, coal and biomass gasification will be used for ammonia processing. For large amounts of nitrogen output, cryogenic air separation is often used. Electricity or heat for the operation, cooling water, waste heat and equipment for the air separation plant are included in the life cycle assessment of nitrogen generation. SMR is a mature production method in which hydrogen from a methane source, such as natural gas, is produced using high-temperature steam. In steam-methane reforming, in the presence of a catalyst to create hydrogen, carbon monoxide, and a relatively limited amount of carbon dioxide, methane reacts with steam under 3–25 bar pressure. The reforming of steam is endothermic. In a Haber–Bosch facility, combining the generated hydrogen with nitrogen produces ammonia, which is the process widely used to date. The second most widely used method for processing ammonia is coal gasification. Coal is gasified underground and syngas acquired is sent to the surface to be extracted in cleaning units for syngas. It is mixed with nitrogen to create ammonia after hydrogen is formed. Figure 3.8 represents a simplified life cycle of ammonia

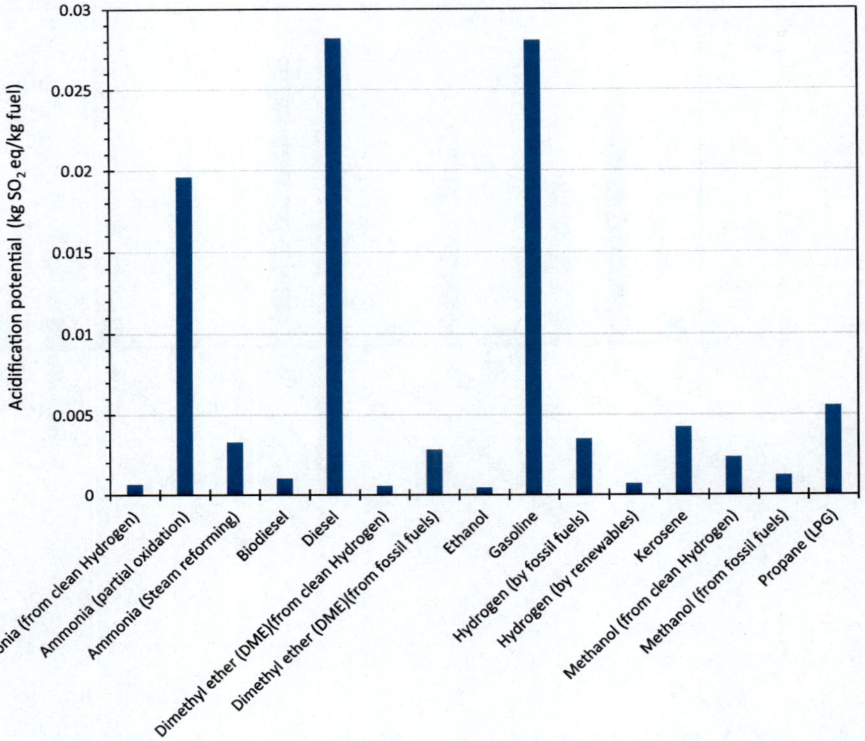

Fig. 3.6 Acidification potential LCA results of various fuels

economy, which describes its production sources, production methods, ammonia storage and distribution methods and its application areas. Moreover, Fig. 3.8 shows the scope and boundaries of the cradle-to-grave LCA of various ammonia production methods and their use in different applications.

A commercial SimaPro software package, the Ecoinvent database, and literature data are utilised to determine LCI data for materials, fuel generation, part manufacturing processes, energy production, distribution, and transportation in both ammonia and other fuel life cycles. 1 kg of ammonia production is the functional unit for the LCA study.

3.3.2 LCA Results and Discussion

The LCA results of various ammonia production technologies are presented in Tables 3.1 and 3.2 in terms of impact categories. The findings of the LCA show that fossil-fuel-based ammonia production technologies perform poorly when compared to renewable ammonia production methods. Ammonia from coal, heavy oil, and

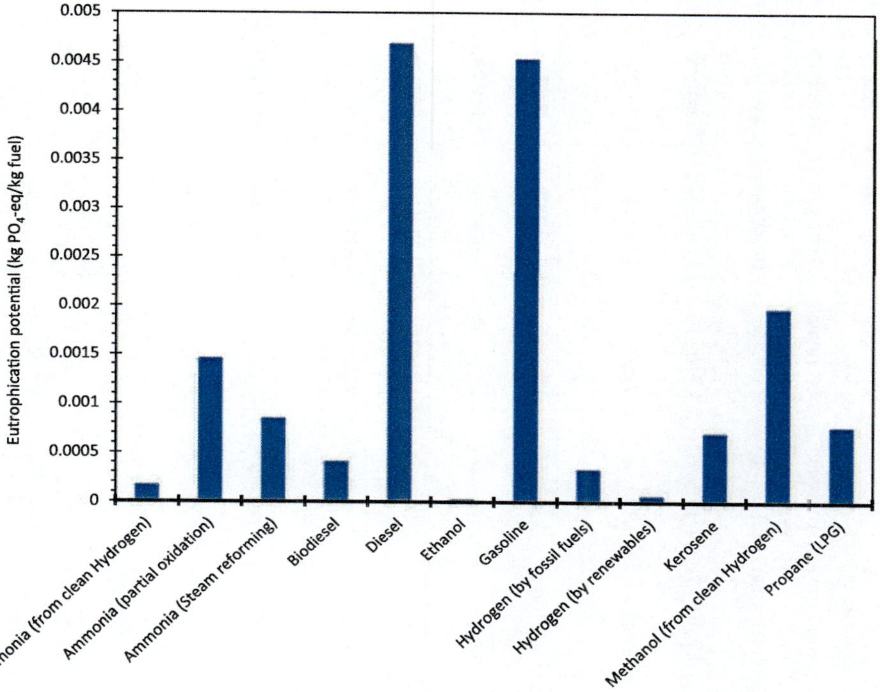

Fig. 3.7 Eutrophication potential LCA results of various fuels

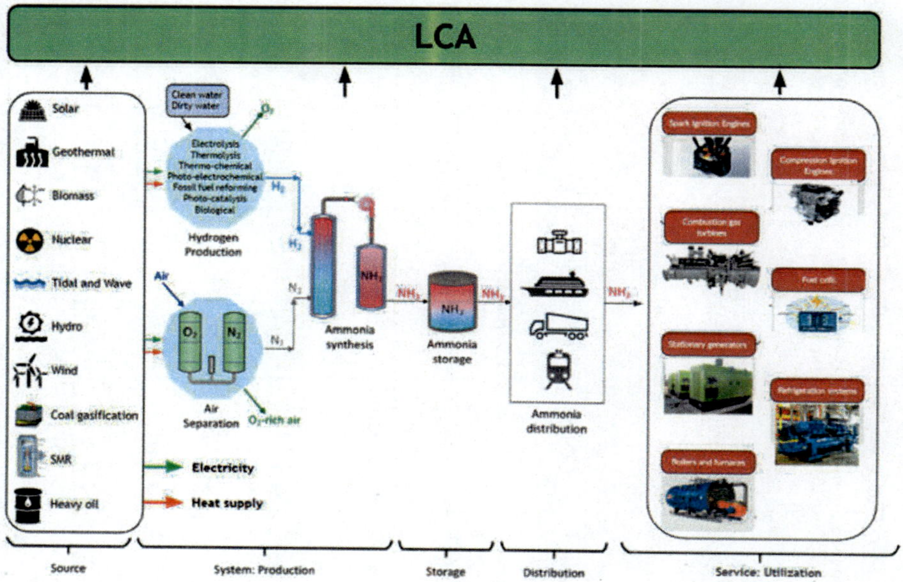

Fig. 3.8 Cradle-to-grave LCA of various ammonia production methods and their use in different applications

Table 3.1 LCA impact assessment results of various ammonia production methods based on impact categories—part 1

Impact category	Unit	Ammonia from biofuels	Ammonia from biomass	Ammonia from coal	Ammonia from geothermal	Ammonia from heavy oil	Ammonia from hydro (pumped storage)	Ammonia from hydro (reservoir)	Ammonia from hydro (run river)
Abiotic depletion	kg Sb eq	7.94E−07	0	1.09E−07	9.76E−07	1.19E−07	1.21E−07	1.26E−08	7.43E−09
Abiotic depletion (fossil fuels)	MJ	0.14240445	0.008981493	3.685782	0.066224	3.834445	1.161723	0.013451	0.010991
Global warming (GWP100a)	kg CO_2 eq	0.003791554	0.012633459	0.325205	0.00719	0.306314	0.079051	0.001908	0.001313
Ozone layer depletion (ODP)	kg CFC-11 eq	1.57E−09	2.64E−14	3.09E−09	9.57E−10	4.91E−08	1.02E−08	1.08E−10	7.80E−11
Human toxicity	kg 1,4-DB eq	0.011199304	0.000775429	0.126243	0.014732	0.071027	0.031723	0.003323	0.002395
Fresh water aquatic ecotoxicity	kg 1,4-DB eq	0.005767032	0.000196906	0.132707	0.007587	0.014566	0.030803	0.001083	0.000698
Marine aquatic ecotoxicity	kg 1,4-DB eq	11.453772	0.7492993	694.0902	14.35719	36.07797	23.86569	1.462164	0.959733
Terrestrial ecotoxicity	kg 1,4-DB eq	5.04E−05	2.26E−08	0.000211	0.000127	0.000214	3.76E−05	3.37E−06	1.81E−06
Photochemical oxidation	kg C_2H_4 eq	6.28E−05	0.006417777	2.49E−05	2.76E−06	8.97E−05	7.89E−06	3.96E−07	3.96E−07
Acidification	kg SO_2 eq	0.000631173	0.000209343	0.000778	2.75E−05	0.002557	9.55E−05	5.14E−06	4.06E−06
Eutrophication	kg PO_4 eq	0.000330044	2.94E−05	0.000669	1.96E−05	0.000181	3.17E−05	2.37E−06	1.90E−06

Table 3.2 LCA impact assessment results of various ammonia production methods based on impact categories—part 2

Impact category	Unit	Ammonia from natural gas	Ammonia from nuclear BWR	Ammonia from nuclear PWR	Ammonia from solar PV	Ammonia from solar tower power plant	Ammonia from wind
Abiotic depletion	kg Sb eq	3.84E−08	2.91E−08	2.84E−08	3.13E−06	3.23E−07	8.16E−08
Abiotic depletion (fossil fuels)	MJ	2.996567	0.032134	0.030102	0.233606	0.174348	0.070124
Global warming (GWP100a)	kg CO_2 eq	0.177529	0.002745	0.00258	0.019815	0.013662	0.006633
Ozone layer depletion (ODP)	kg CFC-11 eq	2.32E−08	2.94E−08	2.87E−08	1.97E−09	9.51E−10	4.00E−10
Human toxicity	kg 1,4-DB eq	0.017851	0.019536	0.018821	0.052618	0.014993	0.036198
Fresh water aquatic ecotox	kg 1,4-DB eq	0.009912	0.007028	0.006778	0.039913	0.010643	0.025924
Marine aquatic ecotoxicity	kg 1,4-DB eq	14.28869	11.38256	10.89174	56.86932	12.31108	19.78811
Terrestrial ecotoxicity	kg 1,4-DB eq	2.22E−05	2.44E−05	2.32E−05	3.42E−05	1.89E−05	2.73E−05
Photochemical oxidation	kg C_2H_4 eq	1.37E−05	7.23E−07	6.85E−07	5.21E−06	3.75E−06	2.76E−06
Acidification	kg SO_2 eq	0.000153	1.23E−05	1.17E−05	6.67E−05	4.54E−05	2.83E−05
Eutrophication	kg PO_4 eq	3.28E−05	1.20E−05	1.13E−05	4.68E−05	1.21E−05	1.63E−05

natural gas are the methods that have highest environmental impacts. However, ammonia from hydropower, as well as ammonia from biomass and biofuels perform much better. As a result, in attempt to lessen ammonia's environmental impact, fossil-based ammonia production methods must be substituted with renewable-based ammonia production methods.

For LCA analyses of ammonia from biofuels the dataset represents a modern sugarcane mill processing of 4 million metric tonnes of sugarcane stalks per year for the production of raw sugar, ethanol (anhydrous and hydrated) and electricity. The plant operates 200 days per season. The industrial process diverts 50% of sugarcane juice for raw sugar production and the other 50%, along with molasses (a by-product of sugar crystallization process), is diverted to ethanol production process. Plant configuration includes reduction of overall energy consumption to maximize of electricity output. It considers high-pressure boilers (65 bar, 485 °C); application of thermal integration strategies reducing 20% of low-pressure steam demand, electrical drives in the crushing mills and molecular sieves for anhydrous ethanol production.

Overall, the process efficiencies assumed were 96% for sugars extraction, 89.5% for ethanol fermentation and 76.5% for sugar recovery as raw sugar. Each 1000 kg of cane stalks yields 51.4 kg of raw sugar, 32.6 L of hydrated ethanol (93% w/w), 22.8 L of anhydrous ethanol (99.6% w/w); and 111.3 kWh of surplus electricity. Included activities start when sugarcane arrives at the industrial area. The data includes the sugarcane conversion into raw sugar, ethanol and surplus electricity. Processing equipment, chemical inputs, industrial area, water use treatment, and emissions to air are included in the inventory. Sugarcane processed arrives with 5% of vegetal impurities (sugarcane straw) and 1% of mineral impurities (dirt). No additional sugarcane straw recovery and industrial use is considered. The activity ends when products are available at factory gate. The inventory is modelled for Brazil. The data represents a modern large-scale sugarcane mill facilitiy existing in Brazil. It includes technologies aiming at maximizing exports of surplus electricity to the grid, while producing raw sugar and ethanol.

Fuel and material extraction, biomass production, biomass gasification power plant, and transportation are all included in LCA assessments of ammonia from biomass. A poplar tree crop with a seven-year growing cycle represents biomass production. The gasification of biomass is followed by a gas turbine to generate electricity. Biomass is produced within the system's borders; hence, it is not included as a fuel source. This is data from start to finish, from the extraction of fuels and raw materials through the generation of energy. There are also infrastructure requirements for biomass production, biomass gasification, and biomass electricity generation. In the original data source, process and fuel emissions are recorded jointly and cannot be separated.

The module describes the operation of an average plant in the Czech Republic for LCA studies of ammonia from coal. The plant is used for intermediate load and operates at full capacity for 4000 h per year. During its lifetime, the plant is expected to run for 150,000 h. A bottom-up strategy was utilised to examine main parameters (LHV, sulphur and ash content of coal, plant efficiency) and criterion emissions

(SO_x, NO_x, particulates, and CO_2). It consists of gathering information on individual plants. To the degree possible, base data from all major UCTE (Union for the Coordination of Transmission of Electricity in Europe) power plants from 1993 have been combined with updated information for the year 2000. The particle size distribution was calculated using German data. The content of the species in the country-specific coal input mix was used to predict halogen emissions, assuming average retention rates. Average UCTE values are included for CO and VOC emissions. For the two instances with and without DeNOx, two average values for N_2O emissions from UCTE plants are evaluated; the country-specific emissions are thus dependent on the percentage of $DeNO_x$.

The literature was used to determine the emission of gaseous radon and K-40. The waste heat releases to air and water were assigned based on the estimated percentage of river-cooled power plants and assumptions about direct air losses. The percentage of recycled ash varies by country. In order to dispose of the residual ash, appropriate disposal modules take into account common country-specific components.

3.3.3 Closing Remarks

The outcomes of the LCA show that fossil-fuel-based ammonia production technologies perform poorly when compared to renewable ammonia production methods. Ammonia from coal, heavy oil, and natural gas are the methods that have greatest environmental impacts. On the other hand, ammonia from hydropower, as well as ammonia from biomass and biofuels perform better. As a result, in attempt to lessen ammonia's environmental impact, fossil-based ammonia production methods must be substituted with renewable-based ammonia production methods.

- The highest global warming values are calculated for coal electrolysis, heavy oil electrolysis, natural gas electrolysis and pumped hydro storage.
- Coal electrolysis corresponds to 0.488 kg CO_2 eq heavy oil corresponds to 0.459 kg CO_2 eq, and natural gas electrolysis gives 0.266 kg CO_2 eq.
- Renewable based methods have minimal impacts in comparison with fossil fuel based ammonia production methods.
- The ozone layer depletion is highest for heavy oil, and followed by nuclear BWR, nuclear PWR, and natural gas.
- The ozone layer depletion values calculated for each method is as follows; for heavy oil 7.37×10^{-7} kg CFC-11 eq, for nuclear BWR 4.42×10^{-8} kg CFC-11 eq, for nuclear PWR 4.3×10^{-8} kg CFC-11 eq, and for natural gas it corresponds to 3.48×10^{-8} kg CFC-11 eq.
- For human toxicity category, coal electrolysis, heavy oil and solar PV electrolysis yield highest environmental impact with values of 0.189 kg 1,4-DB eq, 0.107 kg 1,4-DB eq, and 0.079 kg 1,4-DB eq respectively.

3.4 Case Studies

In this section, numerous case studies are presented to illustrate the benefits, and drawbacks of ammonia utilization in different applications or sectors.

3.4.1 Case Study 1: Utilization of Ammonia as a Fuel in Road Freight-Transportation

In this case study, a comparative LCA of various type of heavy-duty trucks that powered with different fuels is performed. Type of fuels that were considered for analyses are ammonia, propane, CNG, electric and diesel. From raw material extraction to vehicle disposal, all processes are analyzed using the cradle-to-grave LCA methodology. The functional unit for the LCA analysis is 1 t km of freight-transportation by road. The results of this study illustrated different impact in order to reflect the sustainability of the ammonia.

Fuel consumption and the primary emissions (nitrogen oxides, nitrous oxide, methane, non-methane hydrocarbons, ammonia, benzene and lead) are modelled based on information and assumptions on fleet mixes and usage of vehicles in different road and traffic situations. Basic emission data is taken from Handbook Emission Factors for Road Transport (HBEFA) 3.4 with extensive weighted averaging of vehicles in different traffic situations performed. Emission factors for Sulphur dioxide and for carbon dioxide are not taken directly from HBEFA but are calculated using the carbon content and the sulphur content of fuel_type in South Africa and the fuel consumption. Heavy metal emissions to air (other than lead) are extrapolated from transport, freight, lorry 3.5–7.5 metric tonnes. Emissions arising from wear of tyres, brakes and road are extrapolated from wear dataset.

Note that vehicle production and maintenance are described together in the individual global datasets. Vehicle demand per tonne km is calculated based on the lifetime kilometric performance, taken from literature. The average freight load is calculated from the Assumptions and Methodologies in the South African TIMES (SATIM) energy model, weighted to account for the current commercial vehicle fleet in South Africa. Road construction, maintenance and disposal, and road operation are described in the individual global datasets. Road demand per tonne km is calculated based on weighted road usage of each vehicle type of the road network in South Africa.

The expenditures and environmental actions associated with road building have been apportioned based on gross tonne kilometre performance. Expenditures for road infrastructure maintenance and land usage have been allocated depending on annual vehicle kilometre performance. The Ecoinvent v.2 datasets of 540,000 km/vehicle were used to transfer the same assumption on vehicle lifetime performance.

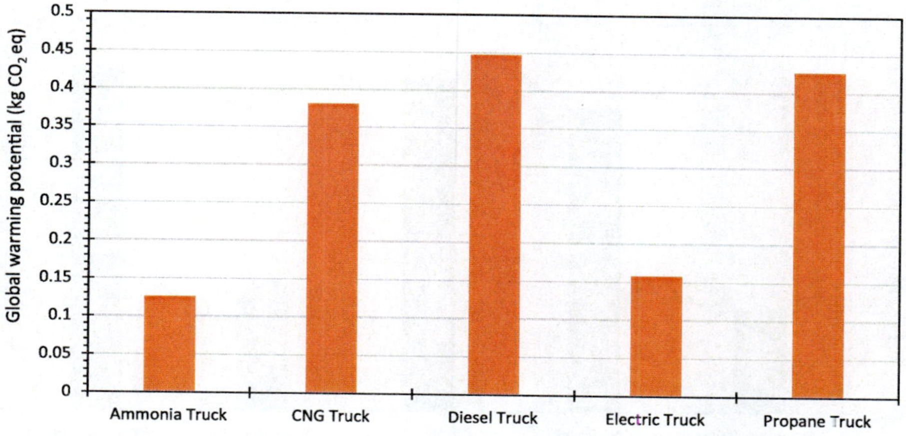

Fig. 3.9 Impact assessment results for global warming potential in road freight-transportation

The operations begin with the combustion of the fuel in the engine. The dataset includes the lorry and road network infrastructure, as well as the materials and labour required to maintain them, as well as the fuel burned by the vehicle during the route.

The activity concludes with a 1 t km transport service and exhaust and non-exhaust emissions into the air, water, and land. Figure 3.9 illustrates Impact assessment results for global warming potential in road freight transportation. The highest global warming values are calculated for the diesel truck and lowest global warming values are calculated for ammonia-fueled truck followed by electric powered truck. Diesel truck corresponds to 0.447 kg CO_2 eq per t km propane fueled truck corresponds to 0.427 kg CO_2 eq/t km and CNG truck yield 0.381 kg CO_2 eq/t km, electric truck produces 0.158 kg CO_2 eq/t km, and ammonia fueled heavy duty truck emits only 0.125 kg CO_2 eq/t km.

Figure 3.10 shows LCA results for abiotic depletion in road freight-transportation for various fuels. In this environmental category, the best performing fuel is ammonia that followed by electric. On the other hand, propane and diesel trucks performs poorly in comparison with ammonia. The impact of ammonia and electric fuels are only about one third of diesel and propane fuels.

Ozone layer depletion potential results of different fuels for road freight-transportation is illustrated in Fig. 3.11. The ozone layer depletion is highest for diesel truck, and followed by propane-fueled truck, and CNG fueled truck. The values calculated for each fuel is as follows; for diesel 9.88×10^{-8} kg CFC-11 eq/t km, for propane 7.25×10^{-8} kg CFC-11 eq/t km, for CNG 6.43×10^{-8} kg CFC-11 eq/t km, and for electric truck it corresponds to 2.79×10^{-8} kg CFC-11 eq/t km. The lowest ozone layer depletion value is obtained for ammonia which corresponds to 2.28×10^{-8} kg CFC-11 eq/t km. Figure 3.12 displays impact assessment results for photochemical oxidation in road freight-transportation. Ammonia is the

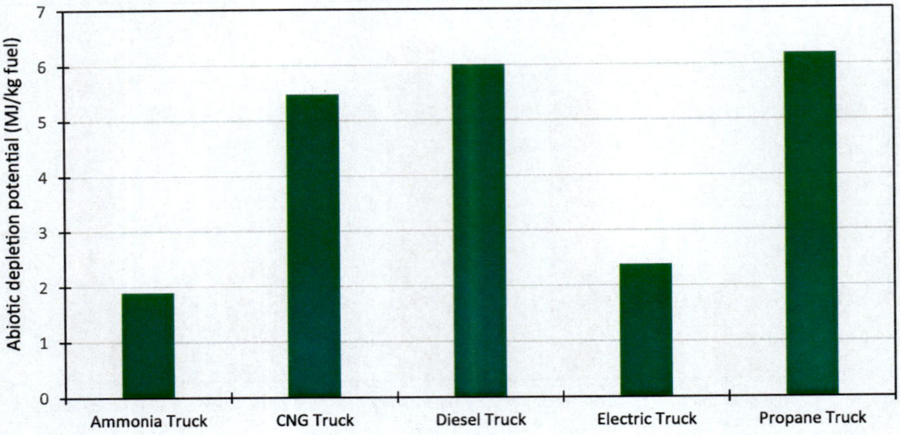

Fig. 3.10 Impact assessment results for abiotic depletion potential in road freight-transportation

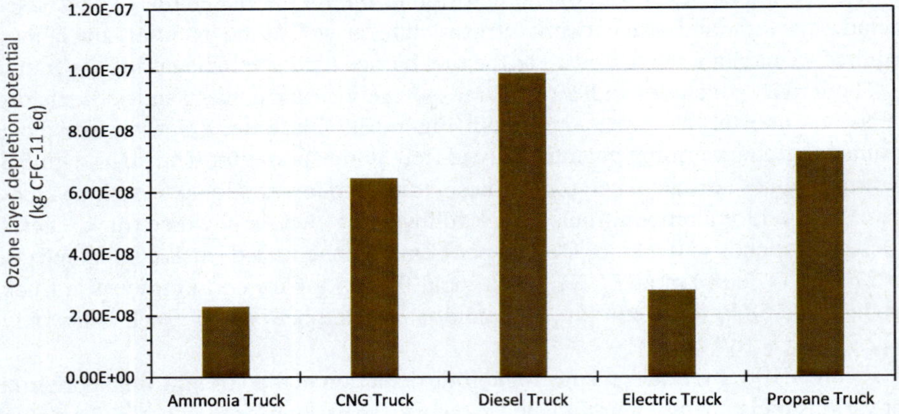

Fig. 3.11 Impact assessment results for ozone layer depletion potential in road freight-transportation

best performing fuel in this impact category and followed by electric powered truck. Diesel and propane emits highest photochemical oxidation emissions.

Figure 3.13 depicts impact assessment results for acidification in road freight-transportation. For acidification impact category propane and diesel yield highest environmental impact with values of 0.0016 and 0.0012 kg SO_2 eq/t km. Thus, ammonia and electric trucks emits lowest acidification values. Figure 3.14 shows LCA results for eutrophication in road freight-transportation. For the eutrophication category, lowest value is obtained for ammonia, and highest value are obtained for propane and diesel fueled trucks.

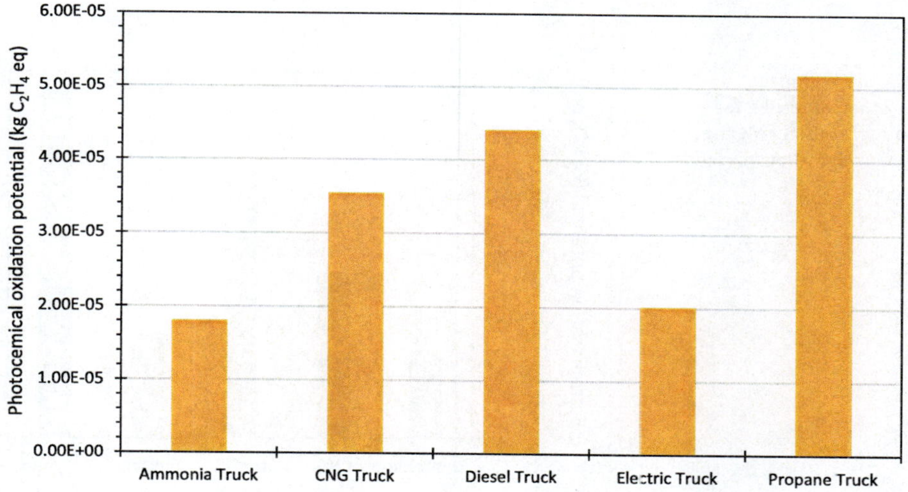

Fig. 3.12 Impact assessment results for photochemical oxidation potential in road freight-transportation

Closing Remarks

The LCA results show that fossil-fuel-based fuels have utmost burden on the environment. On the other hand, ammonia and electric powered trucks have minimal environmental impact. In order to minimize the emission released from road freight-transportation, ammonia fueled trucks and electric trucks must substitute the conventional fuels such as diesel, propane and CNG.

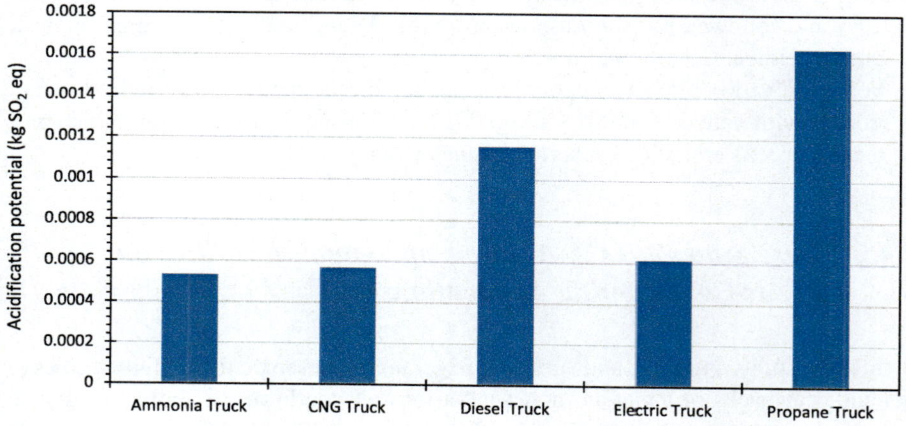

Fig. 3.13 Impact assessment results for acidification potential in road freight-transportation

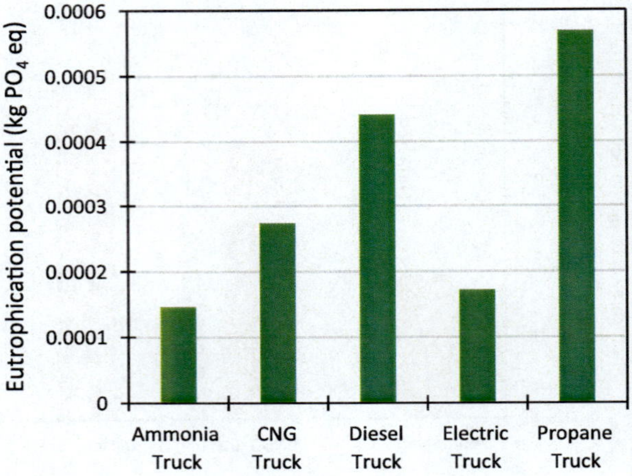

Fig. 3.14 Impact assessment results for eutrophication potential in road freight-transportation

- The highest global warming values are calculated for the diesel truck and lowest global warming values are calculated for ammonia-fueled truck followed by electric powered truck.
- Diesel truck corresponds to 0.447 kg CO_2 eq per t km propane fueled truck corresponds to 0.427 kg CO_2 eq/t km and CNG truck yield 0.381 kg CO_2 eq/t km, electric truck produces 0.158 kg CO_2 eq/t km, and ammonia fueled heavy duty truck emits only 0.125 kg CO_2 eq/t km.
- The ozone layer depletion is highest for diesel truck, and followed by propane-fueled truck, and CNG fueled truck.
- Ammonia is the best performing fuel in photochemical oxidation impact category and followed by electric-powered truck. Diesel and propane emits highest photochemical oxidation emissions.
- For acidification impact category propane and diesel yield highest environmental impact with values of 0.0016 and 0.0012 kg SO_2 eq/t km. Thus, ammonia and electric trucks emits lowest acidification values.

3.4.2 *Case Study 2: LCA Analysis of Ammonia Utilization in Air Conditioning Applications*

In this case study, ammonia and various refrigerants is presented to highlight the benefits and drawbacks of using ammonia in various air conditioning applications. The case study investigates a comparative LCA of ammonia with different refrigerants in air conditioning applications.

Ammonia has been recognised and used as a prominent refrigerant in developed locations due to its good thermal characteristics, negligible ozone depletion, and zero global warming potential. Ammonia provides the highest refrigerating result per unit mass when compared to all other refrigerants, including halocarbons. The following are some of the primary advantages of ammonia over R-134a: reduced average running costs, versatility in meeting complicated and varied cooling requirements, and lower start-up expenditures for multiple applications. Ammonia is the cheapest of all the refrigerants used on a daily basis and can be found practically anywhere. Ammonia has better heat transfer properties than most chemical refrigerants, allowing devices with a smaller heat transfer zone. In addition, because the thermodynamic efficiency in the system often benefits from these characteristics, it also decreases the system's operating costs. The cost of ammonia per mass is far below the cost of HFCs in many countries.

Dairies, ice cream manufacturers, frozen food processing plants, cold storage facilities, fish, poultry, and meat processors, and a number of other industries all employ ammonia as a refrigerant. While the specific ammonia volume is large, the required compressor displacement per tonne of refrigeration is little since tiny compressors are required per tonne of cooling capacity. Finally, this saves a lot of fuel.

For normal conditions around −15 °C in the evaporator, the condenser and evaporator pressures are roughly 2.37 and 11.67 bar, respectively. Because the pressures are not particularly high, lightweight materials can be used in the machinery design. Because the evaporator's pressure is so high, extending the gas to a very low pressure is not a good idea. High compressor suction pressure and a lower compression ratio are also possible. Because the ammonia refrigerant's release temperature is high, it's critical to use water to cool the cylinder heads and compressor cylinders [9].

In this case study, a comparative LCA of ammonia as refrigerant in air conditioning applications will be conducted. Results of the LCA study will be compared with few other refrigerant being used on the market. The extraction of raw materials from the earth to manufacturing, product use, and recycling/disposal at the end are included in LCA analysis.

Ammonia (R-717) Refrigerant

This dataset represents the production of 1 kg of liquid ammonia using the partial oxidation process. Ammonia is a colourless gas with a penetrating, pungent suffocating odour. It is liquid when under pressure. It is hygroscopic and soluble in water [50].

The use of ammonia as a nitrogen source in industry has consumed a rising share of overall ammonia production, now accounting for around 20% of global output. Ammonia makes up about all of the nitrogen utilised in the chemical industry. Plastics and fibres are two of the most common uses of industrial ammonia–nitrogen, which is partly converted into nitric acid. Other notable uses include the production of explosives, hydrazine, amines, amides, nitriles, and other organic nitrogen compounds used as dye and medicinal intermediates. Nitric acid, urea, sodium cyanide, and sodium carbonate are the most important ammonia-based compounds.

Included activities start: The activity starts when the raw materials enter the process.

Ammonia is produced at the end of this process. The manufacturing process, which begins with heavy fuel oil, air, and electricity, as well as auxiliaries, is included in the dataset. Infrastructure and land usage are considered, as well as wastes and emissions into the air and water.

Geography: The inventory is modelled for Europe.

Technology: Process efficiency and plant investment costs are unlikely to improve much. Partial oxidation, on the other hand, will remain interesting in the future due to its feedstock flexibility. To adapt to decreasing residual characteristics or other raw material sources, soot and especially ashes must be separated and disposed of. Steam is provided for the shift conversion in part by a cooler-saturator system and in part by steam. CO_2 is removed with an absorption agent similar to that used in the sulphur removal process. The process gas is next purified with a liquid nitrogen wash to eliminate any remaining amounts of absorption agent and CO_2. Almost all contaminants are eliminated in this unit, and nitrogen is added to achieve the stoichiometric hydrogen to nitrogen ratio. In addition, carbon dioxide, methane, and some soot are produced. The feed's sulphur compounds are transformed to hydrogen sulphide. Mineral components in the feed are converted into distinct ashes. After waste heat recovery, water washing removes particles from the process gas, and the soot is returned into the feed. The ash compounds are drained with the condensate from the process and/or with some of the soot. Heavy metals have been recovered. In a Claus unit, the hydrogen sulphide in the process gas is separated by selective absorption and reprocessed to elementary sulphur.

Refrigerant (R-152a)

R152a, often described as difluoroethane, is a pure fluorinated hydrocarbon having a formula that is nearly identical to R134a. It has the same chemical properties as R-134a and has the same vapour pressure curve, with only 2 K variations, making it compatible with all materials, refrigeration components, thermostatic valves, compressors, and lubricating oils. Thermodynamically, R-152a is superior to R-134a and HFOs. Because R152a has better physical properties than R-134a, the refrigerant's heat transfer coefficient in evaporators is enhanced by about 20%. Because of the lower gas viscosity, the pressure drop in the suction lines will be reduced by 30%. R-152a has a lower molecular weight than R-134a, which means it has a higher latent heat of vaporisation, higher compressor volumetric efficiency, and better COP performance in the refrigeration cycle, as well as a higher discharge temperature of roughly 10 K.

Refrigerant (R-290)

R290 is a refrigerant grade propane that is utilised in new refrigeration and air conditioning systems as an alternative to R404A and R407 series refrigerants. R290 is a hydrocarbon, which is extremely flammable. It is a high-efficiency refrigerant with a low GWP.

Included activities start: The activity begins when crude oil enters the refinery. Wastewater treatment, natural freshwater supply, refinery infrastructure, crude oil and product storage on refinery grounds, and energy supply are all covered. The on-site generation mix is assumed to meet electricity requirements.

Included activities end: Activity ends with refined petroleum products (and surplus electricity and/or hydrogen, where relevant) leaving the refinery.

Refrigerant (R-32)

The R32 refrigerant, often known as difluoromethane, belongs to the HFC refrigerant family. This gas is set to replace other gaseous refrigerants like R-410A and R-407C as the preferred gas due to its lower global warming potential. CH_2F_2 is its chemical formula.

R-410A and R-407C are the refrigerant blends that are used to replace ozone depletion refrigerants such as R22 as a result of the Montreal Protocol 1987, which mandated the phase-out of ozone depletion gases. Both of these gases have 0% ODP (Ozone Depletion Potential). Because difluoromethane is found in significant quantities in both of these gases, its availability on the market is unaffected.

Refrigerant (R-410a)

The chlorofluorination of ethylene is represented in this dataset. One of the most important procedures for producing monochloropentafluoroethane is ethylene chlorofluorination. 1 kg of ethylene is the functional unit. The system encompasses the use of raw resources, energy, infrastructure, and land usage, as well as the discharge of pollutants into the atmosphere and water. Transportation of raw materials is also included. The production during steady operation conditions is not included, as are transient or unstable processes like starting up or shutting down. The final product's storage and shipment are also not covered. Because the manufacturing plants are presumed to be in an urban/industrial location, the emissions are classified as coming from a high population density area. It is anticipated that the emissions into water are emitted into rivers. Monochloropentafluoroethane and hexafluoroethane are co-products of the chlorofluorination of ethylene.

Included activities start: From the reception of ethylene, chlorine and hydrogen fluoride to the factory gate.

Included activities end: This activity ends with the chlorofluorination of ethylene including materials, energy uses, infrastructure and emissions.

Geography: The inventory is modelled for Global.

Technology: Organic fluorine compounds are frequently purified by fractional distillation, fractional crystallisation, or chromatographic techniques. Fluoropolymers and high-boiling per fluorinated oils, on the other hand, necessitate specific precautions, such as the use of exceptionally pure starting ingredients.

In most circumstances, quantitative fluorine determination is attainable through combustion and subsequent measurement of the hydrogen fluoride produced. Fluoride ions are measured using wet chemical techniques. Because organic fluorine compounds are highly volatile, purity may be easily verified using gas chromatography. The structure of organic fluorine compounds can be determined via 19F-nuclear magnetic resonance spectroscopy.

Refrigerant (R-134a)

This information represents 1 kg of tetrafluoroethane manufacturing (R-134a). Tetrafluoroethane is a fluorinated gas with an ether-like odour. It's mostly employed as a refrigerant in "high-temperature" refrigeration systems like refrigerators and air conditioners. Tetrafluoroethane is also utilised in heat pumps, both as a pure gas and as part of common refrigerant combinations (WPZ). Plastic foam blowing, as a solvent for particular cleaning applications, as an aerosol propellant for medical inhalers and as a fire extinguisher in place of halons are all possible applications.

It is assumed that the tetrafluoroethane is produced 50% by hydro fluorination of trichloroethylene and 50% by hydro fluorination of tetrachloroethylene.

Included activities start: From reception of trichloroethylene and tetrachloroethylene at the factory gate.

Included activities end: This activity ends with the production of refrigerant R134a. Emissions during production process are included. This dataset doesn't include an output of hydrochloric acid, however 1.07 respectively 2.2 kg of hydrochloric acid are produced both with the trichloroethylene and with the tetrachloroethylene process.

LCA Results for Ammonia and Various Refrigerants in Air Conditioning Applications

Figure 3.15 shows impact assessment results for abiotic depletion in air conditioning applications for various refrigerants. In this environmental category, the best performing fuel is renewable based ammonia refrigerant that followed by ammonia (R-717) from fossil fuels. On the other hand, refrigerant (R-134a) and Refrigerant (R-410a) performs poorly in comparison with ammonia (R-717) refrigerant. Refrigerant (R-134a) has value of 143.6 MJ while renewable ammonia based refrigerant only has 6.6 MJ value of abiotic depletion impact.

Figure 3.16 illustrates impact assessment results for global warming in air conditioning applications. The highest global warming values are calculated for refrigerant (R-32) and lowest global warming values are calculated for ammonia and R-290 refrigerants. Refrigerant (R-32) corresponds to 65.5 kg CO_2 eq refrigerant (R-134a) corresponds to 19.6 kg CO_2 eq and refrigerant (R-410a) yield 6.84 kg CO_2 eq, refrigerant (R-152a) produces 5.6 kg CO_2 eq, and renewable based ammonia based refrigerant has zero global warming impact.

The ozone layer depletion potential results of different refrigerants for air conditioning applications is illustrated in Fig. 3.17. The ozone layer depletion is highest for refrigerant (R-32), and followed by refrigerant (R-134a). Rest of the refrigerants has zero ozone layer depletion impact including ammonia-based refrigerant.

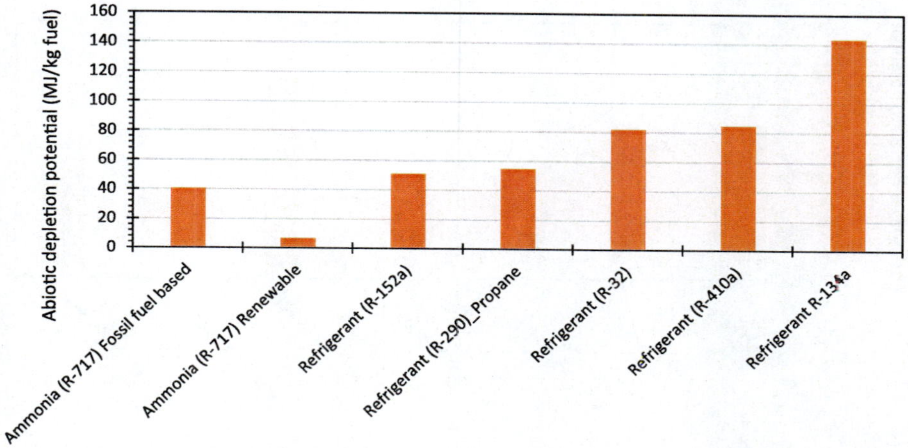

Fig. 3.15 Impact assessment results for abiotic depletion potential in air conditioning applications

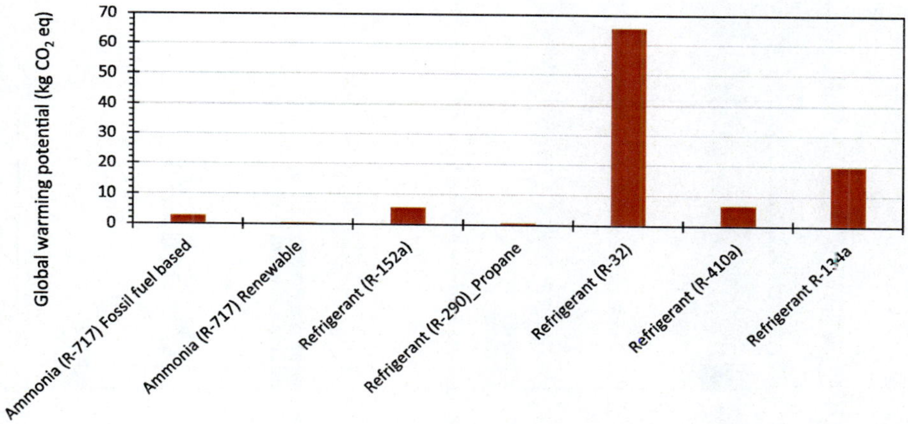

Fig. 3.16 Impact assessment results for global warming potential in air conditioning applications

Figure 3.18 depicts the impact assessment results for human toxicity in air conditioning applications. For human toxicity impact category refrigerant (R-410a) and refrigerant (R-32) yield highest environmental impact with values of 19.6 and 13.2 kg 1,4-DB eq. On the other hand, refrigerant (R-290) propane emits lowest human toxicity impact.

Figure 3.19 portrays impact assessment results for fresh water aquatic ecotoxicity in air conditioning applications. For fresh water aquatic ecotoxicity impact category refrigerant (R-410a) and refrigerant (R-152a) yield highest environmental impact with values of 12.97 and 4.97 kg 1,4-DB eq. On the other hand, refrigerant (R-290) propane and renewable based ammonia emit lowest fresh water aquatic ecotoxicity impact.

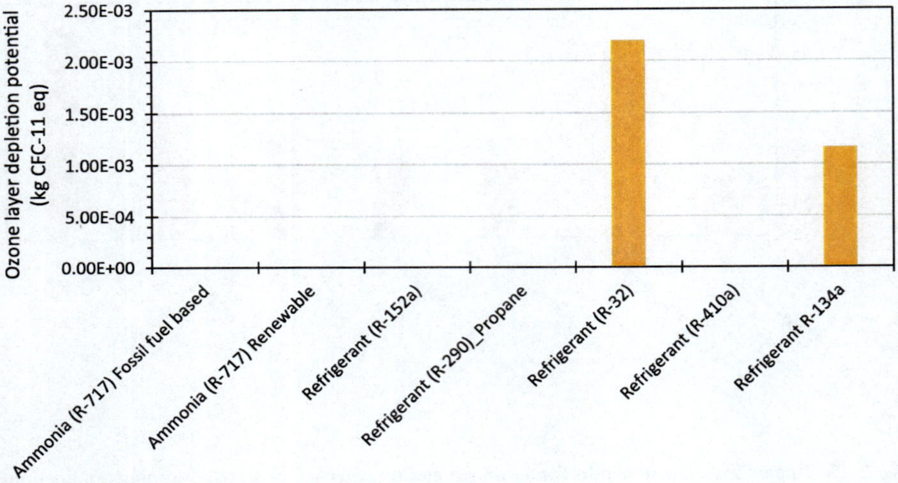

Fig. 3.17 Impact assessment results for ozone layer depletion potential in air conditioning applications

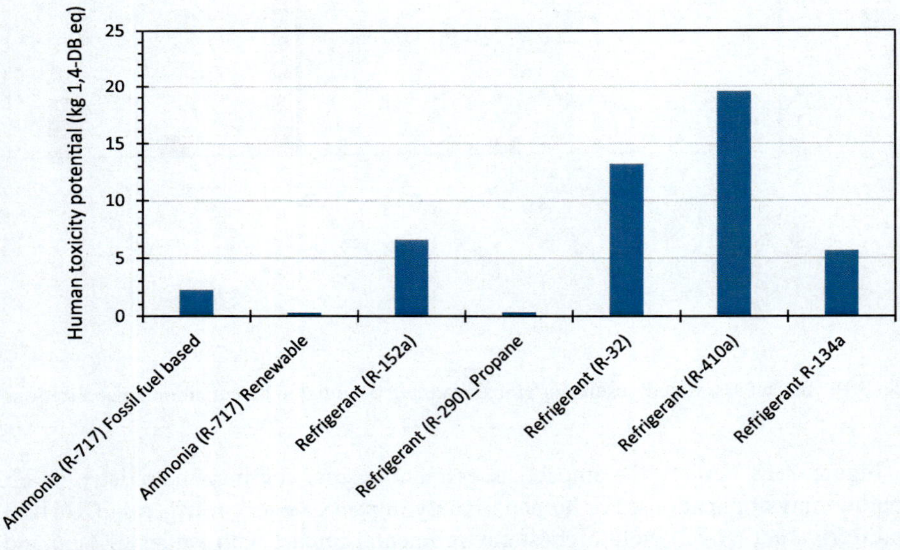

Fig. 3.18 Impact assessment results for human toxicity potential in air conditioning applications

Figure 3.20 represents the impact assessment results for marine aquatic ecotoxicity in air conditioning applications. For marine aquatic ecotoxicity impact category refrigerant (R-32) and refrigerant (R-152a) produce highest environmental impact with values of 128,837 and 11,307 kg 1,4-DB eq. On the other hand, refrigerant (R-290) propane and renewable based ammonia release lowest fresh water aquatic ecotoxicity impact.

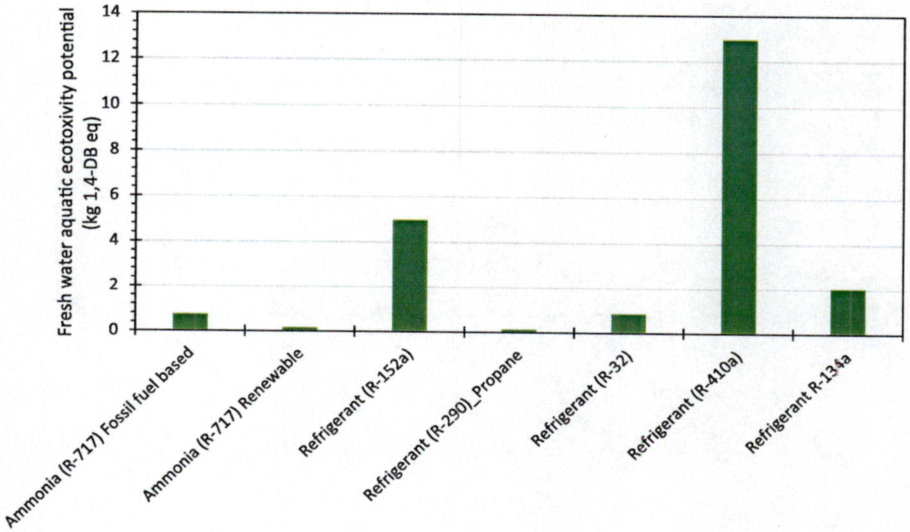

Fig. 3.19 Impact assessment results for fresh water aquatic ecotoxicity potential in air conditioning applications

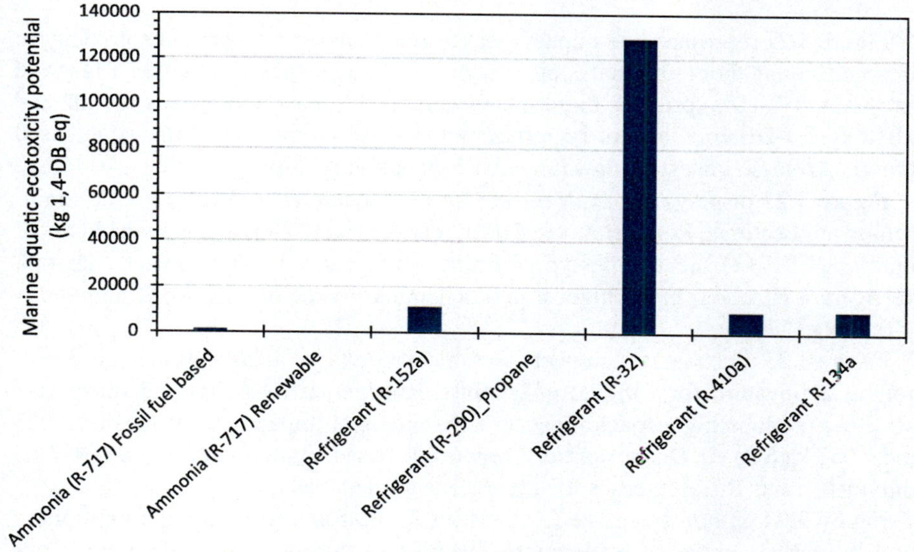

Fig. 3.20 Impact assessment results marine aquatic ecotoxicity potential in air conditioning applications

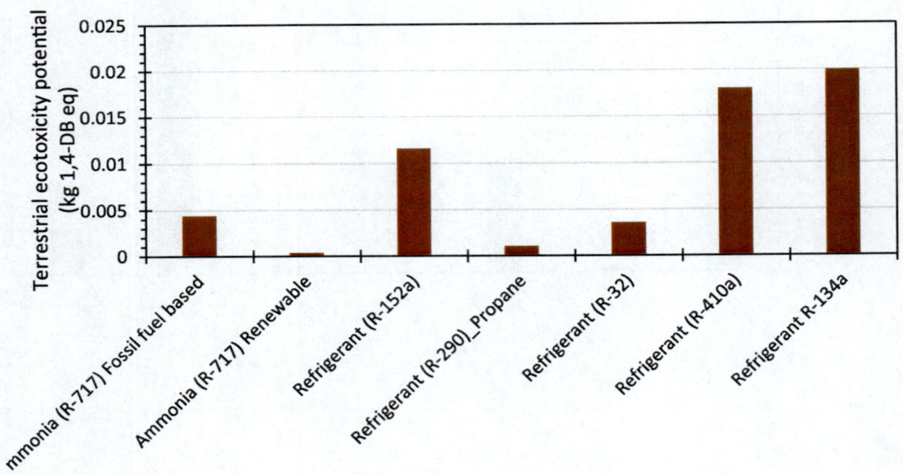

Fig. 3.21 Impact assessment results for terrestrial ecotoxicity potential in air conditioning applications

Figure 3.21 represents the impact assessment results for terrestrial ecotoxicity in air conditioning. For terrestrial ecotoxicity impact category refrigerant (R-134a) and refrigerant (R-410a) produce highest environmental impact with values of 0.02 and 0.018 kg 1,4-DB eq. Conversely, refrigerant (R-290) propane and renewable based ammonia release lowest fresh water aquatic ecotoxicity impact.

Figure 3.22 displays the LCA results for photochemical oxidation in air conditioning applications. Refrigerant (R-410a), refrigerant (R-290) and renewable based ammonia (R-717) are the best performing refrigerants in this impact category. Refrigerant (R-134a) emits highest photochemical oxidation emissions with value of 0.01 kg C_2H_4 eq.

Figure 3.23 illustrates the impact assessment results for acidification in air conditioning application for various refrigerants. Refrigerant (R-410a) and refrigerant (R-134a) produce highest acidification environmental impact with values of 0.063 and 0.033 kg SO_2 eq. On the contrary, renewable based ammonia refrigerant (R-717) and refrigerant (R-32) releases lowest acidification values.

Figure 3.24 demonstrates the LCA results for eutrophication in air conditioning applications for numerous refrigerants. The highest eutrophication values are calculated for refrigerant (R-410a) and lowest eutrophication values are calculated for refrigerant (R-32) followed by ammonia (R-717)) and refrigerant (R-290). Refrigerant (R-410a) corresponds to 0.022 kg PO_4 eq and refrigerant (R-152a) corresponds to 0.009 kg PO_4 eq. On the other hand, renewable based ammonia refrigerant has zero eutrophication impact, and R-290 refrigerant has close to zero eutrophication impact on the environment.

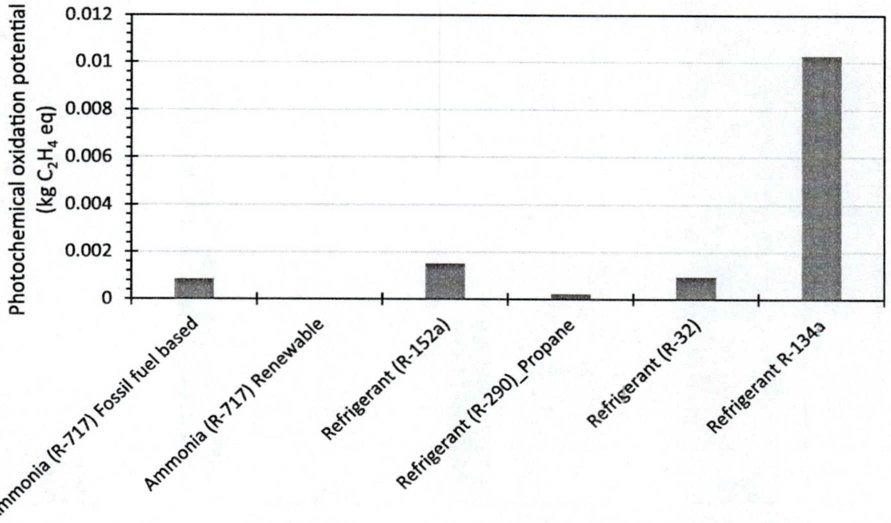

Fig. 3.22 Impact assessment results for photochemical oxidation potential in air conditioning applications for various refrigerants

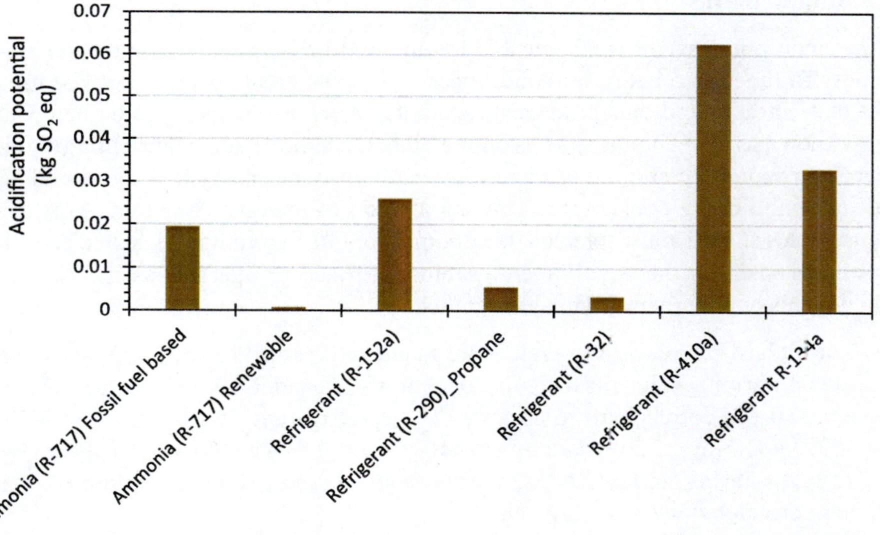

Fig. 3.23 Impact assessment results for acidification potential in air conditioning applications

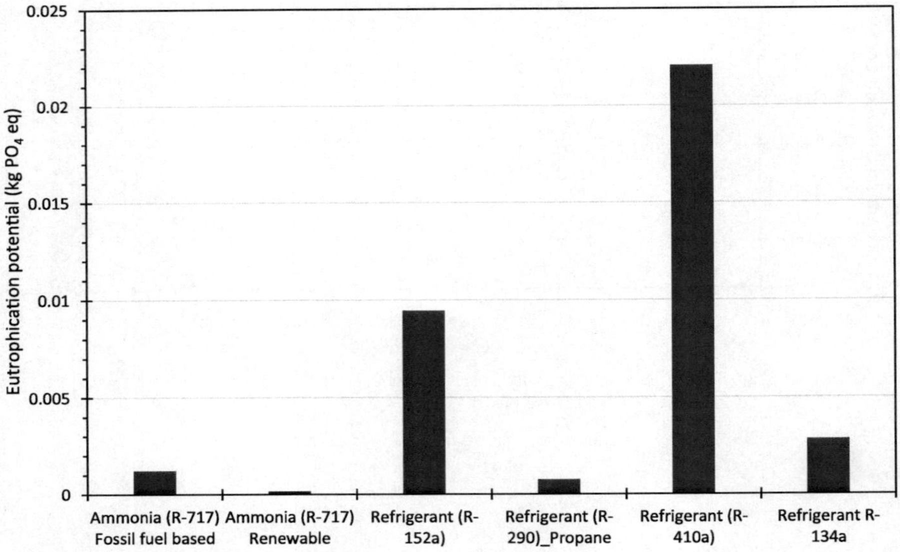

Fig. 3.24 Impact assessment results for eutrophication potential in air conditioning applications

Closing Remarks

Air conditioning sector is responsible for an important part of environmental emissions. To function reliably, air conditioners demand a great deal of energy. It uses a lot of electricity and thus produces pollutants. When fossil fuels are burned, CO_2, also known as a greenhouse gas, is emitted into the atmosphere, contributing significantly to ozone depletion. It is critical to decarbonize the energy that used buildings, as air conditioning consumes a growing amount of energy. The global growth of renewables should allow for such decarbonisation. Refrigerant gas leakage can have serious impact on the environment. Ammonia based refrigerants as if R-717 can reduce this environmental damage significantly.

- The highest global warming values are calculated for refrigerant (R-32) and lowest global warming values are calculated for ammonia and R-290 refrigerants. Refrigerant (R-32) corresponds to 65.5 kg CO_2 eq refrigerant (R-134a) corresponds to 19.6 kg CO_2 eq and refrigerant (R-410a) yield 6.84 kg CO_2 eq, refrigerant (R-152a) produces 5.6 kg CO_2 eq, and renewable based ammonia based refrigerant has zero global warming impact.
- The ozone layer depletion is highest for refrigerant (R-32), and followed by refrigerant (R-134a). Rest of the refrigerants has zero ozone layer depletion impact including ammonia-based refrigerant.
- For human toxicity impact category refrigerant (R-410a) and refrigerant (R-32) yield highest environmental impact with values of 19.6 and 13.2 kg 1,4-DB eq. On the other hand, refrigerant (R-290) propane emits lowest human toxicity impact.

- Refrigerant (R-410a), refrigerant (R-290) and renewable based ammonia (R-717) are the best performing refrigerants in photochemical oxidation.
- Renewable based ammonia refrigerant (R-717) and refrigerant (R-32) produce lowest acidification values.
- Renewable based ammonia refrigerant has zero eutrophication impact.

3.4.3 Case Study 3: LCA Analysis of Ammonia Utilization in Renewable Energy Storage

In this case study, a comparative LCA case study regarding ammonia utilization in renewable energy storage is presented. A comprehensive LCA analysis of ammonia utilization in renewable energy storage is provided. Then results are compared with energy storage options that using rechargeable lithium-ion batteries and thermal energy storage options.

The energy sector has largely focused on two main solutions to the storage of renewable energy: rechargeable batteries and hydrogen. To produce rechargeable batteries of various types and sizes, a range of technologies are used. The benefits of batteries include energy efficiency and long lifetime. Their drawbacks, on the other hand, include their large weight and footprint, their high cost of production, rapid degradation, safety hazards and their disposal issues.

Hydrogen technology is another common trend for energy storage in conjunction with batteries. Such as applications that use hydrogen to drive fuel cells, charging automobiles and mobile machinery, among other revolutionary applications. Using surplus renewable energy, applications such as electrolysis generate clean, green hydrogen from water; this hydrogen can be transported, stored and re-electrified by means of fuel cells. The storage potential of hydrogen energy is much greater than that of batteries. Then why hydrogen has not dominate the market with this technology? The explanation is that hydrogen energy storage is difficult. It is not an easy matter to build a viable hydrogen infrastructure for distribution from manufacturing to consumption sites. Centralized mass production of hydrogen, whether as gas or liquid, entails high delivery and storage costs, whereas dispersed low-volume production is incredibly expensive. Hydrogen distribution is limited by high costs, losses in transit energy quality, problems with hydrogen purity and leakage tendency. To solve these limitations, researchers are working on novel hydrogen carriers. Besides free hydrogen molecules, hydrogen carriers can store hydrogen in a chemical state.

The use of ammonia as an energy carrier is another choice. The benefits are numerous. Ammonia, firstly, is economical. Secondly, 200 million tonnes of ammonia are produced per year, the second most commonly produced chemical in the world. Thirdly, ammonia is simple to store and does not require storage at high pressure. Another main benefit is the chemical compatibility of ammonia with alkaline fuel cells using an alkaline electrolyte, and no harmful contaminants are released as ammonia is cracked in order to produce hydrogen [51]. A schematic of a cradle-to-grave LCA of solar based ammonia for sectoral use is shown in Fig. 3.25. The system boundary of the LCA, and considered components can be seen in Fig. 3.25.

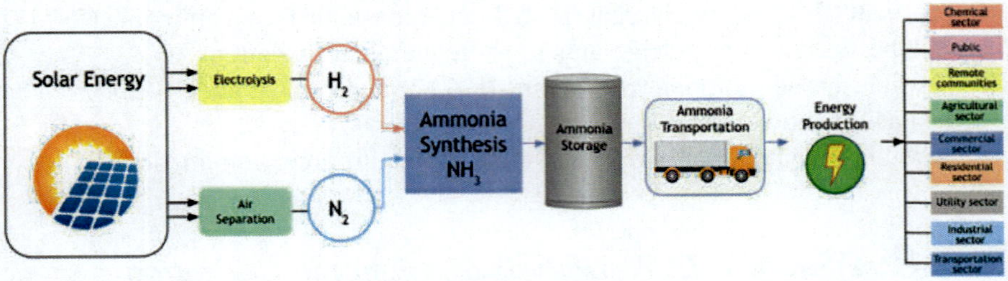

Fig. 3.25 Cradle-to-grave LCA of solar based ammonia for sectoral use

The functional unit of the LCA is set as 1 MJ of solar energy storage. Ecoinvent background database and literature data used for modeling [50].

In this case study, a comparative LCA of renewable energy storage system will be carried out. A comprehensive LCA analysis of ammonia utilization in renewable energy storage will be provided. Then results will be compared with the storage of renewable energy with rechargeable batteries and thermal energy storage options. The extraction of raw materials from the earth to manufacturing, product use, and recycling/disposal at the end are included in LCA analysis.

Engineers and policymakers are increasingly shifting their focus to energy storage technologies, in response to growing worries about the environmental implications of fossil fuels, as well as the capacity and resilience of energy systems around the world. Indeed, energy storage can assist address the intermittent nature of solar and wind generation; in many circumstances, it can also respond quickly to huge swings in demand, and making the grid more efficient and decreasing the need to build backup energy plants.

LCA Results for Renewable Energy Storage Options

Figure 3.26 depicts the findings of an impact assessment for renewable energy storage intended for utilization in many sectors such as residential, commercial, agriculture, industrial, transportation etc. The top performing energy storage method is ammonia energy storage in terms of environmental impacts. For 1 MJ energy storage with batteries produce 0.00115 kg SO_2 eq acidification impact while ammonia has nearly zero acidification impact. Figure 3.27 demonstrates LCA results for eutrophication in energy storage systems such as thermal energy storage, battery and ammonia energy storage. The highest eutrophication values are calculated for batteries and lowest for thermal energy storage. Thermal energy storage and ammonia energy storage gives values near to zero.

Figure 3.28 illustrates the impact assessment results for global warming in renewable energy storage systems. The highest global warming values are calculated for battery storage and lowest global warming values are calculated for solar energy storage with ammonia. Battery energy storage corresponds to 0.31 kg CO_2 eq thermal energy storage corresponds to 0.11 kg CO_2 eq and ammonia energy storage emits only 0.011 kg CO_2 eq. Therefore, storing renewable energy with option of ammonia eliminates global warming effect of energy production.

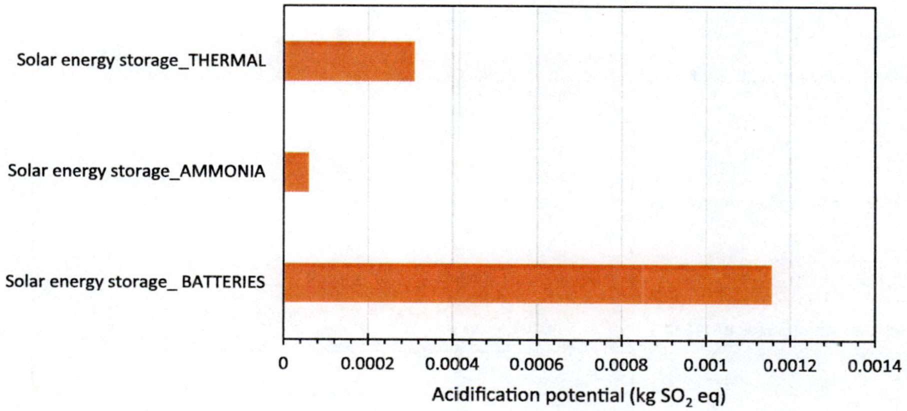

Fig. 3.26 LCA results for acidification potential in renewable energy storage for various applications

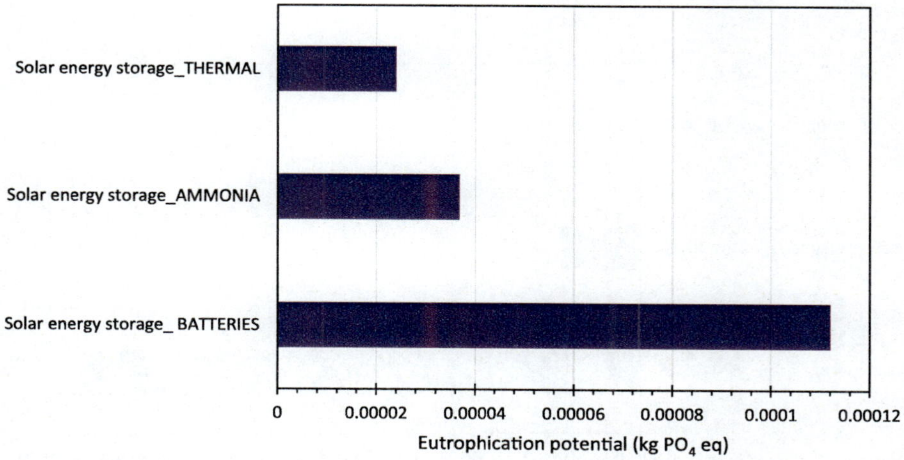

Fig. 3.27 LCA results for eutrophication potential in renewable energy storage for various applications

The LCA impact assessment results for ozone layer depletion in renewable energy storage for various sectors is shown in Fig. 3.29. Although battery energy storage is the most common energy storage system used nowadays, batteries have highest load on the environment. Ammonia energy storage and thermal energy storage are better options for the environment. While world is switching from fossil-based energy options to renewable based energy options, emission-free storage systems or more cleaner systems are needed. Therefore, ammonia can be the solution of this problem.

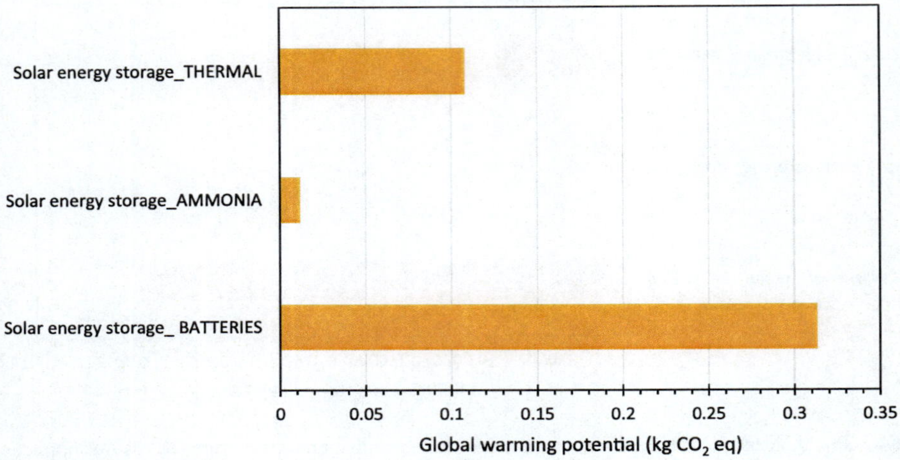

Fig. 3.28 LCA results for global warming potential in renewable energy storage for various applications

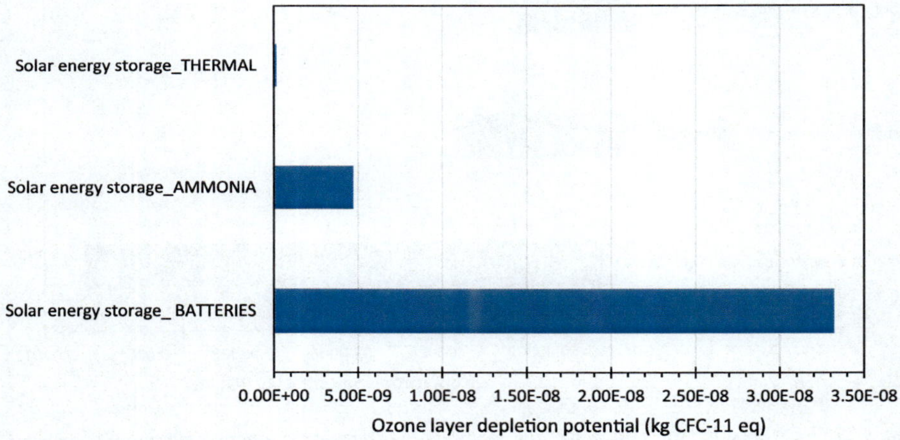

Fig. 3.29 LCA results for ozone layer depletion potential in renewable energy storage for various applications

Figure 3.30 depicts the impact assessment results for human toxicity in energy storage applications. For human toxicity impact category, the highest human toxicity emissions are calculated for battery storage and lowest emissions are calculated for thermal energy storage system. Whereas ammonia is highly toxic, batteries are constructed from a variety of materials. Acid, lead, nickel, lithium, cadmium, alkaline, mercury, and nickel metal hydride are among these materials. When batteries are not properly disposed of, the shell disintegrates and the dangerous chemicals within seep into the environment. Leaking materials can contaminate the land and water, and some chemicals can accumulate in wildlife and humans. LCA results for photochemical oxidation in renewable energy storage is demonstrated in Fig. 3.31. Ammonia is the best performing storage method in this impact category.

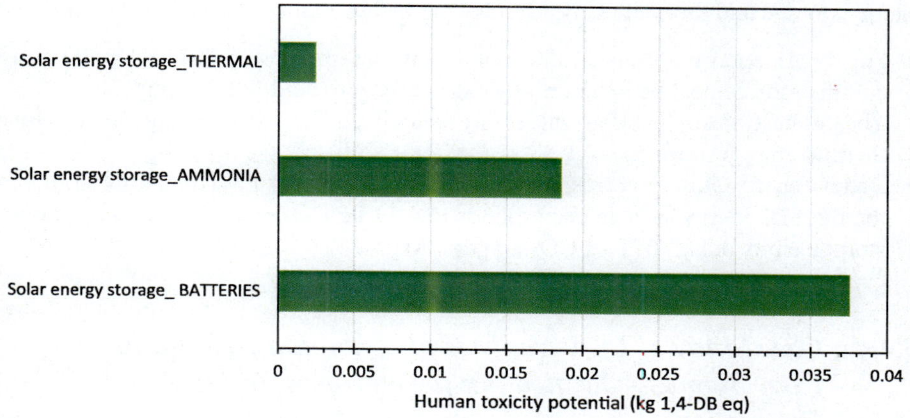

Fig. 3.30 LCA results for human toxicity potential in renewable energy storage for various applications

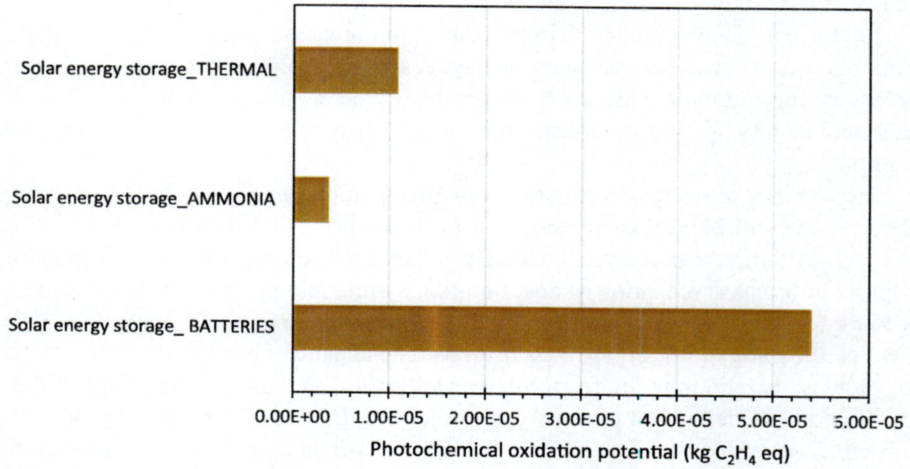

Fig. 3.31 LCA results for photochemical oxidation potential in renewable energy storage for various applications

Closing Remarks

In response to growing worries about the environmental implications of fossil fuels, as well as the capacity and resilience of energy systems around the world. Without a doubt, energy storage can assist address the intermittent nature of solar energy and other renewable energy options. The results of the case study show that ammonia energy storage and thermal energy storage are better options for the environment.

- For 1 MJ energy storage, lithium-ion batteries produce 0.00115 kg SO_2 eq acidification impact while ammonia has nearly zero acidification impact.
- The highest eutrophication values are calculated for batteries and lowest for thermal energy storage.
- Battery energy storage corresponds to 0.31 kg CO_2 eq per 1 MJ of energy storage, the thermal energy storage corresponds to 0.11 kg CO_2 eq and ammonia energy storage emits only 0.011 kg CO_2 eq per 1 MJ.

3.4.4 *Case Study 4: LCA Analysis of Ammonia Generation from Biomass-Gasification for Nitrogen Fertilizer Production*

In this section, a LCA case study of biomass-based ammonia usage for fertilizer production is presented. The aim of this study is to investigate the environmental impact of all aspect of biomass-based ammonia for fertilizer production. A comprehensive LCA analysis will be provided using different impact categories. Gasification and pyrolysis are suitable thermal conversion processes for creating hydrogen from biomass, in which a hydrogen-rich syngas might replace the natural gas input necessary for ammonia generation. Given the correct conditions, a biomass-based ammonia facility might accomplish a production capacity comparable to natural gas facilities.

The viability of manufacturing ammonia from biomass gasification (Crop straw gasification) will be evaluated using life cycle analysis, and then its impacts will be assed in fertilizer production. SimaPro 9.2 is used to assess the environmental impacts of biomass scenarios against traditional ammonia production. Ecoinvent is used as a resource for the data from the life cycle inventory and the impact assessment methodology [50]. Figure 3.32 displays a schematic of a cradle-to-grave LCA for biomass-based ammonia for fertilizer production. The system boundary of the LCA, and considered components can be seen in Fig. 3.32. The functional unit chosen for the LCA is 1 kg of nitrogen fertilizer production from biomass-based ammonia. Production of heat by burning straw, in a furnace of 300 kW capacity for heat production. The infrastructure, and the electricity inputs were obtained from the dataset wood chips, hard wood, burned in furnace 300 kW. The emissions were obtained from literature. Production volume is 9517.4 kg. Included activities starts from reception of straw at the furnace and included activities ends at the combustion of straw. Included are the infrastructure, the straw requirements, the emissions to air,

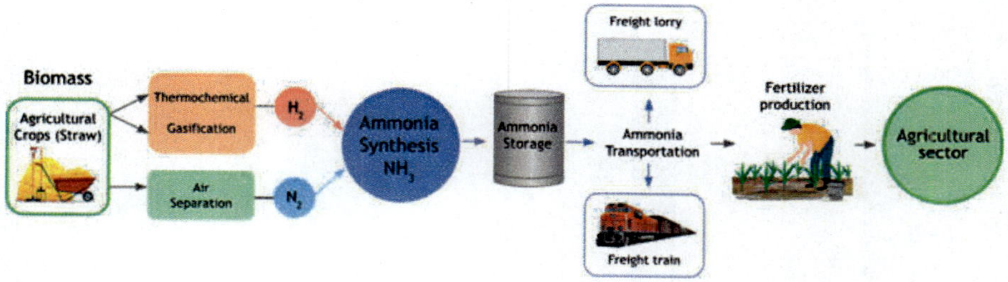

Fig. 3.32 Cradle-to-grave LCA of biomass-based ammonia for fertilizer production

the electricity needed for operation, and the disposal of the ashes. 54 kg of slag are produced per 1 tonne of straw burned. The slag is returned to farmland. The slag produced during straw combustion contains 0.8–2.1% P_2O_5 and 0.05–0.08% N [2]. The transportation dataset depicts a 1 t km ammonia transport service in a lorry with a size class of 32 metric tonnes gross vehicle weight and Euro III emissions. The transport databases include the complete transportation life cycle, including vehicle and road infrastructure creation, operation, maintenance, and disposal. Fuel consumption and emissions are based on average European travel, not a specific transportation scenario.

Due to the sheer importance of ammonia as a fertilizer in agriculture and its bright possibilities as an energy carrier, several studies have recently attempted to identify the best environmentally friendly, energy efficient, and economically feasible ammonia synthesis process. The Haber–Bosch process is the most extensively used ammonia production technology. The disadvantage of this technique is the high level of greenhouse gas emissions. The biomass-gasification process is a dependable and credible alternative, as well as one of the most rapidly increasing renewable technologies. As a source of energy, biomass can be utilised directly through combustion to generate heat, or indirectly through conversion to various forms of biofuel. Converting biomass to biofuel can be accomplished by a variety of processes, which are classified as thermal, chemical, and biological [3]. Biomass is processed before being fed to a gasifier and pressure swing adsorption for hydrogen synthesis in a biomass-based ammonia processing facility. Hydrogen produced can be mixed with nitrogen using either the Haber–Bosch process or innovative electrochemical synthesis methods. Biomass cultivation, gasification and syngas cleaning, syngas conditioning, and ammonia manufacturing are all part of the biomass gasification system. By eliminating the requirement for fossil fuels, biomass gasification will alleviate the environmental implications of ammonia manufacturing. As a result, the global warming potential and ozone layer depletion (ODP) should be reduced.

Figure 3.33 portrays the results of an impact assessment for abiotic depletion category for biomass-based ammonia for fertilizer production. The results show straw production and supply has highest abiotic depletion value, and second highest value calculated for combustion of mixed wood chips. On the other hand, nitrogen fertilizer production, electricity from biomass has near zero abiotic depletion impact.

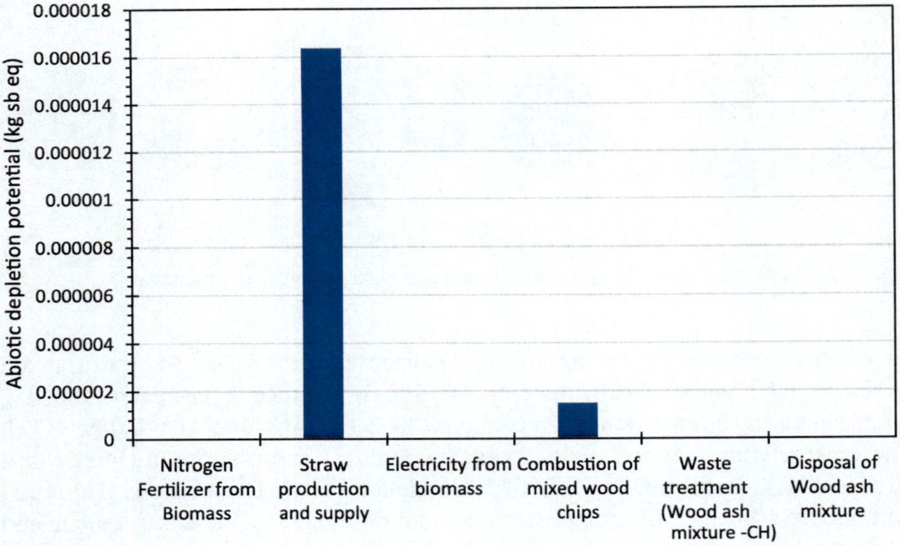

Fig. 3.33 LCA results for abiotic depletion potential for biomass-based ammonia for fertilizer production

The impact assessment results of global warming for biomass-based ammonia in fertilizer production is depicted in Fig. 3.34. The highest global warming values are calculated for straw production and supply, which corresponds to 0.431 kg CO_2 equivalent per kg of nitrogen fertilizer production. The other included activities such as electricity utilization from biomass, combustion of mixed wood and chips has a very low contribution to global warming. As a result, production of nitrogen fertilizer with option of biomass gasification eliminates global warming effect of nitrogen fertilizer production.

Figure 3.35 displays the LCA process flowchart for global warming of nitrogen fertilizer production from biomass gasification, which shows contribution of global warming between each process and shows hot spots in the system. As it can be clearly seen that straw production and transportation has largest contribution to global warming in nitrogen fertilizer production. In other words, straw production and transportation account for 88% of CO_2 emissions. Included operations begin once the previous crop has been harvested. Seeds, mineral fertilisers, and insecticides are all taken into account. No farm manure is applied, it is presumed. All machine activities, as well as machine infrastructure and sheds, are included in the dataset. Soil cultivation, planting, fertilising, weed control, pest and pathogen management, combine harvest, transportation from field to regional processing centre (15 km), and grain drying are all examples of machine operations. The direct field emissions are also mentioned. This task is completed once the grains have been dried at the regional processing plant. Figure 3.36 shows the LCA results for ozone layer depletion for biomass-based ammonia for nitrogen fertilizer production. Similar to other impact

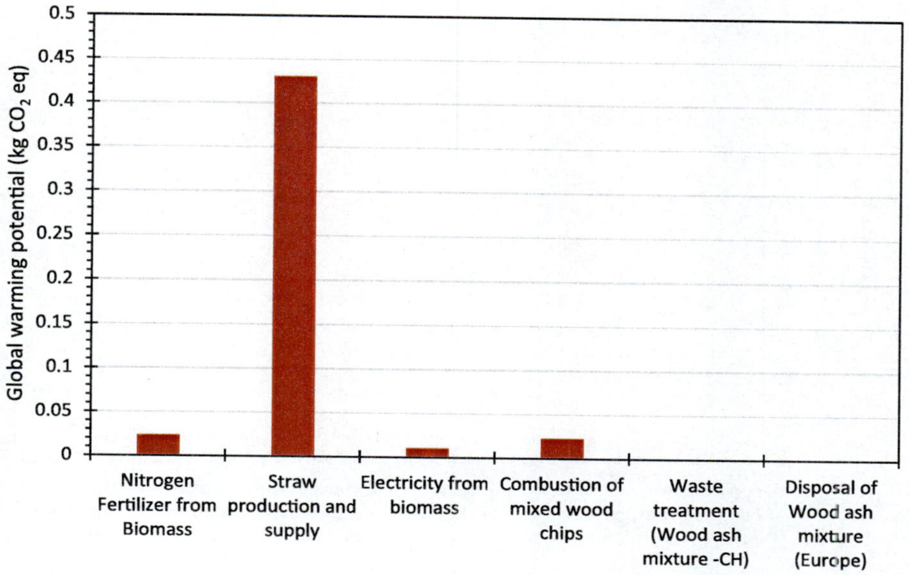

Fig. 3.34 LCA results for global warming potential for biomass-based ammonia for fertilizer production

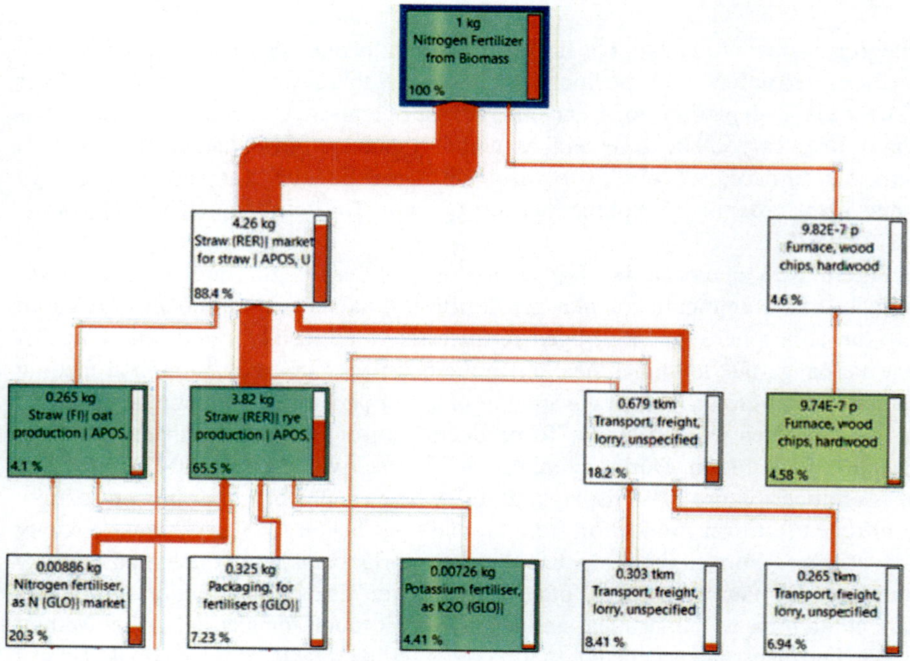

Fig. 3.35 LCA process flowchart for global warming potential of nitrogen fertilizer production from biomass gasification (cut-off 4%)

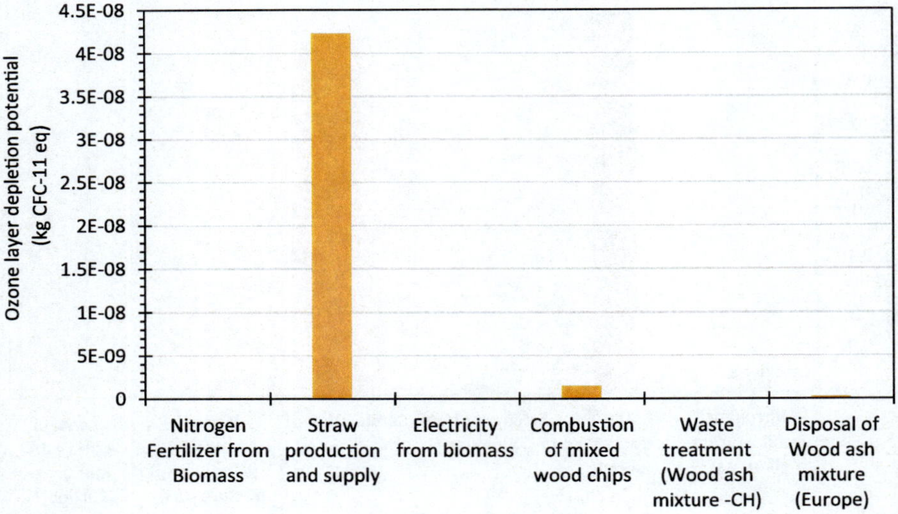

Fig. 3.36 LCA results for ozone layer depletion potential for biomass-based ammonia for nitrogen fertilizer production

categories straw production has highest portion in ozone layer depletion emissions. As illustrated in Fig. 3.37, the thick red lines shows that straw production has 96% of all ozone layer depletion alone, and other 3.34% of emission is sourced from combustion of mixed wood chips. General assumption, obtained from dataset "wood chips burned in furnace, 300 kW". It is assumed that wood-chips furnace needs energy inputs in the order of 1.5% of the fuel energy input. The conversion from MJ to kWh is 3.6 kWh/MJ.

Figure 3.38 demonstrates impact assessment results for human toxicity for biomass-based ammonia for nitrogen fertilizer production. Ammonia is released into the atmosphere when nitrogen fertiliser is volatilized, forming fine particles that are dangerous to human health. As the popularity of confined animal feeding operations has grown, so has the amount of ammonia emitted, which maybe traced back to nitrogen in feed crops. To produce 1 kg of nitrogen fertilizer, 0.292 kg 1,4-DB eq is emitted. During combustion of mixed wood chips 0.042 kg 1,4-DB eq is emitted. Figure 3.39 displays the LCA process flowchart for human toxicity of nitrogen fertilizer production from biomass gasification. 53% of human toxicity impact comes from straw production and transportation, 47% of the human toxicity emissions are sourced from all other activities for nitrogen production. Fertilizers may be harmful to humans and animals by direct touch, inhalation, or inadvertent intake. Fertilizers are acceptable to use on non-edible plants, but one should always exercise caution while handling and storing synthetic fertilizers.

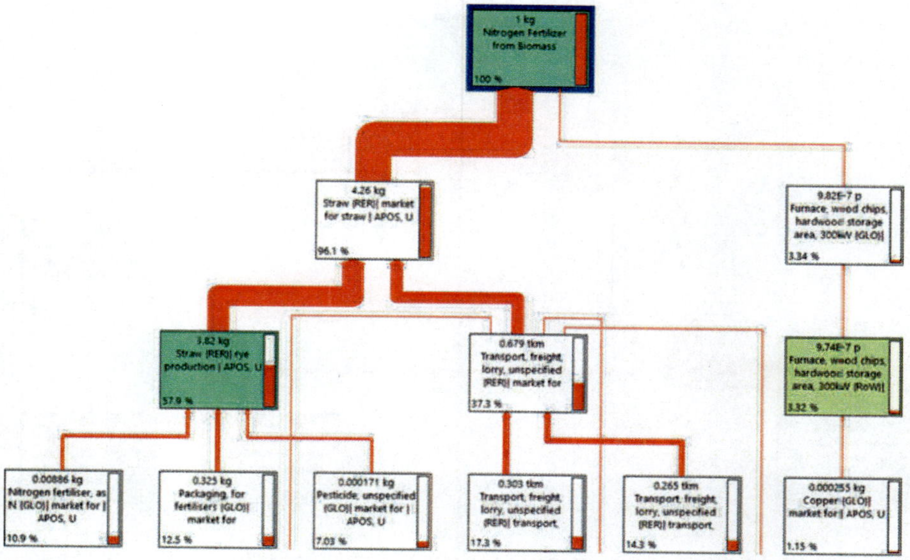

Fig. 3.37 LCA process flowchart for ozone layer depletion potential of nitrogen fertilizer production from biomass gasification (cut-off 6.9%)

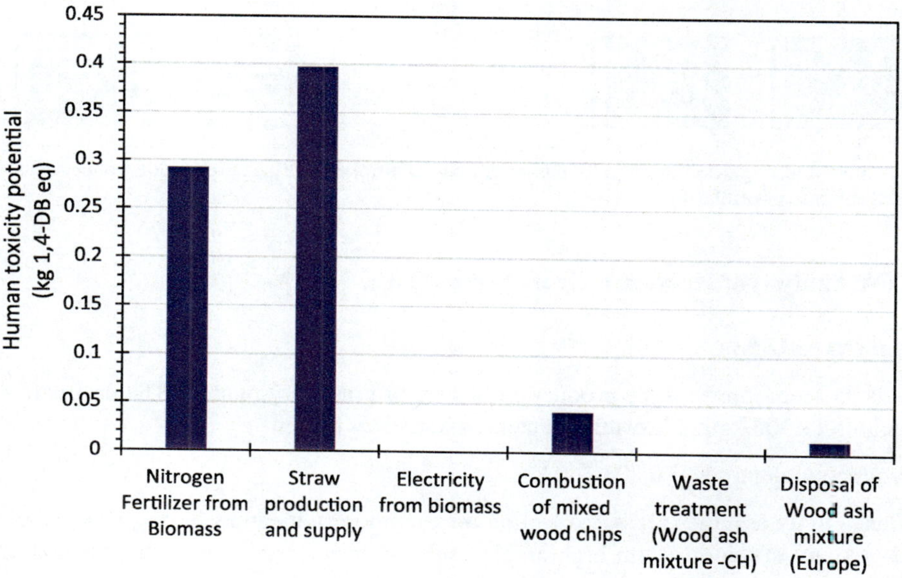

Fig. 3.38 LCA results for human toxicity potential for biomass-based ammonia for nitrogen fertilizer production

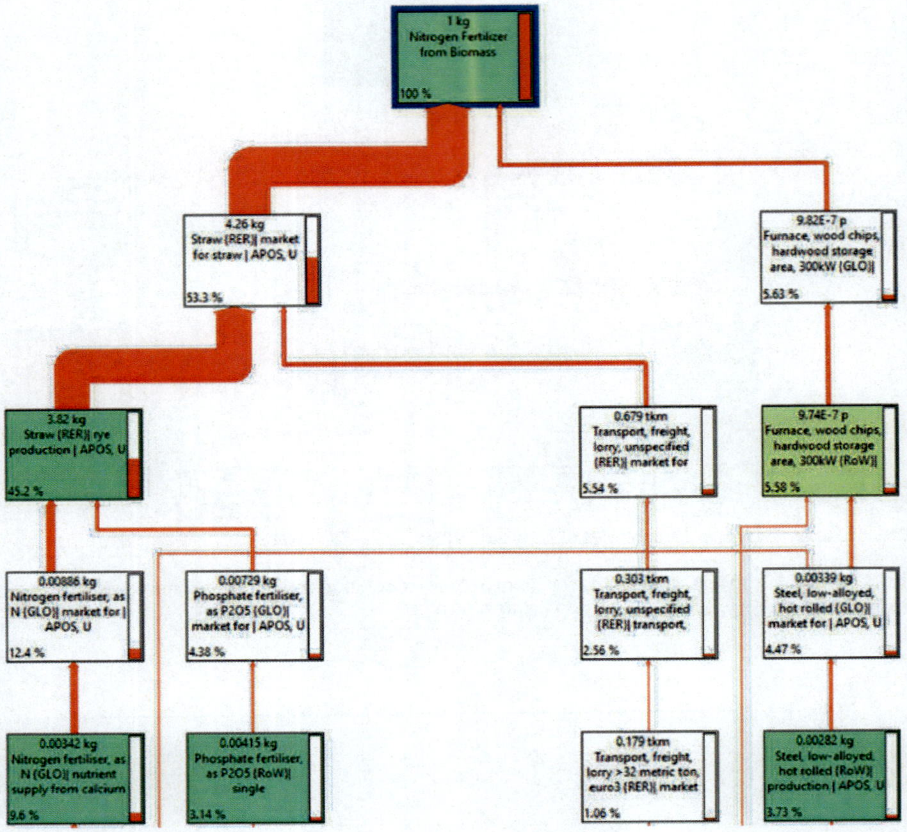

Fig. 3.39 LCA process flowchart for human toxicity of nitrogen fertilizer production from biomass gasification (cut-off 2.8%)

LCA Analysis of Harvested Grass Versus Cattle Feed

Harvested Grass

This dataset represents the production of 1 kg of grass (dry matter). The dry matter yield is 11,500 kg/ha. Moisture content at harvest is 85%.

Production volume: 170,703,360 kg.

This activity assumes grass production on a temporary meadow during three years (2–3 years in regions with high arable use, 3–4 years in regions with low arable use) and represents an average of the three years; it starts with the soil cultivation and sowing after the harvest of the previous crop. The input of seeds is considered. Farm manure as an organic fertilizer is only accounted for in terms of direct field emissions; all pre-processes are included in the animal production systems.

The dataset includes all machine operations and corresponding machine infrastructure and sheds. Machine operations are: soil cultivation, sowing, fertilization, mowing and loading for transport. This activity ends after loading of the grass at the farm gate. The inventory is modelled for Switzerland.

Cattle Feed (Grazed in Pasture)

This process describes the average grass grazed in the pasture, which is on a hectare on a typical beef farm in Ireland. The system boundaries of this process are from farm gate to farm gate. No activities considered, besides the grazing of the animals. The elementary flows include field emissions to the air and water. The amount of nitrogen and phosphorous emissions from manure are calculated based on the net result of the inputs and outputs of nitrogen and phosphorous for the system. In this calculation the amount of hectares (14.4 ha) and the time the animals spend in the field (214 days) for a typical Irish beef farm are taken into account.

Figure 3.40 shows the LCA results for global warming, human toxicity and acidification for the effect of harvesting grass versus cattle feed. Impact assessment results shows that harvested grass has much more environmental impacts than grass feed by grazing of cattle. This is true for all environmental impact categories considered for this LCA study. For global warming, harvesting produces 0.137 kg of CO_2 emission per 1 kg of grass. On the other hand, grazing produces only 0.052 of kg of CO_2 emission per 1 kg of grass. For human toxicity, grass harvesting produces 0.095 kg 1,4-DB eq emission per 1 kg of grass. Grazing produces only 0.0005 kg 1,4-DB eq emission per 1 kg of grass. For acidification, harvested grass emits 0.007 kg SO_2 kg eq and grazing emits 0.0008 kg SO_2 eq per 1 kg of grass. Thus, grass grazing has lowest emission in all categories in comparison with harvesting grass.

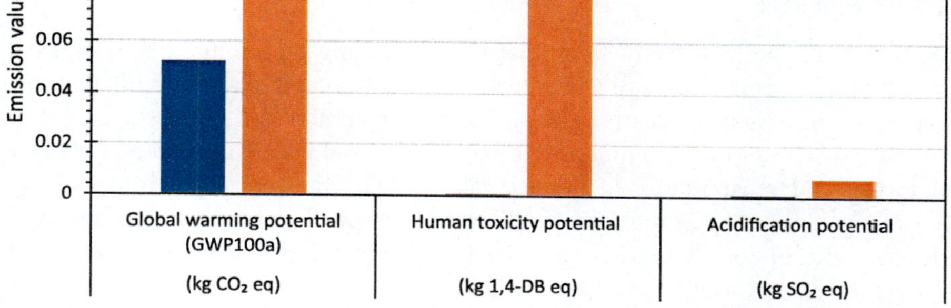

Fig. 3.40 LCA results for global warming potential, human toxicity potential and acidification potential for the effect of harvesting grass versus cattle feed

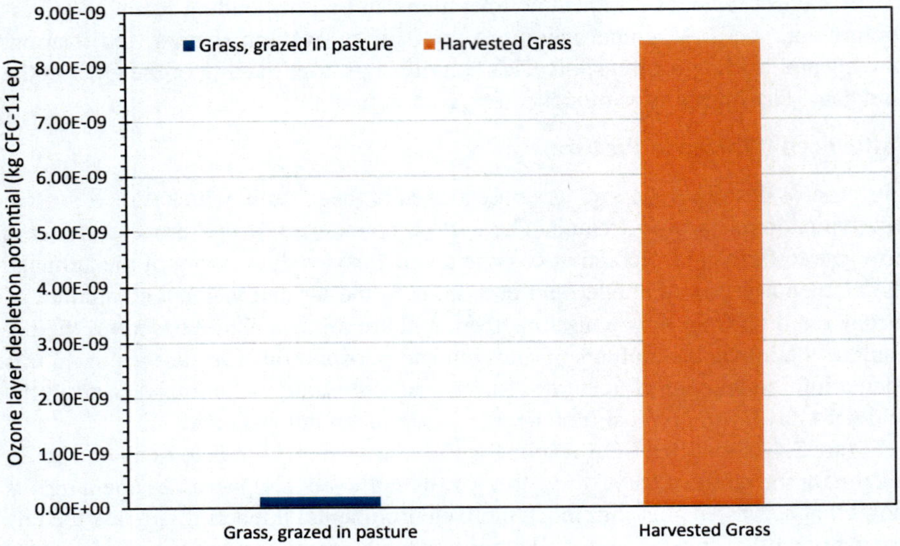

Fig. 3.41 LCA results for ozone layer depletion potential for the effect of harvesting grass versus cattle feed

Figure 3.41 represents the LCA results for ozone layer depletion for the effect of harvesting grass versus cattle feed. While grass grazing has near zero ozone layer depletion effect, harvested grass has 8.43×10^{-9} kg CFC-11 eq.

Figure 3.42 shows the LCA results for photochemical oxidation for the effect of harvesting grass versus cattle feed. The photochemical oxidation emissions are much higher in harvested grass with value of 1.85×10^{-5}; however, grass grazing has 1.99×10^{-6} kg C_2H_4 equivalent per 1 kg of grass. Figure 3.43 demonstrates the LCA results for eutrophication potential for the effect of harvesting grass versus cattle feed. Similar to other impact categories, grass grazing have much lower eutrophication emission then harvested grass. The emission values are 0.0009 kg PO_4 and 0.0019 kg PO_4 respectively.

Closing Remarks

The results of the case study show that by eliminating the requirement for fossil fuels, biomass gasification will alleviate the environmental implications of ammonia manufacturing. As a result, the global warming potential and ozone layer depletion should be reduced. The highest global warming values are calculated for straw production and transportation in biomass gasification, which corresponds to 0.431 kg CO_2 equivalent per kg of nitrogen fertilizer production. The production of nitrogen fertilizer with option of biomass gasification eliminates global warming effect of nitrogen fertilizer production.

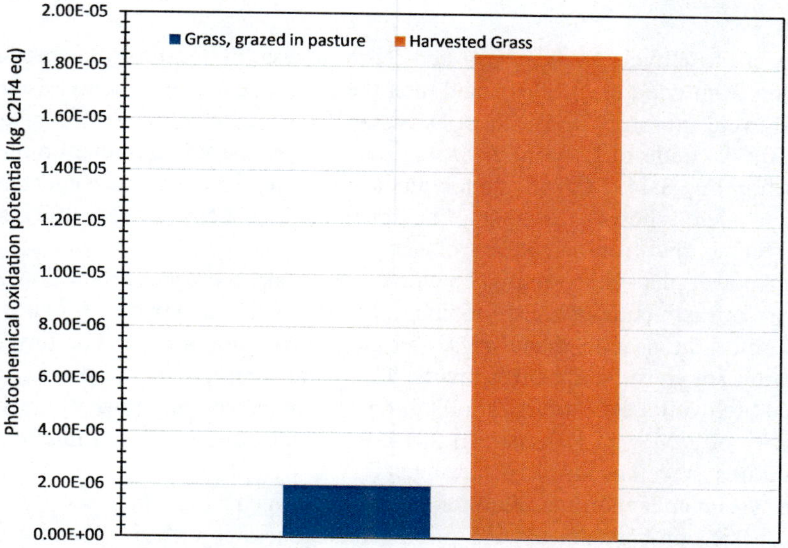

Fig. 3.42 LCA results for photochemical oxidation potential for the effect of harvesting grass versus cattle feed

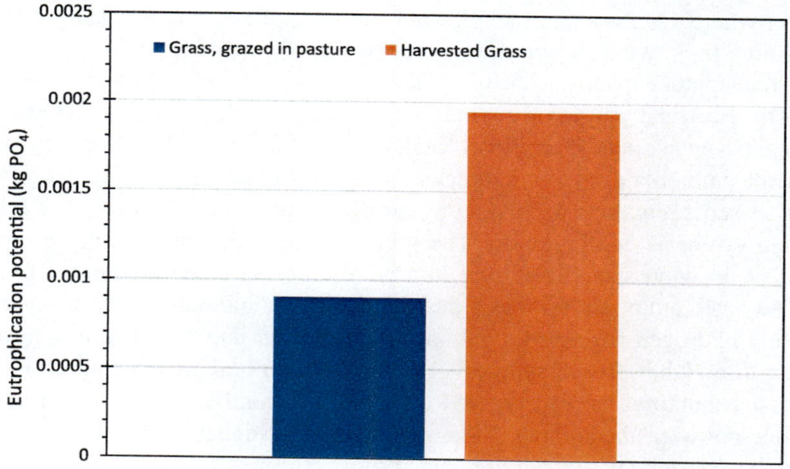

Fig. 3.43 LCA results for eutrophication potential for the effect of harvesting grass versus cattle feed

3.4.5 Case Study 5: LCA Analysis of Ammonia Production from Wind Electrolysis for Sectoral Use

In this case study, a life cycle assessment study of ammonia production from wind electrolysis for its usage in various sectors is presented. The aim of this study is to examine the environmental impact of all phases of wind electrolysis-based ammonia production. Life cycle analysis will be used to assess the feasibility of producing ammonia from wind electrolysis, and then the consequences will be assessed. For the life cycle inventory and impact assessment technique, Ecoinvent database is used [50]. A cradle-to-grave LCA of wind electrolysis-based ammonia production is shown in Fig. 3.44. LCA's system boundaries as well as the components that are being considered shown in Fig. 3.44. The functional unit chosen for this study is 1 kg of ammonia output from wind electrolysis.

Wind power plant is approximated with a 2 MW onshore wind turbine consisting out of the infrastructure datasets for "moving parts" and "fixed parts". Moving parts are the rotor, the nacelle, electronics, and the network connection. The tower and foundation are considered as fixed parts. This dataset represents the production of certified high voltage electricity at offshore grid-connected wind power plants with a capacity of 2 MW in Denmark in 2014. It includes operation and maintenance expenditures as well as infrastructure inputs.

The system under consideration for creating hydrogen from wind energy consists of two basic components: a wind turbine that generates electricity and a water electrolysis unit that generates hydrogen. Wind turbines convert wind energy into mechanical work, which is subsequently converted to AC electricity, and it is sent to the power grid. The effectiveness of wind turbines is determined by their placement. In places where there is a lot of wind, wind energy applications are usually selected. For onshore uses, wind turbines are commercially available in various sizes up to 8 MW nameplate capacity, with much larger units available for offshore applications. The electrical output of a wind turbine is highly dependent on wind speed, resulting in a wide range of electrical energy generation. In a Haber–Bosch synthesis reactor for ammonia production, an electrolyser for creating hydrogen from water and an air separation unit for extracting nitrogen from air are combined in a basic ammonia synthesis configuration. The wind turbine's electrical output is greatly dependent on wind speed, resulting in a wide range of electrical energy production. The basic ammonia synthesis design involves combining an electrolyser for producing hydrogen from water and an air separation unit for obtaining nitrogen from air in a Haber–Bosch synthesis reactor for ammonia manufacture. Offshore wind-based ammonia is also an environmentally favorable method of producing ammonia; however, the unit cost of ammonia remains higher.

The electrolyzer's power requirement is simply the result of the operating voltage and current. The number of electrolyzer cells in series determines the operating voltage, while the number of electrolyzer cells in parallel determines the operating current. For varying hydrogen outputs, most alkaline electrolyzers have roughly equal power profiles. As a liquefied gas, ammonia is commonly stored in huge amounts. In an insulated double containment tank, the liquid is kept at −33 °C at atmospheric pressure. Boil-off of ammonia happens because of heat transfer from the outside to the tank. The boil-off rate is usually around 0.04% every day [11, 14, 16].

A compression-refrigeration loop is established to condense the ammonia vapour and return it to the storage tank in order to recover the ammonia vapours escaping from the storage tanks. Before transferring the ammonia vapour to the condenser, two intercooled compressors raise the pressure to around 13 bar. Prior to being flashed

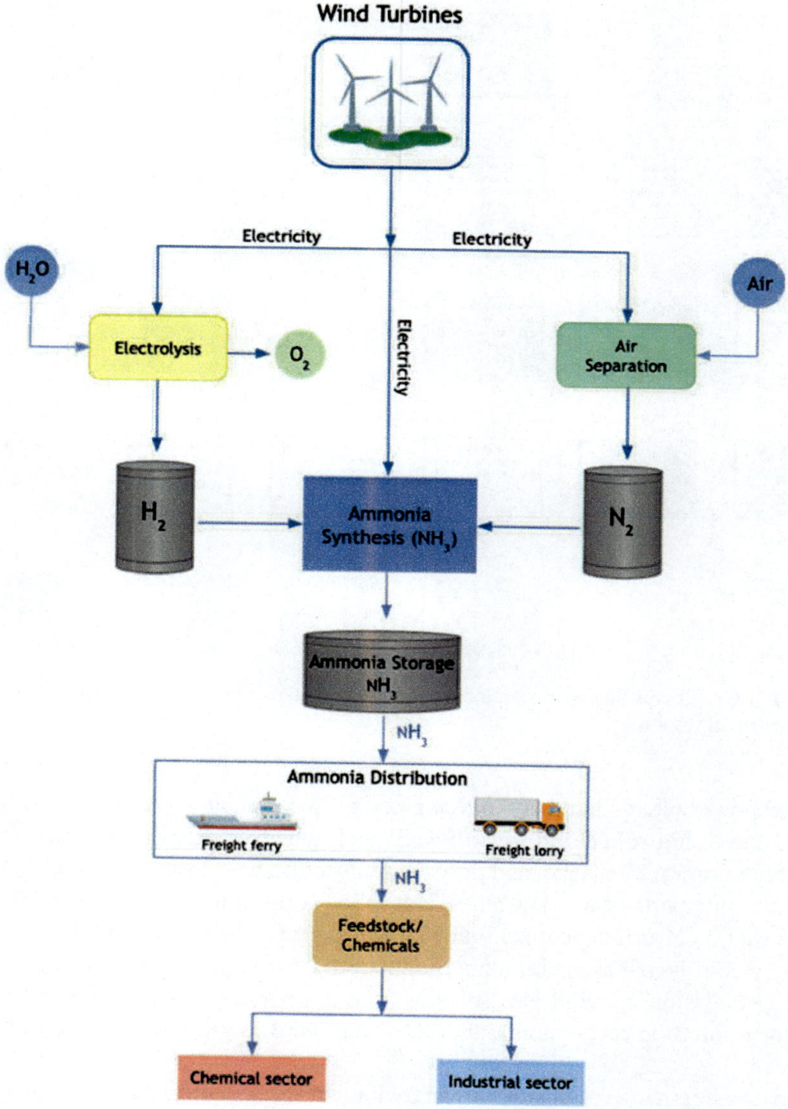

Fig. 3.44 Cradle-to-grave LCA of wind electrolysis-based ammonia production

in an intercooler, the ammonia is condensed with a tiny evaporator or cooling tower. The heat transmission into the vessel, which is proportional to the surface area and heat transfer coefficients, is directly connected to the compression work.

Figure 3.45 shows an LCA process flowchart for global warming of ammonia production from wind electrolysis, which illustrates the contribution of global warming between each process and highlights hotspots in the system. As it shown in Fig. 3.45, the transportation of ammonia via road has highest contribution followed

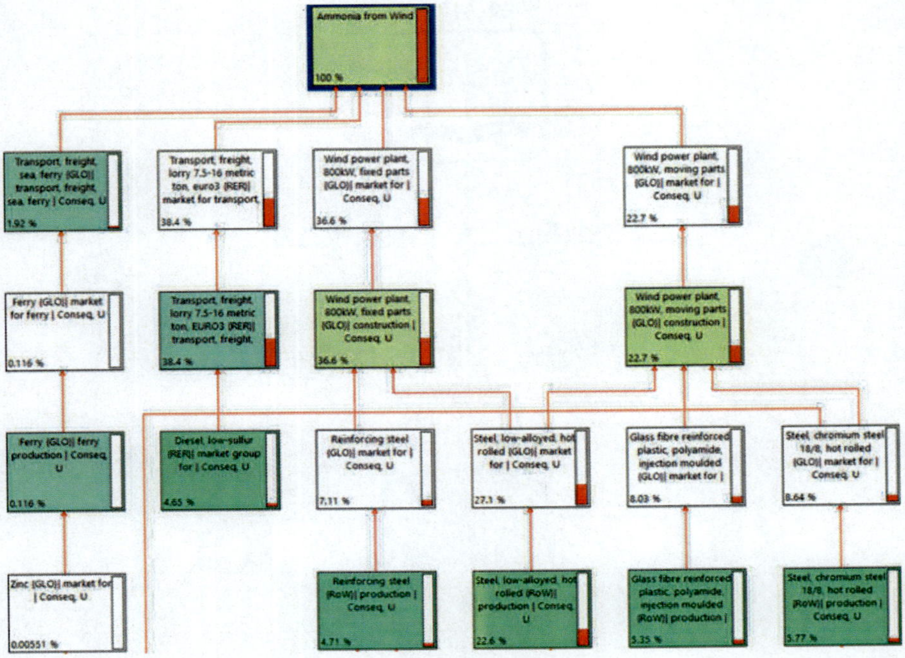

Fig. 3.45 LCA process flowchart for global warming potential of ammonia production from wind electrolysis (cut-off 4.6%)

by wind power plant's fixed and moving parts. Power plant's fixed parts has higher impact due to hot rolled and low-alloyed steel, which is used in the construction. Glass fiber reinforced plastic, and polyamide injection have high impact in construction of moving parts of wind turbines. Hot rolled chromium steel is another hot spot in construction of moving parts. Figure 3.46 displays LCA results for global warming of wind electrolysis-based ammonia production. As it seen in Fig. 3.46, total global warming emission of wind electrolysis-based ammonia production comes mostly from transportation of ammonia and parts that were used in construction of wind turbines.

Impact assessment results for human toxicity of wind electrolysis-based ammonia production is presented in Fig. 3.47. Human toxicity impact is mainly created by wind power plant's fixed and moving parts. While transportation by road has higher impact, ammonia transportation by sea appears more environmentally friendly. A more detailed analysis of human toxicity of wind electrolysis-based ammonia production is provided in Fig. 3.48.

Figure 3.48 shows the process flowchart for human toxicity of ammonia production from wind electrolysis. Each process shown in the figure has some contribution to ammonia production by wind electrolysis. Moving parts contributes to human toxicity emissions by 68.9%, which is highest impact that is followed by fixed parts with 24.6%. Copper production and steel production are the main reasons for high

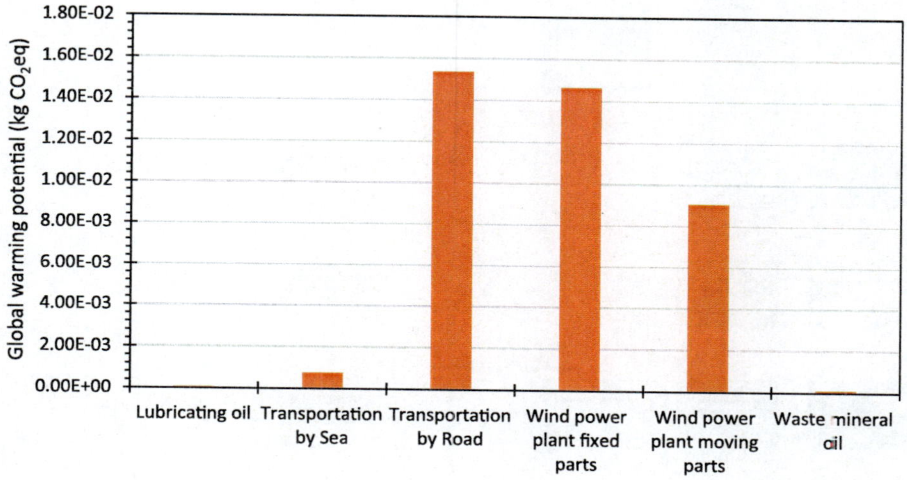

Fig. 3.46 LCA results for global warming potential of wind electrolysis-based ammonia production

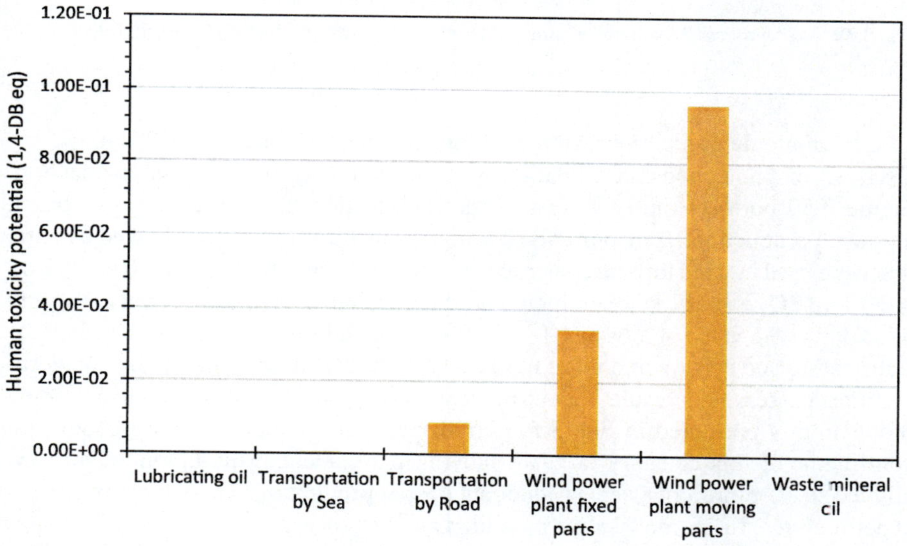

Fig. 3.47 LCA results for human toxicity potential of wind electrolysis-based ammonia production

human toxicity emissions in wind power plant's moving parts. Copper has 37.6% contribution and steel has 32.4% contribution in wind power plant moving parts construction process.

Figure 3.49 displays the impacts assessment results for photochemical oxidation of wind electrolysis-based ammonia production. Fixed parts emits 6.49×10^{-6} kg C_2H_4 eq of photochemical oxidation potential per 1 kg of ammonia production. Moving parts emits 3.43×10^{-6} kg C_2H_4 eq of photochemical oxidation potential per

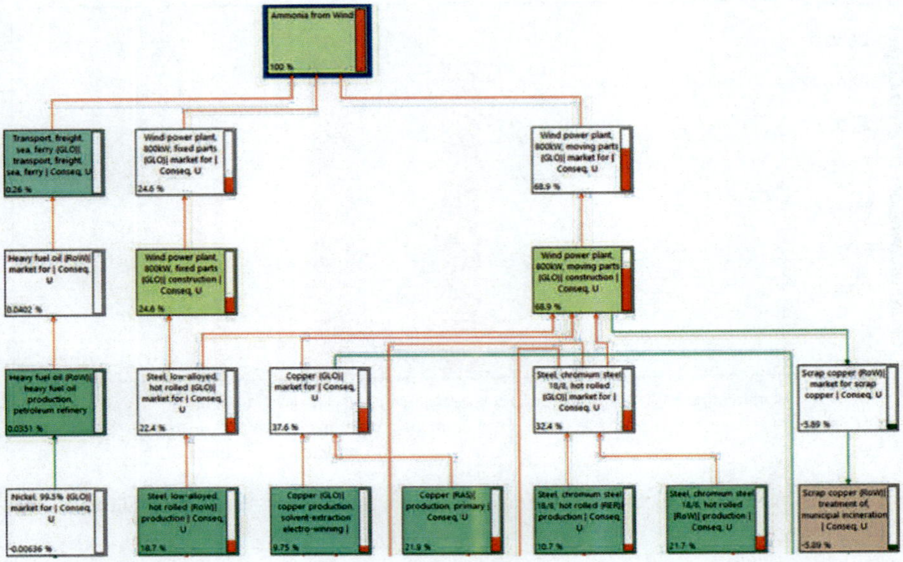

Fig. 3.48 LCA process flowchart for human toxicity of ammonia production from wind electrolysis (cut-off 7%)

1 kg of ammonia production. Ammonia transportation emits roughly 2.4×10^{-6} kg C_2H_4 eq of photochemical oxidation potential per 1 kg of ammonia production. Figure 3.50 portrays the LCA results for acidification of wind electrolysis-based ammonia production. Transportation has highest acidification load followed by parts that were used in wind turbine construction. Ammonia distribution by road emits 7.07 $\times 10^{-5}$ kg SO_2 eq per 1 kg of ammonia production, which is followed by moving and fixed parts with value of 5.64×10^{-5} kg SO_2 eq and 4.53×10^{-5} kg SO_2 respectively. Lubricating and waste mineral oil has impact of near zero in acidification category.

Impact assessment results for eutrophication of wind electrolysis-based ammonia production is presented in Fig. 3.51. Similar to other impact categories, the main contributor of this category is again moving parts and transportation stage. Low alloyed steel, reinforcing steel production, copper production, heavy fuel production at petroleum refinery, nickel mine operation are the processes that mainly contributes in eutrophication emissions.

Closing Remarks

The findings of this study tells that global warming potential emission of wind electrolysis-based ammonia production comes mostly from transportation of ammonia and the parts that were used in construction of wind turbines. Copper production and steel production used in moving parts are the main reasons for high human toxicity emissions in wind-based ammonia production.

- Ammonia transportation emits roughly 2.4×10^{-6} kg C_2H_4 eq of photochemical oxidation potential per 1 kg of ammonia production.

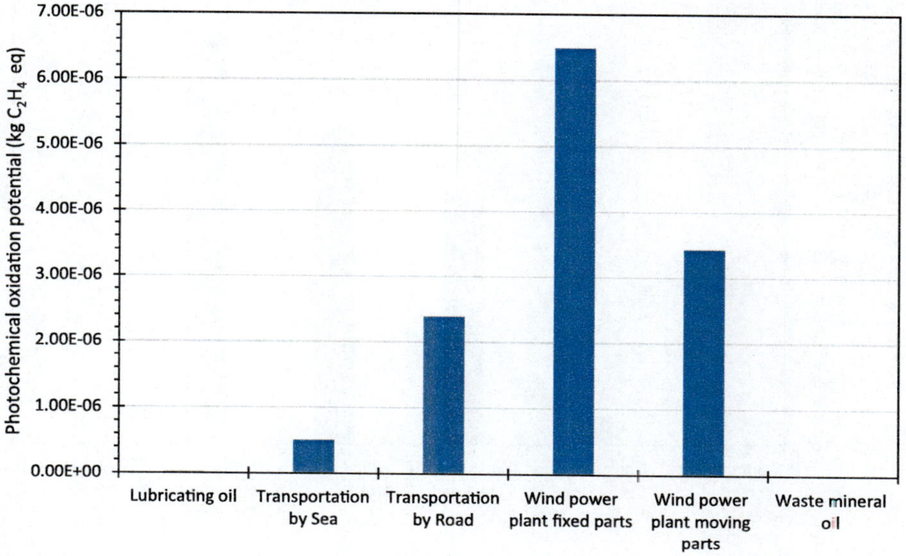

Fig. 3.49 LCA results for photochemical oxidation potential of wind electrolysis-based ammonia production

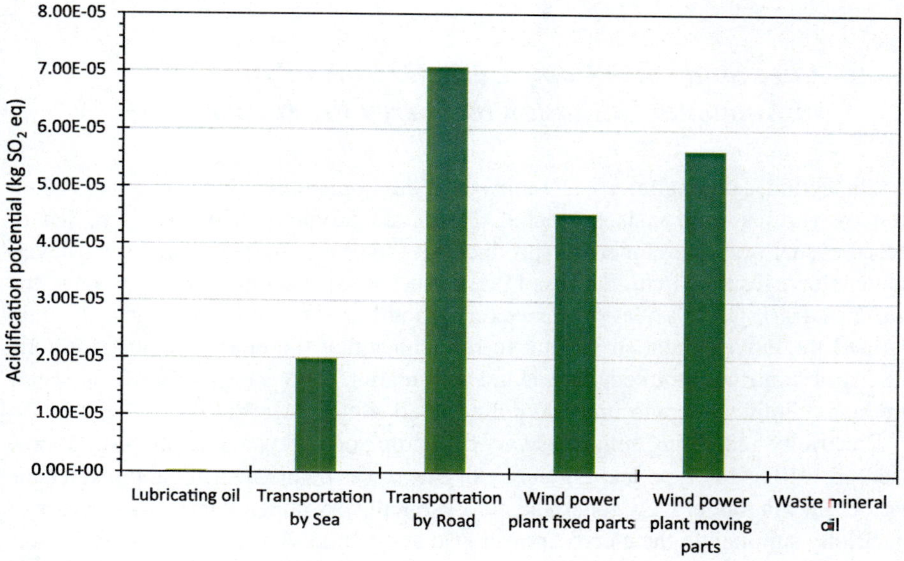

Fig. 3.50 LCA results for acidification potential of wind electrolysis-based ammonia production

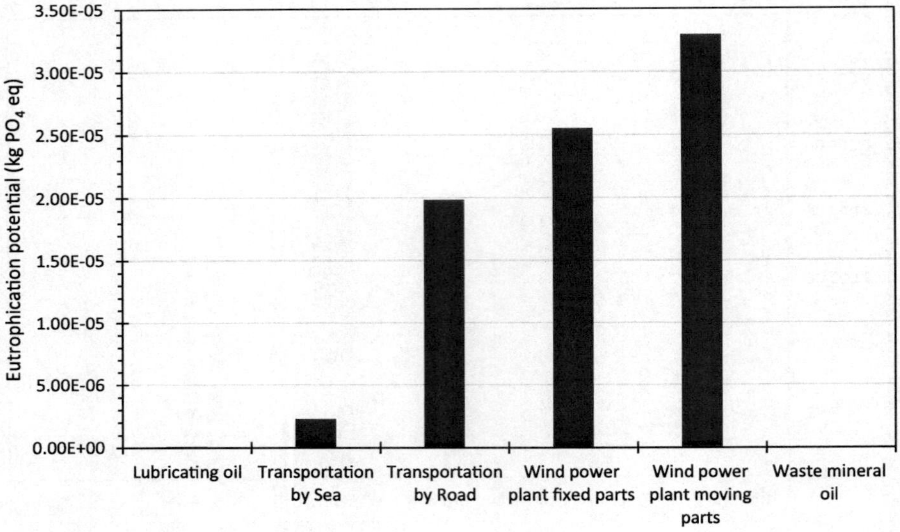

Fig. 3.51 LCA results for eutrophication potential of wind electrolysis-based ammonia production

- Ammonia distribution by road emits 7.07×10^{-5} kg SO_2 eq acidification per 1 kg of ammonia production.

3.4.6 Case Study 6: A Comparative LCA Analysis of Ammonia Utilization in Energy Production

In this section, a comparative LCA analysis of ammonia utilization in energy production by various sources is presented. Ammonia production by extensive list of different energy sources for energy production is analyzed. Increased industrial development have resulted in higher fossil fuel use across all sectors of any economy. The current infrastructure's heavy reliance on carbon-based fossil fuels has significantly harmed the environment. It is vital to remember that the energy sector is among the major contributors to environmental degradation. The energy production sector produces significant environmental pollution all over the world [52].

Electricity generating emissions vary depending on the type of fuel/energy source used, as well as the type and efficiency of the power plants. During any given time period, the amount of CO_2 generated per kWh will vary depending on the sources of electricity supplied to the electric power grid at the time. As a result, CO_2 emissions from energy and CO_2 emission factors will fluctuate hourly, daily, monthly, and annually.

It is difficult to pick the optimal energy source. Advocates for and against different sources use a variety of criteria to determine what is 'better.' Some argue for production consistency. Others place a greater emphasis on economics, such as through a cost–benefit analysis. The public health and environmental implications of diverse energy sources, on the other hand, are arguably the most disputed and crucial. Unfortunately, due to the differences in the stages required to transfer energy to the end consumer, comparisons are typically difficult. For example, based only on exhaust pipe emissions, an electric car may appear to be far cleaner than a gasoline-powered car. When the primary source of electricity is coal, which releases substantial levels of pollutants when burned from a fixed source, such as a power plant, the comparison gets more problematic. Following that, any effective option must take into account the effects of extraction (coalmine versus oil well), transportation (rail versus pipeline), and so on. As a result, using a life cycle analysis to incorporate the effects of energy production, distribution, and transportation is the best way to go.

One of the most important needs for an energy system is that it be long-term. A sustainable solution is one that ensures that an environmental goal is fulfilled continually for the near future, rather than just meeting it once. As a result, engineering and science are becoming more system-focused. As early as possible in the life cycle, proper design must consider all conceivable scenarios. These include the raw materials that will be required under various scenarios, as well as the energy, labor, and other costs, benefits, and risks associated with each choice prior to material extraction from the ground. Costs and risks must be maintained as the project progresses, including during manufacturing, operation, decommissioning, and recycling. The LCA is a crucial tool for any energy decision since it reveals what a site is worth beyond its perceived utility. Every source of energy has downstream consequences. That is, all elements that must be removed from the earth in order for a system to create energy must be addressed. Coal and uranium must certainly be mined, as well as crude oil and natural gas drilling and tree harvesting. This, however, includes 'non-emitting' sources. Materials taken from the soil must be used to manufacture solar panels (e.g. silica and crude oil to make plastics) [53].

The goal and scope definition step of a life cycle assessment includes specifying the study objectives, system boundaries, functional unit, and so on. Quantifying the inputs to the system boundary as well as the outputs from the boundaries is the next phase in the inventory analysis. In addition, the data gathered during the inventory analysis process is used to conduct the impact assessment. This stage entails turning environmental emissions into environmental consequences throughout the life cycle. The environmental consequences are calculated using a specialized impact assessment approach. Furthermore, each stage of the research is intertwined with interpretation. This step mostly entails making assumptions and making recommendations based on the received data. In an LCA, different phases of a product's life cycle are considered, and their associated inputs and outputs are calculated. Raw material extraction is the first stage in the life cycle, followed by material processing, manufacturing, distribution, use, and disposal. The life cycle stages covered in this study are represented in Fig. 3.52.

CML-IA baseline impact assessment method is used for the calculations. For this study the chosen functional unit is 1 MJ of electricity production. The environmental impact categories chosen for the LCA study are listed below:

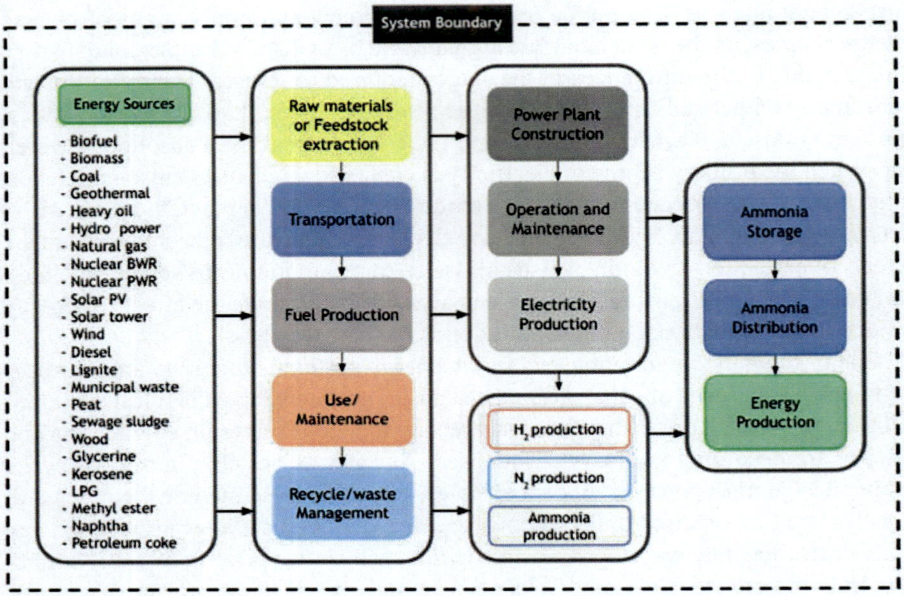

Fig. 3.52 The system boundary and major life cycle stages considered in this study

- Global warming potential
- Ozone layer depletion potential
- Human toxicity
- Photochemical oxidation
- Acidification
- Eutrophication

For this LCA study, following stages are included, and assumptions are listed below:

- Cradle-to-grave life cycle assessment methodology is used.
- The power plant is utilised for moderate load coal, with 4000 h of full capacity operation per year. During its lifetime, the plant is expected to run for 150,000 h.
- Fuel and material extraction, fuel production, power plant construction, transportation, hydrogen production, nitrogen production, ammonia production energy production, and disposal/waste management are included.
- For biofuels, the dataset includes the sugarcane conversion into raw sugar, ethanol and surplus electricity. Processing equipment, chemical inputs, industrial area, water use treatment, and emissions to air are included in the inventory. Sugarcane processed arrives with 5% of vegetal impurities (sugarcane straw) and 1% of mineral impurities (dirt). No additional sugarcane straw recovery and industrial use is considered.
- The inventory is modelled for North America and Europe.

- For geothermal energy, the dataset represents electricity production with a deep geothermal enhanced geothermal systems plant. The power generation unit is run on an Organic Rankine Cycle with benzene as working fluid.
- The dataset includes all energy use, use of chemicals, emissions to air and water including treatment of flue gasses and effluents. In an oil power plant, water is needed for cooling, feed water for the steam circuit, for the flue gas treatment, or other cleaning. The dataset only considers water use (evaporation) for cooling and steam circuit.
- For hydro (pumped storage), the dataset includes the use of pump energy (electricity, high voltage), lubricant oil, the volume of the reservoir, the mass of water passing through the turbines and an estimation of greenhouse gas emissions out of the water reservoir.
- For hydro (Reservoir), lifetime is assumed 150 years for the structural part and 80 years for the turbines. Net average electrical efficiency, including pipe losses is 78%. The data set includes the area/volume occupied for the barrage, power station, and the lake, preliminary estimation of emissions of methane and dinitrogen oxide. It also includes operation and maintenance activities and materials of the power plants, such as lubricating oil or mass of water passing through the turbines.
- For natural gas, the plant is used for middle load with 5000 h of operation at full capacity per year. The plant is assumed to operate 180,000 h during its lifetime.
- The fuel elements, chemicals, and diesel requirements, as well as the necessary transport requirements, are all included in the module for nuclear power plants. Cooling with fresh and decarbonized water is supplied. Spent fuel for reprocessing and conditioning; operational low active waste for conditioning in the intermediate repository; and contaminated waste from disassembly are all considered radioactive waste streams. Non-radioactive wastes are taken into account.
- For solar tower power plant, the technology is 20 MW solar tower power plant, air cooling, collector field area is 215,520 m^2, thermal heat storage capacity: 440 MWh, hot and cold two stage nitrate salt thermal storage system, steam mass flow rate is 37 kg/s.
- For wind power plant, the dataset represents the production of high voltage electricity at offshore grid-connected wind power plants with a capacity of less than 1 MW. It includes operation and maintenance expenditures as well as infrastructure inputs. Wind load hours have been adapted to local conditions.
- For lignite fuel, the dataset represents the production of high voltage electricity in an average lignite power plant in Alberta, Canada. 'Lignite' is used as represent for 'brown coal', which includes sub-bituminous coal and lignite according to the definition of the IEA electricity information 2014.
- For energy from municipal waste, the inventory is modelled for Ontario in year 2012. Production volume is 1 kWh.
- For peat, the dataset represents the production of high voltage electricity in an average peat power plant. The dataset has been extrapolated from a global 2008 dataset representing electricity production with lignite due to the lack of specific peat data.

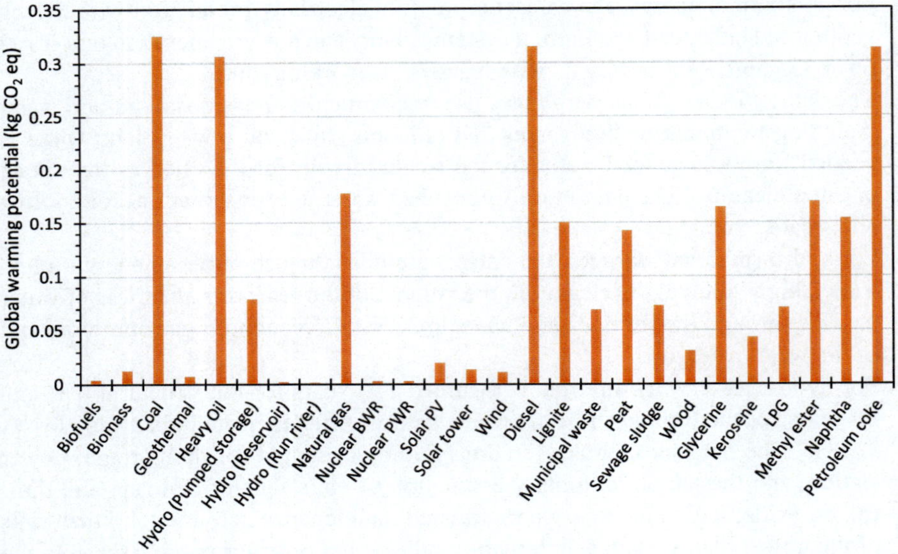

Fig. 3.53 Comparison of LCA results for global warming potential of ammonia utilization in energy production by various sources

- For sewage sludge, the dataset represents the production of electricity and heat from a biogas mix from different sources (bio-waste, sewage sludge) when burning it in a cogeneration unit with gas engine. It is intended to represent the production of grid-connected electricity with biogas. The main product is then considered electricity at high voltage, while heat is produced as a co-product. The cogeneration unit has a capacity of 160 kW; the degrees of efficiency are as follows: electricity: 0.37 and heat: 0.53. A mix of biogas is treated in this dataset with an average lower heating value of 22.73 MJ/N m^3.
- The conveyance of wood to the pulp factory is the first step in the process. Transports to the pulp mill, wood handling, chemical pulping and bleaching, drying process, on-site energy production, chemical recovery cycles, and internal waste water treatment are all part of the process.
- For glycerin, included activities starts from reception of purified cooking oil and chemicals at factory gate. This activity ends at esterification plant with the produced methyl ester and glycerine.
- The production of kerosene begins when crude oil enters a petroleum refinery. Wastewater treatment, drinking water supply, refinery infrastructure, crude oil and product storage on refinery grounds, and electricity generation are all covered.

Despite the fact that electricity is a usually clean and safe form of energy when consumed, its production and transmission have an influence on the environment. Almost all electric generating plants have an impact on the environment, albeit some have a higher impact than others [54]. For this study, the chosen functional unit is 1 MJ of electricity production. A comparative LCA result of global warming for

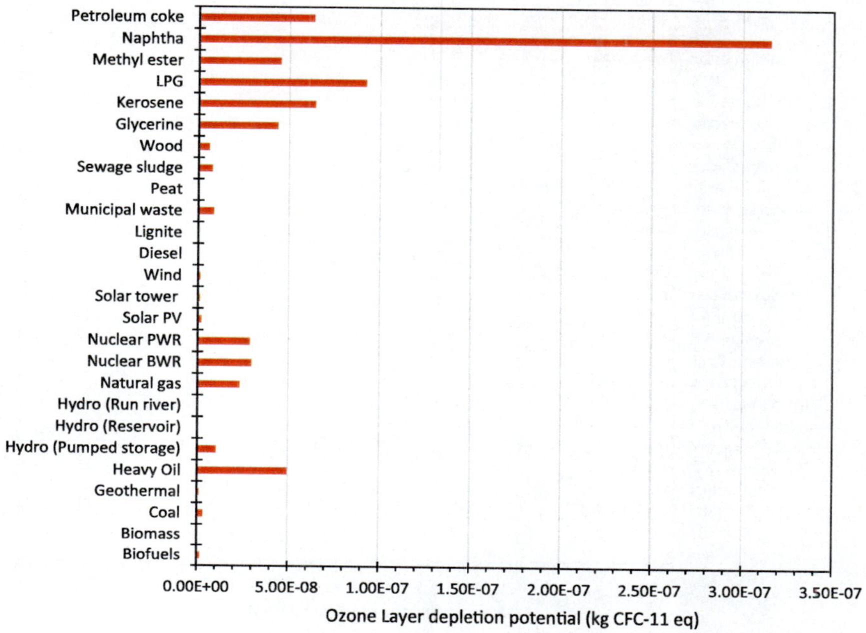

Fig. 3.54 Comparison of LCA results for ozone layer depletion potential of ammonia utilization in energy production by various sources

ammonia utilization in energy production by various sources is shown in Fig. 3.53. Life cycle impact assessment results reveal that fossil fuel-based ammonia energy sources emit highest CO_2 emissions when they are consumed in energy production. Coal, petroleum coke, diesel and heavy oil are damaging the environment most. However, other fossil fuel-based fuels such as LPG, kerosene, naphtha, peat and lignite are comparatively perform better in terms of environmental impacts. For instance, while coal emits 0.325 kg CO_2 eq per MJ of energy production, kerosene emits only 0.044 kg CO_2 equivalent per MJ of energy production. Apart from this, all renewable-based sources emit a very small amount of CO_2 emissions during energy production. For example, hydroelectric, nuclear power, biofuels, geothermal and wind sources have nearly zero global warming impact as shown on Fig. 3.53. Physical footprints are present in all power plants based on where they are located. The footprint of some power plants is quite modest because they are positioned inside, or adjacent to an existing building. The construction of most big power plants necessitates land removal. For fuel supply, electrical transmission lines, as well as cooling water deliveries, some power plants may require road access, railroads, and pipelines. Solid-fuel power plants may feature storage areas for the combustion waste. Ash is the solid residue left over after solid fuels like coal, biomass, and municipal solid waste are burned. Bottom ash refers to the largest particles that settle at the bottom of the combustion chamber in power plant boilers. Fly ash refers to the smaller and lighter particulates that gather in air emission control systems. In most

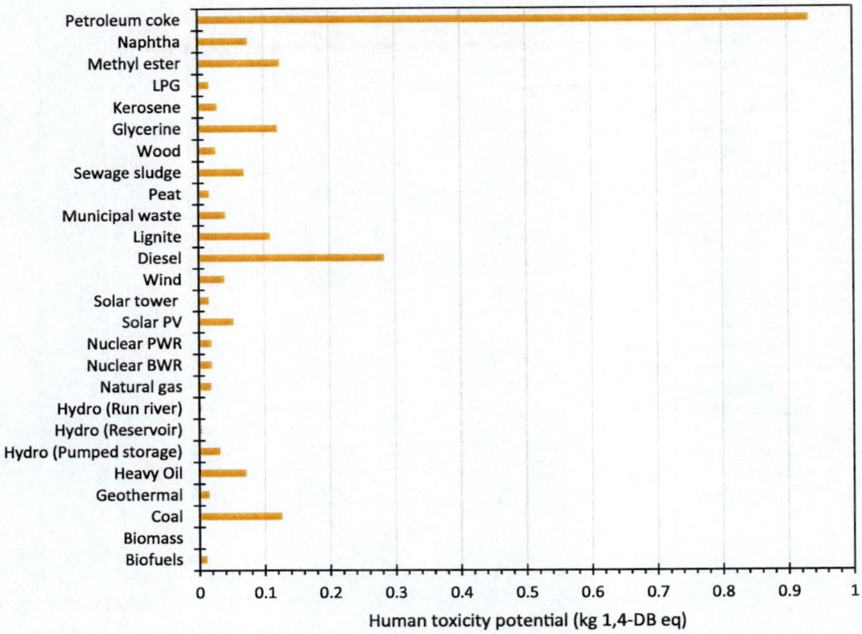

Fig. 3.55 Comparison of LCA results for human toxicity potential of ammonia utilization in energy production by various sources

cases, fly ash is combined with bottom ash. All of the harmful elements captured by pollution control equipment are contained in the ash. In retention ponds, many coal-fired power plants store ash sludge (ash combined with water). A number of these reservoirs have burst, causing significant damage and pollutants downstream. Some coal-fired power facilities dump ash in landfills, sell it for use in concrete, or asphalt production.

Figure 3.54 compares LCA results for ozone layer depletion of ammonia utilization in energy production by various sources. The calculations show that naphtha has highest ozone layer depletion due to onshore petroleum production and platinum group metal extraction and refinery. Also, transportation of petroleum in tankers freight by sea. LPG, petroleum coke, kerosene and heavy oil fuels have highest ozone layer depletion values after naphtha. As it presented with Fig. 3.54 renewable-based energy production sources give much better results. As a result, employing clean energy production techniques is critical for achieving lower life cycle environmental impacts.

Comparison of LCA results for human toxicity of ammonia utilization in energy production by numerous sources is depicted in Fig. 3.55. The first notable fuel in the human toxicity category is petroleum coke. Petroleum coke emits 0.933 kg 1,4-DB equivalent per MJ of energy production, and this due to coking during production, it is followed by diesel with 0.283 kg 1,4-DB equivalent per MJ of energy production.

Besides that, all selected renewable sources for this study have minor human toxicity impact.

Figure 3.56 shows LCA results for photochemical oxidation of ammonia utilization in energy production by various sources. Except biomass and petroleum coke, all selected fuels resulted in low photochemical oxidation impact. Photochemical smog, a complicated mixture of primary and secondary air pollutants, contains photochemical oxidants. Photochemical oxidants are secondary air pollutants created when sunlight interacts with nitrogen oxides and their precursors, reactive hydrocarbons. Ozone and peroxyacetyl nitrate are the most prominent phytotoxic components produced by these atmospheric photochemical processes. Other products include peroxy compounds, aldehydes, ketones, organic and inorganic acids, aerosols, and nitrogen dioxide.

Figure 3.57 demonstrates the LCA results for acidification of ammonia utilization in energy production by various sources. As it observed here, all fossil fuel based-energy sources have high acidification impact. The leading fuels are heavy oil, naphtha, petroleum coke, coal and diesel, which are most commonly, used fuel for energy production and transportation. On the other hand, Renewable-based sources have minimal acidification impact as expected. Acidification Potential is a measurement of the drop in the pH of precipitation and fog, which has the consequence

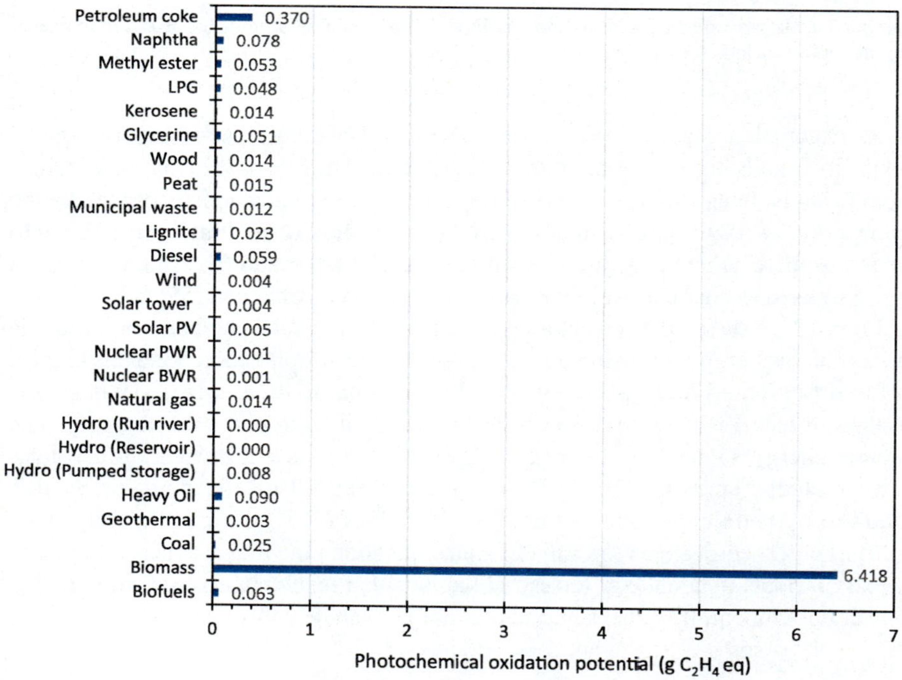

Fig. 3.56 Comparison of LCA results for photochemical oxidation potential of ammonia utilization in energy production by various sources

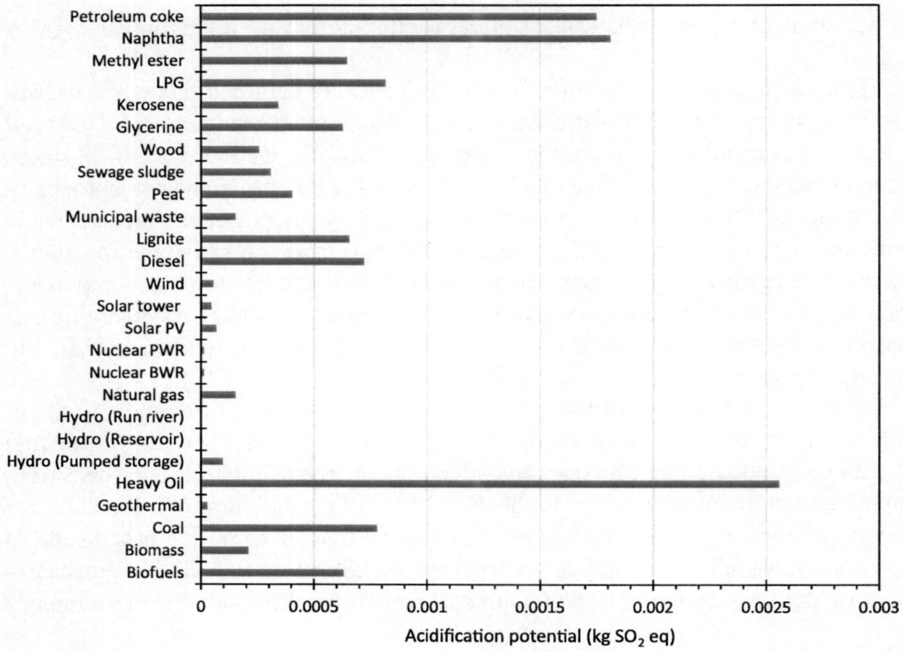

Fig. 3.57 Comparison of LCA results for acidification potential of ammonia utilization in energy production by various sources

of ecosystem degradation due to nutrition being flushed out of soils and trace metal solubility in soils, for example. Acidification potential is measured in mass of sulphur-dioxide equivalents and has a regional impact. Acidification is mostly caused by the combustion of fossil fuels, which releases sulphur-dioxide and nitrogen-oxide into the atmosphere, where they dissolve with condensed water and fall as rain. Acid rain is a term used to characterize severe instances of this mechanism [55].

Figure 3.58 shows the impact assessment results for eutrophication of ammonia utilization in energy production by various sources. Eutrophication impact is highest in fossil fuel-based fuels as seen in Fig. 3.58. Similar to other assessed categories, renewable-based fuels are most promising sources in eutrophication category. The notable energy sources are sewage sludge and lignite in eutrophication category. Sewage sludge emits 0.0014 kg PO_4 equivalent per MJ of electricity production followed by petroleum coke and lignite with values of 0.0013 kg PO_4 equivalent, 0.0011 kg PO_4 equivalent respectively. Eutrophication potential is a measurement of excess nutrients in aquatic or terrestrial habitats that results in ecological harm due to over enrichment. It is quantified in mass of phosphate equivalents.

Closing Remarks

LCA results of this study tells that fossil fuel-based ammonia energy sources emit highest CO_2 emissions when they are consumed in energy production. Coal, petroleum coke, diesel and heavy oil are most notable fuels in high CO_2 emissions.

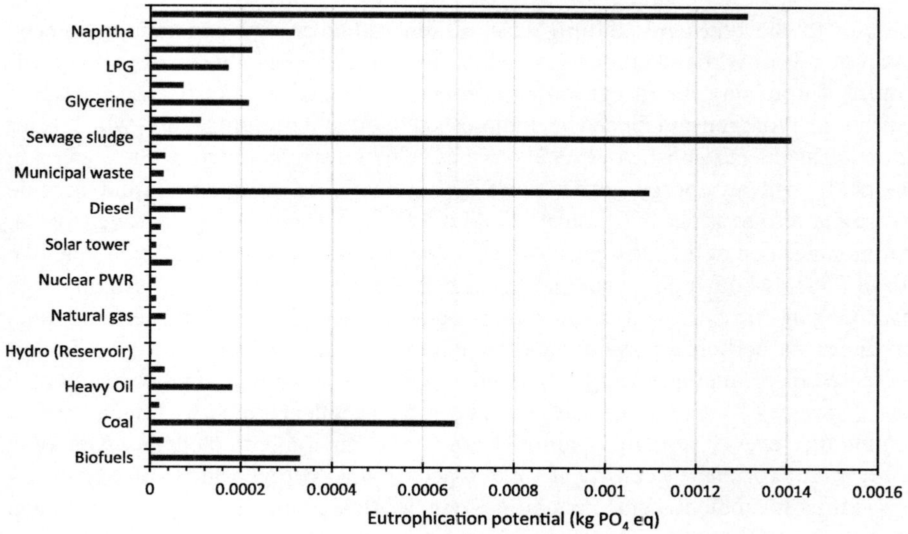

Fig. 3.58 Comparison of LCA results for eutrophication of ammonia utilization in energy production by various sources

Environmentally friendly routes of energy production can be achieved by renewable based ammonia production.

- Coal emits 0.325 kg CO_2 eq per MJ of energy production, kerosene emits 0.044 kg CO_2 equivalent per MJ of energy production. Apart from this, all renewable-based sources emit a very small amount of CO_2 emissions during energy production.
- The calculations show that naphtha has highest ozone layer depletion potential.
- In human toxicity category, petroleum coke emits 0.933 kg 1,4-DB equivalent per MJ of energy production, and it is followed by diesel with 0.283 kg 1,4-DB equivalent per MJ of energy production.
- The notable energy sources are sewage sludge and lignite in eutrophication category. Sewage sludge emits 0.0014 kg PO_4 equivalent per MJ of electricity production followed by petroleum coke and lignite with values of 0.0013 kg PO_4 equivalent, 0.0011 kg PO_4 equivalent respectively.

3.4.7 *Case Study 7: A Comparative LCA Analysis of Ammonia Production Using Spectrum of Hydrogen Colors*

In this case study, a comparative LCA analysis of ammonia production is presented by using spectrum of hydrogen colors. The main ammonia colors are investigated that vary depending on the energy source and the hydrogen production process.

Due to the constantly falling cost of renewable energy generation, "green" hydrogen (i.e. hydrogen produced via electrolysis using renewable energy) is considered as a promising energy carrier for a variety of applications. For storage and transportation, hydrogen can indeed be compressed or liquefied, a procedure, which is far from ideal due to substantial safety issues. Chemically converting the hydrogen molecule into an energy carrier such as ammonia, methanol, or liquid organic hydrogen carriers is one of the most suitable methods of transporting hydrogen by sea. Ammonia is one of the most promising energy carriers among these. Ammonia, like fossil fuels, is both a chemical energy carrier and a potential fuel, releasing energy by changing the chemical composition. Furthermore, if used as a fuel, ammonia produces no carbon emissions, and its environmental credentials can be further increased if sustainable energy is utilized to operate ammonia production. Ammonia has a variety of characteristics that make it an excellent green hydrogen carrier: Ammonia's energy storage capabilities are very comparable to methane. Ammonia has a volumetric energy density of 150% that of liquid hydrogen and hydrogen densities similar to ambient storage conditions can be attained. In contrast, pure hydrogen requires high pressures or low temperatures to attain effective volumetric hydrogen density. In comparison to pure hydrogen, ammonia possesses lower explosive limitations in air. As a result, hydrogen storage is more complicated, energy-intensive, and costly than ammonia storage. Because of its widespread use as a source for inorganic nitrogen-containing fertilizers and a range of other industrial chemicals, the ammonia supply chain has reached a high level of maturity. Ammonia has a well-established global infrastructure for storage and trading. In comparison to other potential hydrogen transporters, it has a high hydrogen density [56]. The major ammonia colors differ depending on the energy source and the hydrogen synthesis process, as shown in Fig. 3.59.

The steam reforming of natural gas (SMR) provides 72% of ammonia production. Naphtha, heavy fuel oil, coal, natural gas coke oven gas, and refinery gas are being used as feedstock in ammonia synthesis when fossil fuels are used. Because coal is China's primary source of ammonia, the country's energy consumption and greenhouse gas emissions are higher than the rest of the globe. Natural gas prices account for almost 70–90% of the cost of producing ammonia using the steam methane reforming process. Since the SMR method of ammonia production relies on natural gas, increased natural gas prices result in higher ammonia production costs. Even though there are numerous methods for producing ammonia, the Haber–Bosch process and the solid-state ammonia synthesis process are the most common methods. An air separation process in both operations provides nitrogen. Cryogenic air separation is the most efficient and cost-effective method for creating large quantities of oxygen, nitrogen, and argon.

At a low added cost, the cryogenic process can also produce high-purity nitrogen as a beneficial by-product stream. Cryogenic air separation is the most well established and industrialised form of air separation. Because ammonia is produced in large numbers, needed nitrogen must be produced at a cheap cost and with high efficiency. The necessary electricity for air separation could come from either conventional or

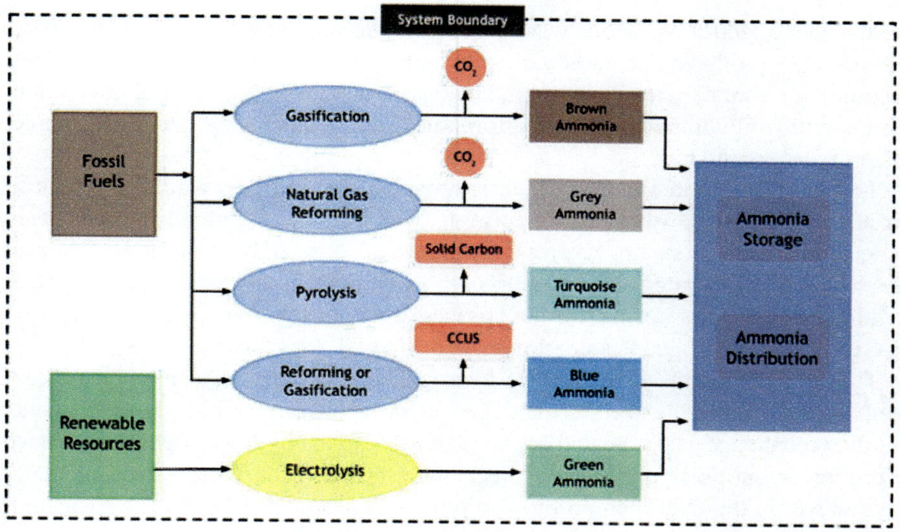

Fig. 3.59 The system boundary and major life cycle stages considered in this study

renewable sources. The most prevalent method for producing ammonia is the Haber–Bosch process. An exothermic process produces ammonia by mixing hydrogen and nitrogen in a ratio of 3:1. The catalyst helps the reaction, and the ideal temperature range is 450–600 °C. Each of the major issues affecting the world, such as population growth, food security, and the use of biomass to generate energy, will raise the demand for nitrogen fertilizer. If fossil fuels are used to produce ammonia and other fertilizers, Earth will become a world with dirty air, greater human health risks, diminished biodiversity, and more GHGs. The transformation of feed gases to ammonia is thermodynamically limited, resulting in several Haber–Bosch process drawbacks. Furthermore, the environment is severely contaminated, and energy consumption is excessive. Modifications to the catalyst and heat recovery are the most common ways to improve Haber–Bosch [10, 12, 29, 57].

Reducing the carbon footprint of ammonia synthesis is a problem that must be met if sustainable goods are to be developed that have minimal environmental consequences throughout their entire life cycle. There are essentially two ways to reduce CO_2 emissions: the first is to absorb and sequester CO_2 after it is produced, which leads to "blue" products; the second is to completely eliminate CO_2 production by using renewable energy and feedstock, which leads to "green" products. Different colors are attributed to hydrogen based on the kind of production. However, because there is no common naming convention, color definitions can alter over time and even between countries [56].

Green hydrogen is the color of hydrogen produced with no damaging GHGs in the spectrum of hydrogen colors. Green hydrogen is created by electrolyzing water with clean electricity generated from surplus renewable energy sources such as solar or wind power. Electrolysis divide water into hydrogen and oxygen via an electrochemical reaction, producing no carbon dioxide in the process. Green hydrogen currently accounts for a small percentage of total hydrogen due to the high cost of generation. Green hydrogen will become more affordable as it becomes more widely available, just as wind energy has.

The most common source of blue hydrogen is natural gas, which is produced through a process known as steam reforming, which combines natural gas and heated water in the form of steam. Hydrogen is produced, but carbon dioxide is produced as a by-product. To catch and store this carbon, carbon capture and storage (CCS) is required. Because the steam reforming process does not actually avoid the generation of GHGs, blue hydrogen is sometimes referred to as "low-carbon hydrogen".

This is the most used method of hydrogen production at the moment. Steam methane reformation is used to make grey hydrogen from natural gas or methane, but the greenhouse gases created are not captured. In the hydrogen spectrum, brown hydrogen is the polar opposite of green hydrogen and the most environmentally destructive. In the hydrogen production process, black coal or lignite (brown coal) are used.

The hydrogen colour wheel has a new addition, and scale manufacturing has yet to be confirmed. The methane pyrolysis method, which creates hydrogen and solid carbon, is used to make turquoise hydrogen. In the future, turquoise hydrogen could be considered a low-emission hydrogen if the heating process is powered by renewable energy and the carbon is permanently stored or used [58].

In this particular study, the CML-IA-baseline impact assessment method is used for the LCA calculations. The chosen functional unit is 1 kg of ammonia production for the analyses and assessments. The environmental impact categories chosen for the LCA study are listed below:

- Global warming potential (GWP)
- Ozone layer depletion potential (ODP)
- Human toxicity potential (HTP)
- Photochemical oxidation potential (POP)
- Acidification potential (AP)
- Eutrophication potential (EP).

The module illustrates how an ordinary ammonia plant in Europe works. The plant is used for intermediate load and operates at full capacity for 4000 h per year. During its lifetime, the plant is expected to run for 150,000 h. A bottom-up approach was utilised to examine the key characteristics of brown ammonia from coal (LHV, sulphur and ash content of coal, plant efficiency) and criterion emissions (SO_x, NO_x, particulates, and CO_2). It consists of gathering information on individual plants. The particle size distribution was calculated using German data. Halogen emissions were calculated using the species content in the country-specific coal input mix and average retention rates. The module illustrates how an ordinary ammonia plant in Europe works. The plant is used for intermediate load and operates at full capacity for 4000 h per year. During its lifetime, the plant is expected to run for 150,000 h. A bottom-up approach was utilised to examine the key characteristics of brown ammonia from coal (LHV, sulphur and ash content of coal, plant efficiency) and criterion emissions (SO_x, NO_x, particulates, and CO_2). It consists of gathering information on individual plants. The particle size distribution was calculated using German data. Halogen emissions were calculated using the species content in the country-specific coal input mix and average retention rates.

A comparison of LCA results for global warming of ammonia production methods using spectrum of hydrogen colors is presented with Fig. 3.60. Impact assessment result displays five different ammonia colors based on the hydrogen production sources and processes. The results shows that the brown ammonia is opposite of green ammonia in the hydrogen spectrum and it is the most environmentally harmful ammonia type. Brown ammonia is produced using black coal or lignite (brown coal) in the hydrogen-making process. Brown ammonia emits 0.325 kg CO_2 eq per 1 kg of ammonia production while green ammonia produce only 0.0072 kg CO_2 eq per 1 kg of ammonia production. Grey ammonia is the second most harmful ammonia type after brown ammonia with value of 0.178 kg CO_2 eq per 1 kg of ammonia production followed by turquoise ammonia with value of 0.079 kg CO_2 eq per 1 kg of ammonia production. Blue ammonia is the most environmentally friendly option among the fossil-based ammonia production options. Because hydrogen is produced, but carbon dioxide is produced as a by-product. To catch and store this carbon, CCS is required. Because the steam reforming process does not actually avoid the generation of GHGs.

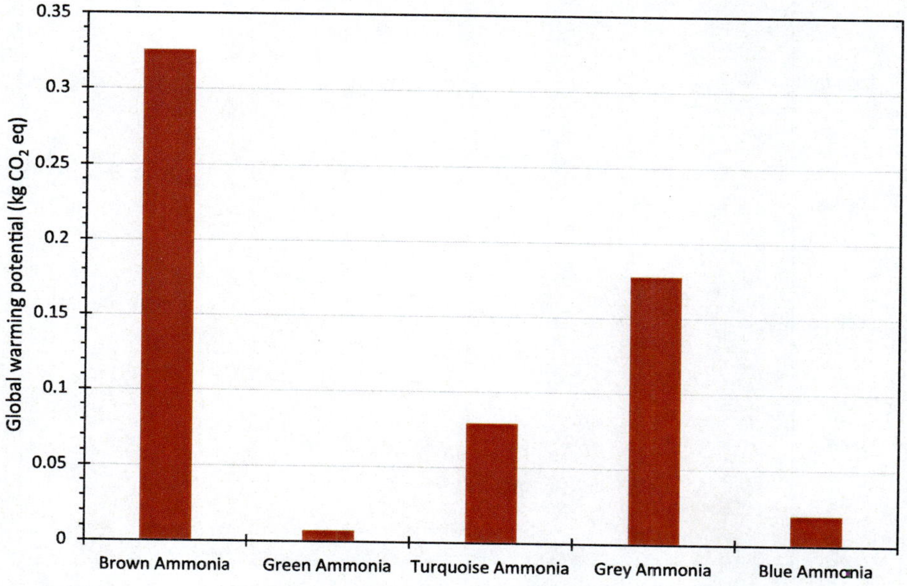

Fig. 3.60 Comparison of LCA results for global warming potential of ammonia production methods using spectrum of hydrogen colors

Figure 3.61 compares LCA results for ozone layer depletion potential of ammonia production methods using spectrum of hydrogen colors. The impact assessment results shows that grey ammonia has the highest ozone layer depletion impact among the ammonia production methods. Grey ammonia from grey hydrogen is manufactured from natural gas, or methane, by steam methane reformation, but without capturing the greenhouse gases produced. Absence of capturing the greenhouse gases in the blue ammonia production method results in higher emission values. The depletion of the ozone layer is primarily caused by nitrogenous chemicals such as NO_2, NO, and N_2O. Turquoise ammonia production has the second highest ozone layer depletion value, and then followed by brown ammonia and blue ammonia. Green ammonia production technique emerge as a most promising and sustainable ammonia production method. As a result, using clean energy production techniques is crucial for achieving lower environmental impacts over the long term.

Figure 3.62 compares LCA results for human toxicity of ammonia production methods using spectrum of hydrogen colors. The LCA calculation results indicate that brown ammonia production technique produce highest human toxicity during ammonia production process. The production method of blue ammonia emits the second-highest level of human toxicity among the ammonia production methods. This is due to using gasification of coal or steam methane reforming process during

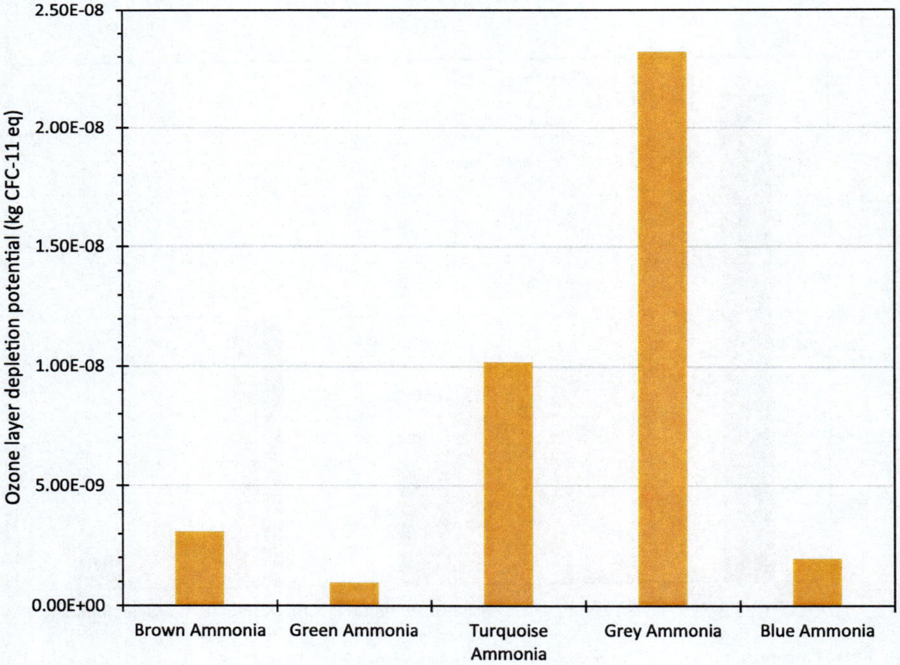

Fig. 3.61 Comparison of LCA results for ozone layer depletion potential of ammonia production methods using spectrum of hydrogen colors

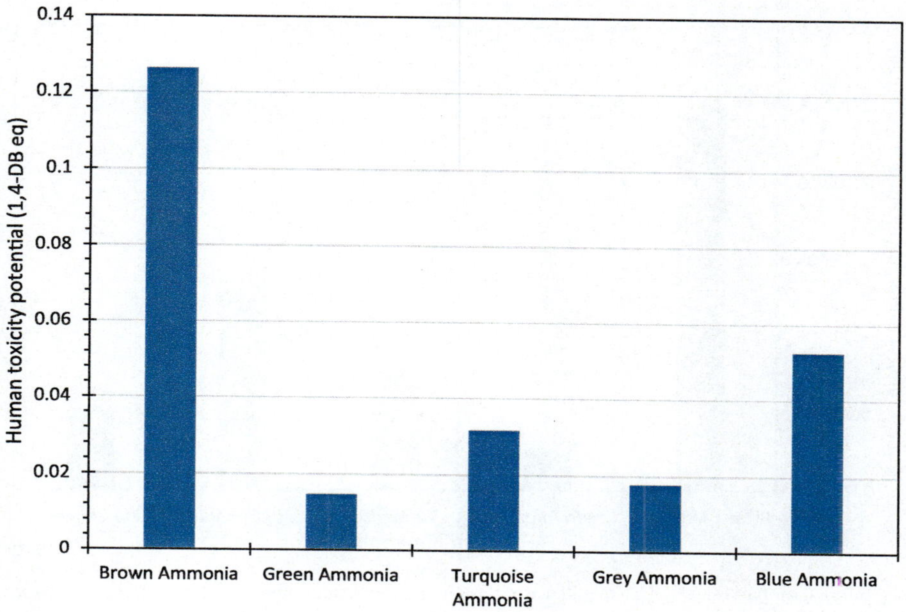

Fig. 3.62 Comparison of LCA results for human toxicity potential of ammonia production methods using spectrum of hydrogen colors

hydrogen production. Green ammonia production technique is the most environmentally benign option in human toxicity impact category. Coal burning emits a complex mixture of gaseous components and particles with different sizes, morphologies, chemical, physical, and biological properties into the atmosphere. Pollutants from power plants are released into the environment, potentially causing health problems. It is not only coal-burning that is harmful to one's health. During the extraction, transport, and handling of coal, large volumes of coal dust are emitted, which reach employees and nearby neighborhoods and create xenobiotic consequences [59].

A comparison of LCA results for photochemical oxidation of ammonia production methods using spectrum of hydrogen colors is illustrated in Fig. 3.63. Impact assessment results shows that brown ammonia production technique has highest photochemical oxidation emission values followed by grey ammonia production method and turquoise ammonia production. When sunlight interacts with nitrogen oxides and their precursors, reactive hydrocarbons, photochemical oxidants are produced as secondary air pollutants. The most notable phytotoxic components produced by these atmospheric photochemical processes are ozone and peroxyacetyl nitrate. Figure 3.64 shows comparison of LCA results for acidification of ammonia production methods using spectrum of hydrogen colors. Brown ammonia and grey ammonia have highest acidification potential impact. On the other hand, green ammonia and blue ammonia comes up as best options for producing ammonia with minimal environmental impact. Acidification potential is a measurement of the drop in pH of precipitation and fog,

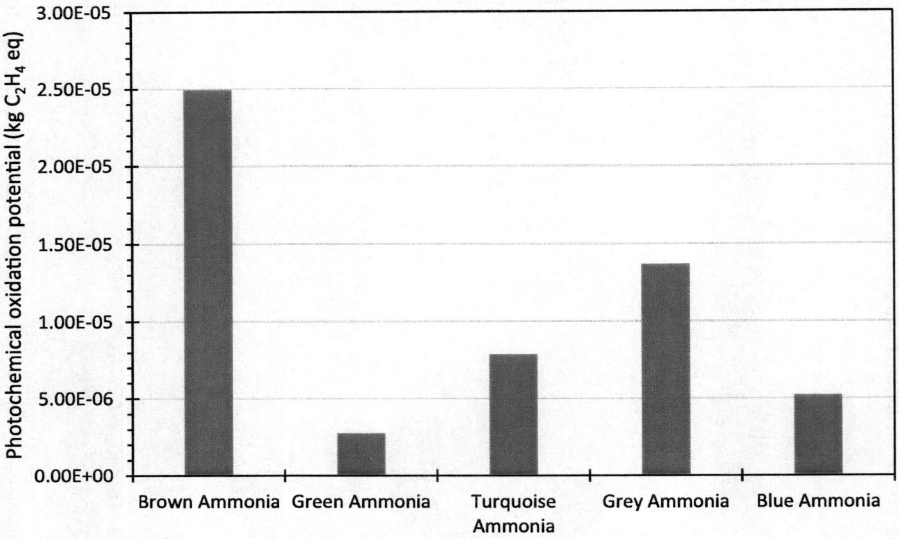

Fig. 3.63 Comparison of LCA results for photochemical oxidation potential of ammonia production methods using spectrum of hydrogen colors

which has the effect of degrading ecosystems by flushing nutrients out of soils and increasing trace metal solubility in soils, among other things. The potential for acidification is assessed in mass of sulphur-dioxide equivalents, and it has a regional influence.

Eutrophication potential is a measurement of excess nutrients in aquatic or terrestrial habitats that results in ecological harm due to over enrichment. It is quantified in mass of phosphate equivalents. Figure 3.65 demonstrates a comparison of LCA results for eutrophication of ammonia production methods using spectrum of hydrogen colors. Brown ammonia has highest eutrophication values compared to all other ammonia production techniques with value of 0.00067 kg PO_4 equivalent per 1 kg of ammonia production. Rest of the ammonia production techniques have low eutrophication values compared to brown ammonia. However, green ammonia is the most promising ammonia production technique in terms of environmental impacts.

Closing Remarks

The brown ammonia is the most environmentally hazardous ammonia form, as it is the absolute opposite of green ammonia in the hydrogen spectrum. Blue ammonia is the most environmentally friendly option among the fossil-based ammonia production options. Green hydrogen is the color of hydrogen produced with no damaging GHGs in the spectrum of hydrogen colors, and it is the most environmentally friendly ammonia production method.

- Brown ammonia emits 0.325 kg CO_2 eq per 1 kg of ammonia production while green ammonia produce only 0.0072 kg CO_2 eq per 1 kg of ammonia production.

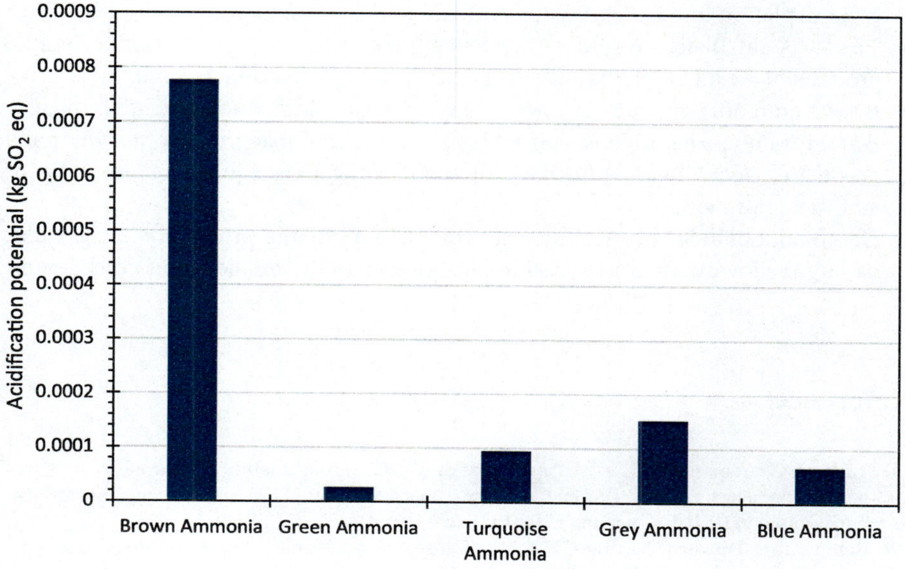

Fig. 3.64 Comparison of LCA results for acidification potential of ammonia production methods using spectrum of hydrogen colors

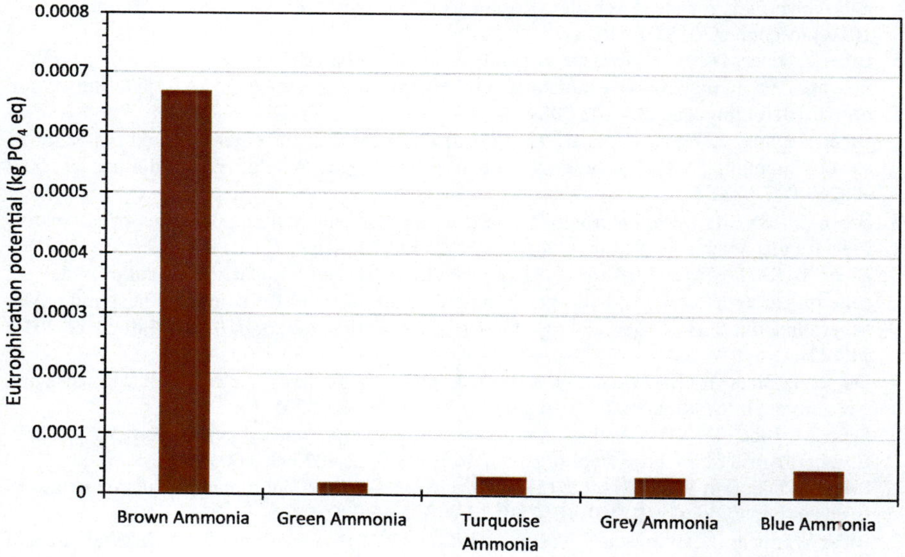

Fig. 3.65 Comparison of LCA results for eutrophication potential of ammonia production methods using spectrum of hydrogen colors

- The impact assessment results shows that grey ammonia has the highest ozone layer depletion impact among the ammonia production methods.
- The LCA calculation results indicate that brown ammonia production technique produce highest human toxicity during ammonia production process.
- Brown ammonia and grey ammonia have highest acidification potential impact.
- Brown ammonia has highest eutrophication values compares to all other ammonia production techniques with value of 0.00067 kg PO_4 equivalent per 1 kg of ammonia production.
- Green ammonia is one of the most promising ammonia production techniques, having the lowest environmental impact in each of the impact areas considered.

References

1. Turkson C, Acquaye A, Liu W, Papadopoulos T (2020) Sustainability assessment of energy production: a critical review of methods, measures and issues. J Environ Manage 264:110464. https://doi.org/10.1016/j.jenvman.2020.110464
2. Baniasadi E, Dincer I, Naterer GF (2013) Causes, impacts and solutions to global warming. https://doi.org/10.1007/978-1-4614-7588-0_23
3. Acar C (2018) A comprehensive evaluation of energy storage options for better sustainability. Int J Energy Res 42:3732–3746. https://doi.org/10.1002/er.4102
4. Bicer Y, Dincer I (2018) Life cycle environmental impact assessments and comparisons of alternative fuels for clean vehicles. Resour Conserv Recycl 132:141–157. https://doi.org/10.1016/j.resconrec.2018.01.036
5. Bicer Y, Dincer I (2017) Life cycle assessment of nuclear-based hydrogen and ammonia production options: a comparative evaluation. Int J Hydrogen Energy 42:21559–21570. https://doi.org/10.1016/j.ijhydene.2017.02.002
6. Bicer Y, Dincer I, Vezina G, Raso F (2017) Impact assessment and environmental evaluation of various ammonia production processes. Environ Manage 59:842–855. https://doi.org/10.1007/s00267-017-0831-6
7. Bicer (2018) Life cycle environmental impact assessments and comparisions of alternative fuels for clean vehicles. Resour Conserv Recycl 132:141–157
8. Bicer Y, Khalid F (2020) Life cycle environmental impact comparison of solid oxide fuel cells fueled by natural gas, hydrogen, ammonia and methanol for combined heat and power generation. Int J Hydrogen Energy 45:3670–3685. https://doi.org/10.1016/j.ijhydene.2018.11.122
9. Dincer I, Bicer Y (2018) Ammonia. In: Comprehensive energy systems, vol 2–5. Elsevier, Amsterdam. https://doi.org/10.1016/B978-0-12-809597-3.00201-7
10. Dincer I, Bicer Y (2018) Ammonia production. In: Comprehensive energy systems, vol 3–5. Elsevier, Amsterdam. https://doi.org/10.1016/B978-0-12-809597-3.00305-9
11. Singh V, Dincer I, Rosen MA (2018) Life cycle assessment of ammonia production methods. Elsevier. https://doi.org/10.1016/B978-0-12-813734-5.00053-6
12. Bicer Y, Dincer I, Zamfirescu C, Vezina G, Raso F (2016) Comparative life cycle assessment of various ammonia production methods. J Clean Prod 135:1379–1395. https://doi.org/10.1016/j.jclepro.2016.07.023
13. Al-Breiki M, Bicer Y (2021) Comparative life cycle assessment of sustainable energy carriers including production, storage, overseas transport and utilization. J Clean Prod 279:123481. https://doi.org/10.1016/j.jclepro.2020.123481

14. Chisalita DA, Petrescu L, Cormos CC (2020) Environmental evaluation of european ammonia production considering various hydrogen supply chains. Renew Sustain Energy Rev 130:109964. https://doi.org/10.1016/j.rser.2020.109964
15. Beccali M, Cellura M, Longo S, Nocke B, Finocchiaro P (2012) LCA of a solar heating and cooling system equipped with a small water-ammonia absorption chiller. Sol Energy 86:1491–1503. https://doi.org/10.1016/j.solener.2012.02.010
16. Morgan E, Manwell J, McGowan J (2014) Wind-powered ammonia fuel production for remote islands: a case study. Renew Energy 72:51–61. https://doi.org/10.1016/j.renene.2014.06.034
17. Morales Mora MA, Vergara CP, Leiva MA, Martínez Delgadillo SA, Rosa-Domínguez ER (2016) Life cycle assessment of carbon capture and utilization from ammonia process in Mexico. J Environ Manage 183:998–1008. https://doi.org/10.1016/j.jenvman.2016.09.048
18. Kim Y, Lim SR, Jung KA, Park JM (2019) Process-based life cycle CO2 assessment of an ammonia-based carbon capture and storage system. J Ind Eng Chem 76:223–232. https://doi.org/10.1016/j.jiec.2019.03.044
19. Cox B, Treyer K (2015) Environmental and economic assessment of a cracked ammonia fuelled alkaline fuel cell for off-grid power applications. J Power Sources 275:322–335. https://doi.org/10.1016/j.jpowsour.2014.11.023
20. Boggia A, Paolotti L, Antegiovanni P, Fagioli FF, Rocchi L (2019) Managing ammonia emissions using no-litter flooring system for broilers: environmental and economic analysis. Environ Sci Policy 101:331–340. https://doi.org/10.1016/j.envsci.2019.09.005
21. Abbasi R, Setzler BP, Wang J, Zhao Y, Wang T, Gottesfeld S et al (2020) Low-temperature direct ammonia fuel cells: recent developments and remaining challenges. Curr Opin Electrochem 21:335–344. https://doi.org/10.1016/j.coelec.2020.03.021
22. Makhlouf A, Serradj T, Cheniti H (2015) Life cycle impact assessment of ammonia production in Algeria: a comparison with previous studies. Environ Impact Assess Rev 50:35–41. https://doi.org/10.1016/j.eiar.2014.08.003
23. Bicer Y, Dincer I (2018) Life cycle assessment of ammonia utilization in city transportation and power generation. J Clean Prod 170:1594–1601. https://doi.org/10.1016/j.jclepro.2017.09.243
24. Sadeghi S, Ghandehariun S, Rosen MA (2020) Comparative economic and life cycle assessment of solar-based hydrogen production for oil and gas industries. Energy 208:118347. https://doi.org/10.1016/j.energy.2020.118347
25. Young B, Krynock M, Carlson D, Hawkins TR, Marriott J, Morelli B et al (2019) Comparative environmental life cycle assessment of carbon capture for petroleum refining, ammonia production, and thermoelectric power generation in the United States. Int J Greenh Gas Control 91:102821. https://doi.org/10.1016/j.ijggc.2019.102821
26. Innocenzi V, Cantarini F, Zueva S, Amato A, Morico B, Beolchini F et al (2020) Environmental and economic assessment of gasification wastewater treatment by life cycle assessment and life cycle costing approach. Resour Conserv Recycl. https://doi.org/10.1016/j.resconrec.2020.105252
27. Sittichompoo S, Nozari H, Herreros JM, Serhan N, da Silva JAM, York APE et al (2021) Exhaust energy recovery via catalytic ammonia decomposition to hydrogen for low carbon clean vehicles. Fuel 285:119111. https://doi.org/10.1016/j.fuel.2020.119111
28. Korre A, Durucan S, Nie Z (2019) Life cycle environmental impact assessment of coupled underground coal gasification and CO_2 capture and storage: alternative end uses for the UCG product gases. Int J Greenh Gas Control 91:102836. https://doi.org/10.1016/j.ijggc.2019.102836
29. Sadeek S, Chan TL, Ramdath R, Rajkumar A, Guo M, Ward K (2020) Life cycle inventory and assessment datasets on the operational sustainability of the ammonia process. Data Br 30:105593. https://doi.org/10.1016/j.dib.2020.105593
30. Bicer Y, Dincer I (2018) Clean fuel options with hydrogen for sea transportation: a life cycle approach. Int J Hydrogen Energy 43:1179–1193. https://doi.org/10.1016/j.ijhydene.2017.10.157
31. Gomez JR, Baca J, Garzon F (2020) Techno-economic analysis and life cycle assessment for electrochemical ammonia production using proton conducting membrane. Int J Hydrogen Energy 45:721–737. https://doi.org/10.1016/j.ijhydene.2019.10.174

32. Sadeek S, Lee Chan T, Ramdath R, Rajkumar A, Guo M, Ward K (2020) The influence of raw material availability and utility power consumption on the sustainability of the ammonia process. Chem Eng Res Des 158:177–192. https://doi.org/10.1016/j.cherd.2020.03.020
33. Cai T, Zhao D, Li X, Shi B, Li J (2021) Mitigating NOx emissions from an ammonia-fueled micro-power system with a perforated plate implemented. J Hazard Mater 401:123848. https://doi.org/10.1016/j.jhazmat.2020.123848
34. Curran MA (2012) Life cycle assessment handbook: a guide for environmentally sustainable products. https://doi.org/10.1002/9781118528372
35. Curran MA (2006) Life cycle assessment: principles and practice
36. Rebitzer G, Ekvall T, Frischknecht R, Hunkeler D, Norris G, Rydberg T et al (2004) Life cycle assessment part 1: framework, goal and scope definition, inventory analysis, and applications. LCA 30:701–720. https://doi.org/10.1016/j.envint.2003.11.005
37. Schenck R (2005) Why LCA? Build Des Constr 2005:4–5
38. Tillman AM (2000) Significance of decision-making for LCA methodology. Environ Impact Assess Rev 20:113–123. https://doi.org/10.1016/S0195-9255(99)00035-9
39. Finnveden G, Hauschild MZ, Ekvall T, Guinée J, Heijungs R, Hellweg S et al (2009) Recent developments in life cycle assessment. LCA 91:1–21. https://doi.org/10.1016/j.jenvman.2009.06.018
40. McAloone TC, Pigosso DCA (2017) Ecodesign implementation and LCA. Life cycle assess. Theory Pract 545–576. https://doi.org/10.1007/978-3-319-56475-3_23
41. European Environment Agency (2006) Life cycle assessment (LCA)—a guide to approaches, experiences and information sources. Environ Issues Ser 2006:1–6. https://doi.org/10.1016/B978-0-12-801775-3.00024-X
42. Jiménez-Guerrero JF, Gázquez-Abad JC, Ceballos-Santamaría G (2015) Innovation in eco-packaging in private labels. Innov Manag Policy Pract 17:81–90. https://doi.org/10.1080/14479338.2015.1011055
43. Frydendal J, Hansen LE, Bonou A (2017) Environmental labels and declarations. Life cycle assess. Theory Pract 577–604. https://doi.org/10.1007/978-3-319-56475-3_24
44. Strazza C, Del Borghi A, Costamagna P, Traverso A, Santin M (2010) Comparative LCA of methanol-fuelled SOFCs as auxiliary power systems on-board ships. Appl Energy 87:1670–1678. https://doi.org/10.1016/j.apenergy.2009.10.012
45. Firth P (2012) Environmental product declarations. Environ Des Constr 2012:33–39
46. ISO 14044:2006 (2006) Environmental management—life cycle assessment—requirements and guidelines
47. ISO (2006) Environmental management-life cycle assessment-principles and framework. ISO 14040 2006:46. https://doi.org/10.1136/bmj.332.7550.1107
48. Guinée JB, Gorrée M, Heijungs R, Huppes G, Kleijn R, Wegener Sleeswijk A et al (2001) Life cycle assessment: an operational guide to the ISO standards. LCA 692. https://doi.org/10.1300/J082v38n04_05
49. Baumann H, Tillman A-M (2004) The Hitch Hiker's guide to LCA. https://doi.org/10.1065/lca2006.02.008
50. Ecoinvent (2013) Ecoinvent database 3.0. Ecoinvent cent. https://doi.org/10.4018/978-1-59140-342-5.ch003
51. Ultimate Energy Storage: Ammonia | Energy Central (2020) https://energycentral.com/c/cp/ultimate-energy-storage-ammonia. Accessed 12 Feb 2021
52. Siddiqui O, Dincer I (2021) A comparative life cycle assessment of clean aviation fuels. Energy 234:121126. https://doi.org/10.1016/j.energy.2021.121126
53. Vallero DA (2013) Environmental impacts of energy production, distribution and transport. Elsevier Ltd. https://doi.org/10.1016/B978-0-08-099424-6.00025-9
54. Electricity and the Environment (2021) U.S. Energy Information Administration (EIA). https://www.eia.gov/energyexplained/electricity/electricity-and-the-environment.php. Accessed 28 Nov 2021
55. Impact Category Definitions—eTool (2015) https://etoolglobal.com/eblog/software/impact-category-definitions/. Accessed 28 Nov 2021

56. Green and Blue Technologies—Casale SA (n.d.) https://www.casale.ch/green-and-blue-solutions/green-and-blue-technologies. Accessed 14 Dec 2021
57. Bicer Y (2017) Investigation of novel ammonia production options using photoelectrochemical hydrogen
58. The Hydrogen Colour Spectrum | National Grid Group (2022) https://www.nationalgrid.com/stories/energy-explained/hydrogen-colour-spectrum. Accessed 14 Dec 2021
59. Gasparotto J, Da Boit MK (2021) Coal as an energy source and its impacts on human health. Energy Geosci 2:113–120. https://doi.org/10.1016/J.ENGEOS.2020.07.003

Chapter 4
Utilization of Ammonia and Ammonia Blends in Power Generators

Nomenclature

Symbols

I	Electricity current (A)
P	Pressure (kPa, bar) and power (W, kW)
PR	Performance ratio
t	Time (s, minute, h)
T	Temperature (°C)
V	Voltage (V)

Subscripts

c.out	Cylinder outlet
env	Environment
e.out	Exhaust outlet

Abbreviations

AP	Acidification
CNG	Compressed natural gas
EP	Eutrophication
LNG	Liquid natural gas
GWP	Global warming potential
HTP	Human toxicity

© The Author(s), under exclusive license to Springer Nature Switzerland AG 2022

I. Dincer et al., *Ammonia Energy Technologies*, Lecture Notes in Energy 91,
https://doi.org/10.1007/978-3-031-13532-3_4

LPG Liquid petroleum gas
ODP Ozone layer depletion
POP Photochemical oxidation

4.1 Introduction

Not only the depletion of fossil fuels but also the environmental effects of using these fuels are one of the main reasons to turn to alternatives. Emissions resulting from the consumption of fossil fuels harm the environment considerably. Although GHG emissions in Canada increased by 13.1% between 1990 and 2020, It was 672 $MtCO_2$ eq, which is 8.9% lower than one in 2019 (738 $MtCO_2$ eq) [1]. Even though it is an ultimate goal to see a tangible reduction in GHG emissions, there is a long way to reach the target, which is declared by signing the Paris Agreement. Canada has adopted 2005 as the base year and committed to dropping its GHG emissions by 40–45%, particularly below 2005 levels by 2030.

In order to reduce environmental problems caused by energy production and to develop a truly sustainable approach, the systems should be evaluated in detail from source to end use. For this reason, a 3S concept (Source-System-Service) was proposed by Acar and Dincer [2]. This approach proposes a cradle-to-grave methodology for energy systems. By adopting this methodology, three steps should be assessed extensively.

- Source: Choosing a clean, abundant, economically feasible, and available energy source
- System: Choosing the right method for production, efficiency of system, integration possibilities with other systems, multigeneration potential, possible wastes generated in the process and recovery options for those wastes
- Service: Application of the products in various fields, their efficiency, possible reuse options etc.

In order to effectively manage the processes between source-system and system-service, energy storage options should be included in the system. This concept is called the rule of 5S (3S + 2S = 5S). Here, ammonia and hydrogen can be critical as both energy storage medium and fuel. When ammonia and hydrogen are obtained from renewables, they can be used as a carbon-free fuels. Therefore, they are often called as clean fuels as they have no carbon footprint. They are also a promising energy storage medium due to their high energy storage density. Due to environmental concerns, attention to them and their usage are ever-increase. The scientists and technologists have been intensely working on systems using ammonia and hydrogen in order to increase sustainability indexes of systems. Current studies have generally focussed on either developing novel systems used ammonia or hydrogen or converting present conventional systems into systems using ammonia or hydrogen as fuel. In the near future, it is expected that the use of ammonia and hydrogen will increase substantially with energy supply/demand issues and environmental concerns. For instance, in Japan, it has been planned that 1% of electricity consumption will be met by ammonia-fueled systems [3]. Additionally, ammonia-fuelled gas turbine program has been started for power generation in Japan [4]. Additionally, The California Public Utilities Commission is considering replacing the diesel generators with ammonia-driven ones in 2021 [5].

As pointed out by many pioneering researchers, there is a strong need for renewable energy sources and clean fuels and effective energy storage mediums such as ammonia and hydrogen. However, in the near future, ammonia, and hydrogen can be used in systems that were initially designed to be powered by fossil fuels instead of designing brand new systems. Thus, the transition to novel sustainable systems can be achieved cost-effectively and environmentally friendly. This can play a critical role to transition from conventional energy systems to sustainable energy systems and increase attention to these applications.

During the past several decades fossil fueled power generators have been used in many sectors due to numerous compelling reasons. They are be used for both continuing the meeting power demands during power outages and meeting power demands where there is no grid connection. Communication towers in rural areas are one of the common applications of power generator usage as off-grid application. It is one of the only options for generating electricity in remote areas. Using ammonia as a fuel in the power generator is a promising option to continue to generate power in a sustainable way and economically feasible in those areas via exiting power generators with small modifications. Table 4.1 demonstrates the challenges and their possible solutions for transforming the fossil fuel powered generator into ammonia and ammonia blends fueled ones. It is clear that the problems in using ammonia in power generators can be fixed with proper research and development progress.

Although the use of ammonia has advantages to reduce CO_2 and NO_x emissions, there a few critical issues in use of them. Figure 4.1 illustrates a flowchart for improvements and the future of power generator technology. Up to now, traditional power generators and their hybrid applications are used in many sectors. Today, carbon-free applications of power generators have become popular with environmental concerns. Therefore, many countries plan to transform their traditional power generator into carbon-free or low-carbon systems. Here, ammonia plays a significant role in carbon-free power generator technologies since it can be integrated into power generators with a reasonable modification.

This chapter deals with the experimental attempts to convert a gasoline fueled power generator into an ammonia and ammonia blends fueled one. The attempts which are done for converting the generator are presented. The performance of the generator under different conditions are discussed. Performance of the ammonia and ammonia blends fueled generator are studied on thermodynamic efficiencies, and life cycle assessment.

Table 4.1 Challenges and possible solutions map for ammonia-driven generator transformation

Issue	Description	Possible solution
High pressures and temperatures	The auto ignition temperature for ammonia is 651 °C where this value is 256 °C for diesel fuel. As the generator system was designed for diesel fuel the material used for the engine, cylinder block should withstand these high temperature and pressure values (*Ammonia has a lower LHV than diesel fuel so combustion temperature and pressure should be lower than with 100% diesel. Higher compression ratios for NH_3 should benefit combustion and thermal efficiency*)	• Adding agents, substance, or fuel source to lower the combustion temperatures and pressures • Changing the combustion ratios by replacing the cylinders • Changing the whole system work under ammonia combustion condition
Corrosion effect	Using ammonia as a fuel source will result in water propagation, which might corrode the parts (inside the engine block, fuel delivery systems) that get in touch with fuel source (*Ammonia is corrosive to copper, zinc, and brass*)	Corroding parts used in a diesel generator should be replaced with non-corroding ones
Cooling system	The cooling system for the diesel engine is designed to operate around temperature much lower compared to the temperature of the ammonia combustion (*Ammonia's higher knock-resistance could allow higher boost pressures, which would in turn allow more NH_3 fuel input for diesel-NH_3 dual fuel systems*)	After the problem for high pressures and temperature are dealt with the cooling systems should be modified
NO_x emissions	The only adverse effect for ammonia combustion for the environment is NO_x emissions, which should be considered for the conversion	Using a slightly fuel rich mixture and a catalytic converter where one mole of NH_3 is added to each mole of NO_x in the exhaust (*EGR may be beneficial in controlling NO_x in diesel-NH_3 dual fuel engines*)

Source Adapted from [6]

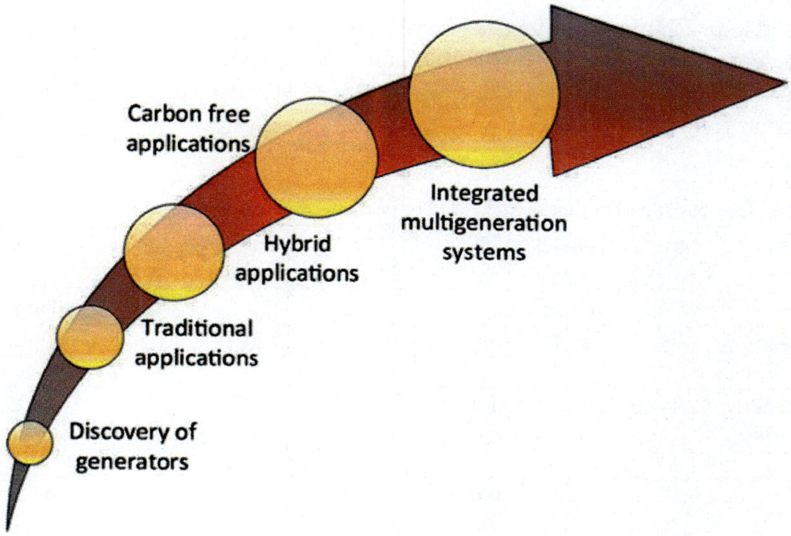

Fig. 4.1 Flowchart of the improvements and the future of power generator technology (modified from [7])

4.2 Challenges in Ammonia Combustion

As mentioned in Chap. 1, ammonia (NH_3) is one of the most produced industrial chemicals in the world since it is used in fertilizer production. It is a compound of nitrogen and hydrogen. It is a pungent-smelling and colorless gas. It can be easily dissolved in water and humidity. It is also hygroscopic. Ammonia is quite corrosive, and therefore requires costly materials in the system using ammonia.

Almost more than 75% of the ammonia produced globally is used in the agriculture sector as fertilizer. Another common use of ammonia is the use as a refrigerant in cooling machines. It is also used in cleaning chemicals, and disinfectants. Despite its common use, almost all ammonia production is performed with inefficient methods, mostly Haber Bosch. Therefore, in addition to the global production and utilization amount of ammonia, when its potential use in various sectors is considered, it has drawn the attention of researchers, scientists, engineers, and technologists. Because of its carbon-free nature, it can be used as a fuel in internal combustion engines. It is also possible to use it in ammonia fuel cells for power generation. Finally, it may be used for hydrogen storage purposes due to its high hydrogen content. Ammonia can be one of the game-changer materials for reducing CO_2 emissions in practical applications. It can play a critical role in solving various challenges related to hydrogen energy options and hydrogen economy as a unique hydrogen storage medium (with three atoms of hydrogen) and transportation and distribution, without changing the infrastructure, with the existing transportation and distribution options where industries currently deploy. In the last decade, the attempts have considerably increased to use ammonia in internal combustion engines and gas turbines. In Table 4.2, the common fuels used in internal combustion engines are given with their combustion properties. Some notable advantages of ammonia as a potential fuel can be listed as follows [8]:

Table 4.2 Comparison of common fuels used in internal combustion engines

Properties	Gasoline	Diesel	LPG	CNG	Gaseous hydrogen	Liquid hydrogen	Ammonia
Formula	C_8H_{18}	$C_{12}H_{23}$	C_3H_8	CH_4	H_2	H_2	NH_3
Lower heating value (MJ/kg)	44.5	43.5	45.7	38.1	120.1	120.1	18.8
Flammability limits, gas in air (vol. %)	1.4–7.6	0.6–5.5	1.81–8.86	5.0–15.0	4–75	4–75	16.25
Flame speed (m/s)	0.58	0.87	0.83	8.45	3.51	3.51	0.15
Autoignition temperature (°C)	300	230	470	450	571	571	651
Minimum ignition energy (MJ)	0.14	N/A	N/A	N/A	0.018	N/A	8
Flash point (°C)	− 42.7	73.8	− 87.7	− 184.4	N/A	N/A	− 33.4
Octane	90–98	N/A	112	107	> 130	> 130	110
Fuel density (kg/m^3)	698.3	838.8	1898	187.2	17.5	71.1	602.8
Energy density (MJ/m^3)	31,074	36,403	86,487	7132	2101	8539	11,333
Latent heat of vaporization (kJ/kg)	71.78	47.86	44.4	104.8	0	N/A	1369
Storage method	Liquid	Liquid	Comp. Liquid	Comp. Gas	Comp. Gas	Comp. Liquid	Comp. Liquid
Storage temperature (°C)	25	25	25	25	25	− 253	25
Storage pressure (kPa)	101.3	101.3	850	24,821	24,821	102	1030
Cost[a] (US$/l)	0.58	0.65	0.72	0.57	0.14	0.18	0.24

[a] Cost data for April 2020
Source Data from [8–10]

- It is carbon free.
- It is environmentally benign.
- Its production, storage, transportation and distribution are much easier and less complicated than many other fuels, especially hydrogen.
- It is cost effective and economically feasible for applications.
- It can be considered a potential carbon-free replacement for gasoline, diesel and kerosene.
- It can be considered for all combustion systems, ranging from small internal combustion engines to gas turbines.
- It can be a potential fuel solution for clean power generation in remote areas.

A comparison of various types of fuels with ammonia is given in Table 4.1. The energy density of gaseous ammonia in the pressurized tank is 2.5 times lower than gasoline. However, the cost of ammonia is half that of gasoline. So, ammonia has a significant advantage in terms of specific energetic cost. Additionally, when ammonia is produced in greener ways, the life cycle cost of the ammonia will be reduced significantly.

Despite ammonia seeming a promising carbon-free fuel, there are a few certain challenges in using it as a fuel. These challenges can be listed as follows:

- High ignition temperature
- Low flame velocity
- Slow chemical kinetics
- Increased NO_x emissions.

Since there are increasing research efforts to minimize the impacts and improve the combustion performance, it is necessary to provide some key remedies to overcome the issues. This chapter focusses on the experimentally solving the issues listed above.

One may look at the combustion reactions of ammonia. Let us first consider a stoichiometric combustion with 100% theoretical air for ammonia combustion, which occurs as a complete combustion, as follows [8]:

$$NH_3 + 0.75(O_2 + 3.76N_2) \rightarrow 1.5H_2O + 3.32N_2 \tag{4.1}$$

When the pure ammonia is ideally combusted, H_2O and N_2 are formed. However, this kind of combustion is practically impossible. Any fuel, including ammonia needs to be combusted with excess air (which will make the theoretical air more than 100%). Let us this time consider a combustion reaction of ammonia with excess air (where X will be percent theoretical air, for example, 50% excess air will make it 150% theoretical air which will bring 1.5 to multiply the air and 100% excess air will make it 200% theoretical air which will bring 2.0 to multiply the air) as follows [8]:

$$\begin{aligned} NH_3 + [X(0.75)](O_2 + 3.76N_2) \rightarrow\ & [1.5]H_2O \\ & + [2.82(X) + 0.5]N_2 + [(X-1)\cdot 0.75]O_2 \end{aligned} \tag{4.2}$$

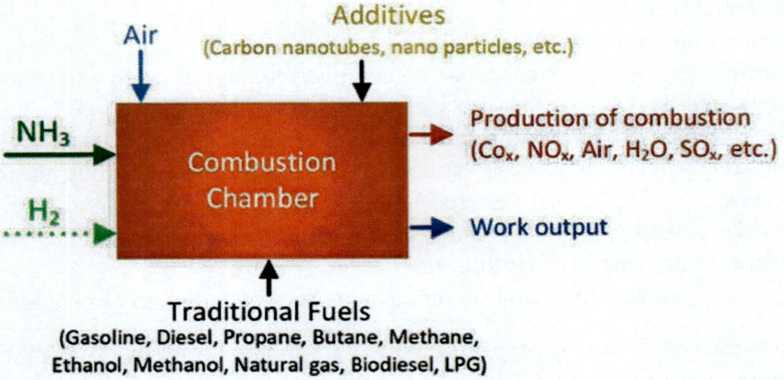

Fig. 4.2 Illustration of ammonia and ammonia blends combustion

On the other hand, under actual conditions, the reaction kinetics of ammonia combustion tend to form nitric oxide. Therefore, ammonia is partially oxidized, and the partial oxidation of ammonia occurs as [8]:

$$\begin{aligned}[Y]NH_3 + [(X)(Y)(0.75)](O_2 + 3.76N_2) \rightarrow & [1 - C \cdot Y]NH_3 + [Z \cdot C \cdot Y]NO_x \\ & + [1.5Y - [1.5 \cdot [1 - C \cdot Y]]]H_2O + [([1 - Z]/2) \cdot C \cdot Y]N_2 \\ & + [[(X \cdot Y \cdot 0.75 \cdot 2) - (Z \cdot C \cdot Y \cdot X) \\ & - [1.5Y - (1.5 \cdot (1 - C \cdot Y))]]/2]O_2 \end{aligned} \tag{4.3}$$

where X is the percent theoretical air, Y is the number of input moles of NH_3, C: the percent conversion of NH_3 and Z is the percent NO_x formation. Figure 4.2 illustrates the ammonia and ammonia blends combustion. The partial oxidation of ammonia reduces the energy output of the reaction by about 30% according to the ideal stoichiometric combustion reaction. Also, as the ignition temperature of ammonia is high, a portion of ammonia does not combust, and it exhausts in the gas form. Therefore, it is critical to provide complete combustion for ammonia. As given in Table 4.2, the autoignition temperature of ammonia is approximately 200 °C higher than gasoline and diesel. Therefore, ignition difficulty is seen in both spark ignition and compression engines. High ignition temperature causes low combustion temperature and power reduction in the engine.

4.2.1 Solutions to Challenges in Ammonia Combustion

The above listed challenges in the ammonia combustion are discussed and possible technical solutions are provided as follows:

(a) **High auto-ignition temperature**

In order to solve the difficult ignition problem of ammonia, it is a common way to mix ammonia with traditional fuels (so-called: promoters) using in internal combustion engines such as gasoline, diesel, LPG, CNG, ethanol, methanol, hydrogen, etc. Ammonia can intake to the engine either with air in its gas form from the air intake manifold or injecting to the cylinder in its liquid form separately from accompanying fuel. Since the traditional fuels will ignite at a lower temperature, it will increase the temperature of the cylinder and it helps to ignite ammonia.

In order to reach an optimum performance for an internal combustion engine, ammonia-fuel blends (promoter) can reduce the need for additional devices or modifications in engines. Thus, the transition to the hydrogen economy can be achieved cost-effectively in internal combustion engines. The common way to send the ammonia into the engine is to send with air in gas form from the air intake manifold. In order to achieve as possible as maximum power and as possible as the lowest missions, the ammonia, fuel, air mixture should adjust carefully. This requires a comprehensive blender design and automatic control system. Therefore, an optimum blender should design for each conventional fuel to be mixed with ammonia to adjust the optimum mixing ratio for ammonia, fuel, and air. Ammonia can also be intake by injecting ammonia and fuel separately into the intake manifold in the liquid phase. In this method, the flow rates of ammonia and fuel should be well adjusted. Also, effective ammonia injectors should be developed to prevent ammonia slip.

In order to enhance/improve the ammonia combustion performance and hence power output, some potential methods to be used are listed as follows:

- Promoters can be included in the fuel mixture.
- A supercharger can be integrated into the engine, especially in diesel engines.
- The compression-ratio of the engine can be increased.
- Preheating for fuel mixture or air can be integrated.
- High performance sparkplugs can be used.
- Oxygen-rich air can be used.

(b) **Low flame velocity**

The second challenge for the use of ammonia in an internal combustion engine is the lower flame speed of ammonia according to other traditional fuels. The lower flame speed restrains temperature diffusion in the cylinder during the combustion stroke and causes the power reduction. This problem is seen in both spark ignition and compression engines. Consequently, the use of ammonia in a conventional engine is not possible to use without power losses. However, when it is taken into consideration that being carbon-free fuel and its potential to reduce carbon emissions, ammonia is still a worthwhile and significant alternative fuel for internal combustion engines. All solutions described above can also help in resolving low flame velocity. Namely, ammonia-fuel blend, preheated ammonia, and higher compression ratios can improve flame velocity.

Ammonia can be used in spark ignition engines by injecting ammonia and gasoline separately into the intake manifold in liquid phase. As ammonia combusts almost 1/5 time slower than gasoline, the spark timing requires special arrangement of the crank angle/piston position. When 70% of gasoline is substituted into ammonia same amount of carbon dioxide is reduced. Ammonia should not be injected cylinder unless pressure is higher that cylinder compression pressure. In the compression ignition engines, ammonia is premixed with the air and introduced through the intake manifold and small quantity of diesel fuel or other promoter is injected inside the cylinder to have the ammonia-air mixture ignited. As mentioned before, here, it is important to adjust and auto control mixture ratios of ammonia, fuel and air is a significant issue.

(c) **Slow chemical kinetics**

The chemical reaction rate is the speed at which a chemical reaction takes place. It is also defined as the speed at which reactant converted into the products. When ammonia is used in internal combustion engines as fuel, the chemical reaction rate is slow than traditional fuels due to its high ignition temperature and low flame velocity. This slow chemical reaction rate causes ammonia to be discharged from the exhaust without burning. The common way to enhance the chemical reaction rate of ammonia combustion is to use promoter in ammonia-air mixture. Traditional fuels and hydrogen are commonly used as promoter in ammonia fueled engines, as expressed in the previous section. Ammonia-fuel blends can increase the chemical reaction rate of ammonia combustion. Also, NaCl, $BaCL_2$ and NaF are used for a catalyst to enhance chemical kinetics of ammonia combustion.

4.3 Experimental Works on Ammonia and Ammonia Blends Fueled Power Generators

This section introduces the attempts to convert a gasoline-fueled generator into an ammonia-driven one. For this purpose, first, the experimental setup is presented. Then, the tests with gasoline, ammonia and ammonia blends are performed with the standard power generator, which does not have any modifications. This is followed by the experiments which have modifications to be improved the generator performance. The following attempts are implemented to the generator to improve its performance:

1. Preheated air
2. Preheated ammonia
3. Using propane and ammonia blends
4. Using propane and ammonia blends, and preheated air
5. Using ethanol and ammonia blends
6. Using methanol and ammonia blends
7. Using different sparkplugs
8. Using hydrogen and ammonia blends
9. Using oxygen and oxygen-rich air
10. Using HHO kit.

4.3.1 *Details of the Power Generator*

A commercially available power generator, which is FIRMAN 3650 is used in the experimental work. Figure 4.3 shows the photo of the power generator. The generator has a 208 cc, single-cylinder, four-stroke gasoline engine. The specification of the engine is given in Table 4.3.

The specification of the generator is given in Table 4.4. It has a brushed alternator. It can generate 120 V and 30 A of electricity at the 60 Hz of frequency. While its starting power is 4550 W, its running power is 3650 W. It generates the single phase electricity at a 120 V of voltage and 30 A of current.

Fig. 4.3 The photo of the power generator used in the experiments

Table 4.3 Specification of the engine

Engine type	Single cylinder, four-stroke, OHV air cooled
Displacement	208 cc
Engine power	3650 W
Ignition system	Spark ignition
Starting system	Recoil
Fuel	Unleaded gasoline (min 87 octane)
Carburetor type	Float
Shaft rotation	Counter clockwise
Idle speed	3000 RPM
Compression ratio	7:1

Table 4.4 Specification of the generator

Rated AC voltage	120 V
Rated current	30 A
Rated frequency	60 Hz
Starting power	4550 W
Running power	3650 W
Alternator type	Brushed
Voltage regulator	AVR
Phase	Single phase

4.3.2 Experimental Setup

In order to convert the gasoline-fueled power generator, shown in Fig. 4.3, the attempts listed in Sect. 4.3 are implemented sequentially. The details of the experimental setup for each attempt will be introduced in detail in the section where they are discussed. Here, the details of the devices used in the experimental setup are introduced. The devices introduced in this section, are used in all experiments.

The Omega PT100 thermocouples are used for measuring the temperatures in the system. The temperatures at the cylinder outlet, exhaust outlet and ambient temperature are measured. Volumetric flowrate of ammonia is measured with Sierra Smart-Trak 2. Alicat Scientific MS-5S is used for the volumetric flowrate of hydrogen. The datalogger system consists of National Instruments NIcDAQ-9174 for base module, two NI 9211 module for thermocouples, an NI9215 module for flowmeters and pressure sensors, and NI 9203 module for analog inputs have been used. Electrical information generated by thermocouples, flowmeters and pressure sensors are processed via DAQ Express software. Figure 4.4 demonstrates the photo of measurement tools and datalogger. The EMS Model 5003 Exhaust Gas Analyzer is used to analyze the exhaust gases. The gas analyzer is placed inside the exhaust muffler. The probe is placed 15 cm inside from the outlet. A multimeter is used for measuring the electricity voltage and current generated.

The view of the experimental setup from a case is shown in Fig. 4.5. In order to start the experiments, the generator is modified, and all sensors are inserted into the experimental setup. First, the gasoline tank is raised from the base in order to apply the modifications easily and control throttle valves. All thermocouples, flowmeters and gas analyzer prob which are tested and calibrated previously are installed. The thermocouples are installed to cylinder and exhaust outlets. The exhaust outlet temperature is measured from the gas analyzer probe, as illustrated in green in Fig. 4.6. For the cylinder outlet (it is also the exhaust muffler inlet), the thermocouples are mounted after machining works. The connection pipe has been drilled and threaded. Then, the thermocouple case is threaded. The thermocouple is mounted on the pipe by tightening. The detailed view of the thermocouple at the cylinder outlet is shown on the right-hand of Fig. 4.6. Both thermocouples allow gas flow over them. The

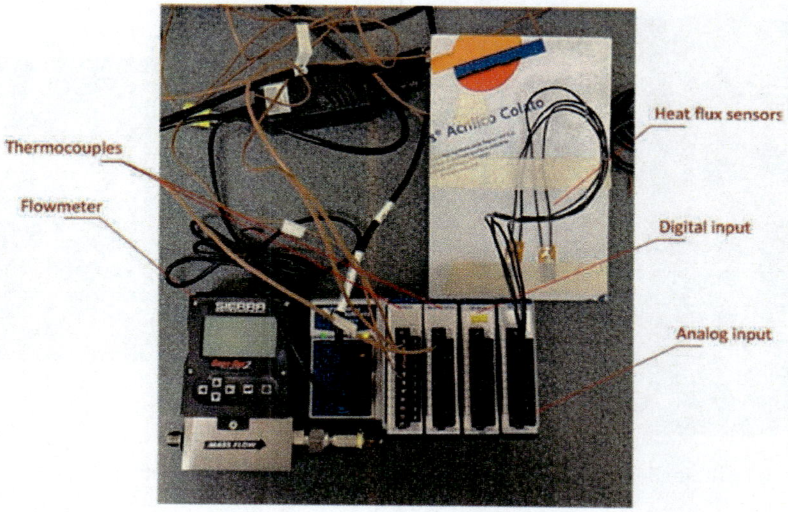

Fig. 4.4 The view of the measurement equipment and datalogger

Fig. 4.5 A photo of the experimental setup

temperature measurements at the exhaust inlet and outlet will provide information about combustion effectiveness, and whether it is to be used or not in the pre-heater as a heat source. The specific details of the experimental setup during the tests will be presented in the following sections.

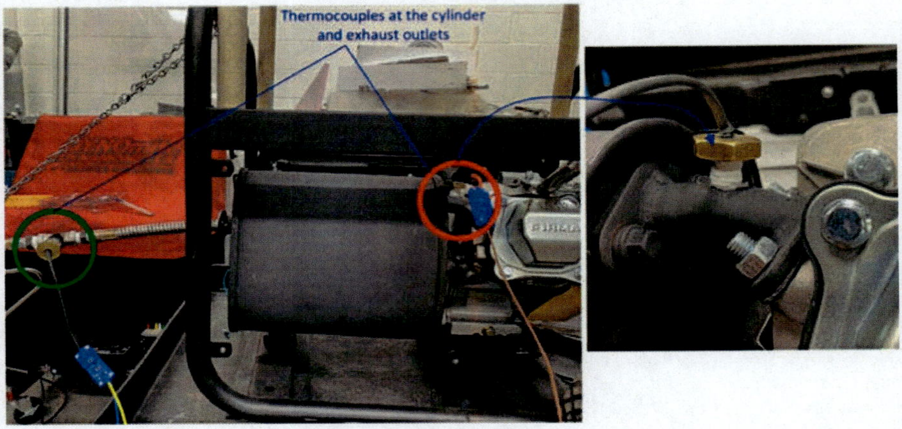

Fig. 4.6 The illustration of thermocouples at the cylinder and exhaust outlets

4.3.3 Performance Assessment

The results which have ammonia-based options are compared with the gasoline-powered generator. For this purpose, performance ratio (PR) is introduced. PR can be defined the ratio of actual power output achieved to nominal power output. It is written as follows:

$$PR = \frac{Actual\ power\ output\ achieved}{Nominal\ power\ output} \tag{4.4}$$

For each case studied, PR is calculated to determine how the techniques applied are reached in the ideal case. Nominal power output is taken as the generator's power output with gasoline, which is 3650 W. Additionally, the emission values are compared with the standard case to determine how emissions changes. In order to investigate the combustion effectiveness, the temperature at the cylinder exit is measured and compared.

4.4 Experimental Works with Gasoline and Ammonia Blends

The main objective of this set of experiments is to develop a fully functional, ammonia-fueled power generator by transforming a gasoline generator with the minimum modifications and cost-effective techniques. All techniques applied to the generator are done by considering economic and easy applicability conditions. Here, it should be noted that despite there are many attempts to use ammonia as a fuel in the internal combustion engines, many of them is not applicable to the existing engines. Because they are complicated and expensive methods.

Although ammonia has a significant advantage since it is a carbon-free fuel, the biggest obstacle to use it is its higher auto-ignition temperature, as mentioned in the previous section. In order to solve this issue, first, the engine is run with gasoline to heat it up for a while until the engine reaches a certain value. When it is heated up, ammonia and ammonia blends are used as the fuel in the engine. The preheating of engine is practically used in the LPG-fueled engines. This system is called the sequential operated LPG system. In the sequential operated LPG system, the engine starts to run with gasoline and continue to use gasoline until the engine has reaches the certain temperature. After the desired temperature is reached, LPG is sent to engine as the fuel. In the present experiments with ammonia, the engine is run with gasoline for 5 min to reaches the steady temperature condition. Then, ammonia is fed to the engine.

The first attempt to use ammonia in the generator is to send ammonia to the engine with air intake. There is no modification in the engine. Gasoline flow is cut, and ammonia is provided to the engine. The engine has worked for 3–5 min with ammonia. While the voltage and current are 120 ± V and 30 ± 2 A with gasoline, they have been measured to be 120 ± 4 °C and 10 ± 3 A with the ammonia. Namely, there was about 70% power loss in the generator. Four more experiments are done with the same configuration and different ammonia flowrate. It is observed that the increasing ammonia flowrate has improved the electric current while reducing the working period. The temperature at the cylinder exit has reduced dramatically after the first one minute.

The second attempt is to implement the sequential operating strategy to heat the engine up before starting the experiments with ammonia. For this purpose, the generator is run with gasoline for 5 min. Then, gasoline flow is cut, and ammonia is fed to the generator. The flowrate of ammonia is adjusted via a needle valve. The operating period of the generator is extended up to 10 min, thanks to the engine preheating. During the experiments, while voltage values are almost constant with the value of 120 ± 3 V, the current values have varied from 5 to 11 A with the increasing ammonia flowrate. There is no correlation between the operating period and power output. The temperature at the cylinder exit has reduced after 3–4 min after it starts to run. It happens due to inefficient ammonia combustion.

Two attempts studied above show that it is impossible to use ammonia as the fuel in the generator with the current engine configuration and engine preheating. Two potential reasons are causing this problem. The first one is the carburetor may not adjust the intake contents correctly. Another one is the current spark plug may not provide enough energy to ignite the intake configuration. It is observed that the exiting carburetor and sparkplug are not suitable for maintaining ammonia combustion. The current carburetor in the generator engine does not run with ammonia as the carburetor chamber has filled with ammonia. During the compression, combustion (power) and exhaust strokes, ammonia have escaped through the atmosphere. Finally, those attempts have shown that a special carburetor or fuel blender is required.

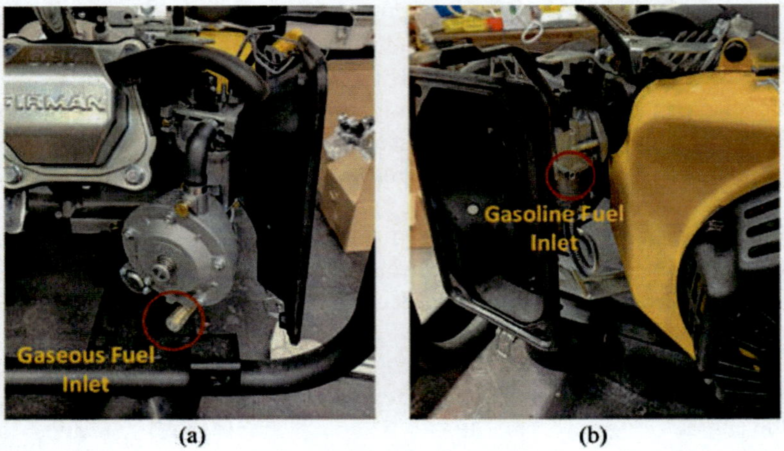

Fig. 4.7 The view of the new carburetor to be used with ammonia: **a** Gaseous fuel inlet and **b** gasoline inlet

In order to blend the ammonia, gasoline and air efficiently, a modified dual fuel carburetor is used. Figure 4.7 shows the modified dual fuel carburetor. The carburetor is designed for using liquid and gaseous fuels. Since the regular carburetor is not suitable for direct mounting to the engine, a new mounting hole is drilled on the new carburetor. Also, the pressure adjustment needle is modified to adjust the ammonia content of the fuel mixture. In addition, the bottom of the part has been machined with a turner. Thus, the pressure of the ammonia will be adjusted more precisely. In order to adjust the mixture content of the fuel, volumetric flowmeters are used in the gasoline and ammonia streams.

The gasoline and ammonia blends experiments are performed with the dual-fuel carburetor given in the previous section (Fig. 4.7). First, the generator has tested with 100% gasoline to determine the reference generator performance. The thermocouples, flowmeters, pressure sensors, gas analysis and datalogger are checked and verified. After the experiments with gasoline, gasoline and ammonia blends are tested. The experimental schedule is given in Table 4.5.

The schematic view of the experimental setup is shown in Fig. 4.8. Figure 4.9 demonstrates the photographic views of the experimental setup. The flow rates of ammonia and gasoline are controlled by needle valves and measured by volumetric flow meters. All temperatures, voltage, current and emissions values are recorded systematically.

The following experimental procedures are implemented:

1. The engine is warmed up for 5 min with gasoline.
2. Flowrates of the ammonia and gasoline is adjusted.
3. Quick check is done for system stability and safety.
4. Experiment is commenced, and data is accused.
5. Experiment is conducted for 45 min.
6. Each experimental configuration is repeated two times for checking data stability.

Table 4.5 Experimental schedule for gasoline/ammonia blends

Experiment	Fuel(s)	Experiment label
1	100% Gasoline	G100
2	90% Gasoline–10% Ammonia	G90–A10
3	80% Gasoline–20% Ammonia	G80–A20
4	70% Gasoline–30% Ammonia	G70–A30
5	60% Gasoline–40% Ammonia	G60–A40
6	50% Gasoline–50% Ammonia	G50–A50
7	40% Gasoline–60% Ammonia	G40–A60
8	30% Gasoline–70% Ammonia	G30–A70
9	20% Gasoline–80% Ammonia	G20–A80
10	10% Gasoline–90% Ammonia	G10–A90
11	100% Ammonia	A100

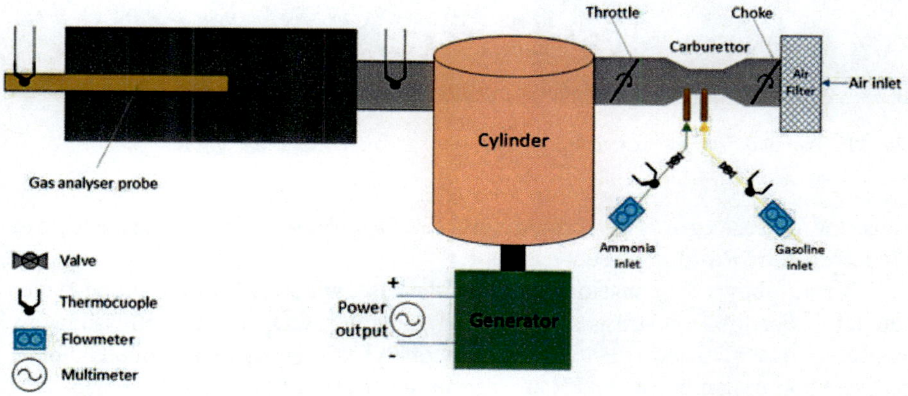

Fig. 4.8 The schematics of the experimental setup

Results of Experiments with Gasoline and Ammonia Blends

100% Gasoline

All experimental data are illustrated in three divisions: temperature changes with time, voltage and current changes with time and the average emission values. Also, the performance ratio (PR) values are calculated for each case studied. In the experimental works, first, the generator is tested with 100% gasoline to determine the reference conditions. Figure 4.10 and Table 4.6 give the experimental data for the case with 100% gasoline-fueled generator. As seen from Fig. 4.10a, the temperature at the cylinder exit has reached 900–980 °C. The exhaust outlet temperature has been measured as about 200 °C. The ambient temperature is stable at the value of 24 °C. The generator generates a 120 $\pm$ 1 V of the potential and a 30.2 A of current, corresponding to a 3650 W of power. The power output is constant during the tests.

Fig. 4.9 Two photos of the experimental setup

Since the present case is performed with 100% gasoline, it is the reference case. Therefore, the PR of the present case is 1.

Table 4.6 shows the emission values for the case, which is 100% gasoline-fueled generator. For the standard gasoline-fueled generator, CO_2 and CO emissions are measured to be 8.70 and 10.34 vol.%, respectively. NO_2 emission is around 37 ppm. HC is 242 ppm. Lambda and AFR are measured to be 0.711 and 10.461, respectively. As the present case will be used as the reference value, the tests have repeated five times in order to verify the data stability. Each trial indicates the similar results.

90% Gasoline and 10% Ammonia

Figure 4.11 and Table 4.7 depict the effect of 10% ammonia adding to the fuel mixture. The temperature at the cylinder exit has slightly reduced. The temperature change with time is similar to the case with 100% gasoline. While there are no changes in the voltage value, the electricity current has reduced to 28.40 A. The generator has produced the power of 3409 W which is corresponds a PR = 0.93. CO_2 and CO values are 8.336 and 9.774 vol.%, respectively. O_2 is measured to be 0.587 vol.% NO_2 and HC are observed to be 232.216 and 74.478 ppm, respectively. Lambda and AFR values are observed 0.711 and 10.461, which are the same with the case with 100% gasoline. When the emission values are compared for the cases studied, adding 10% ammonia in the fuel content has reduced the CO_2 emissions by 1.5%. The CO emission has decreased by 1.8%. On the other hand, it has almost doubled the NO_2 emission.

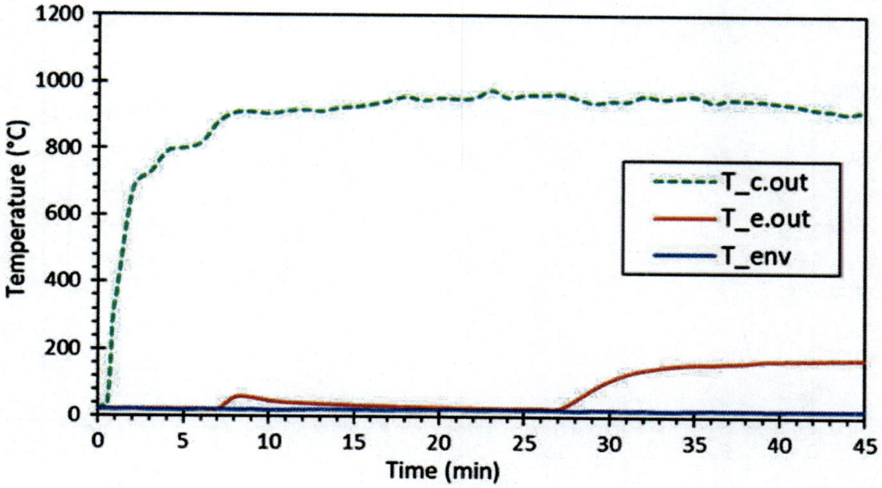

(a)

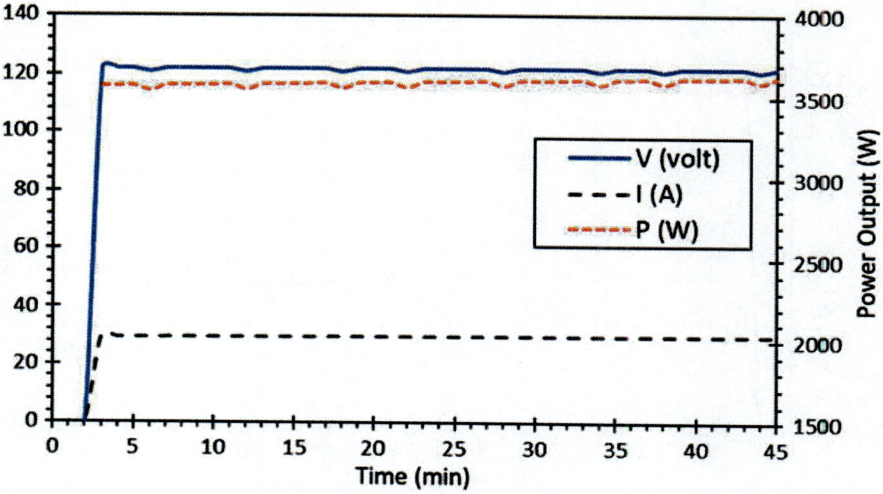

(b)

Fig. 4.10 Experimental data for the case which is 100% gasoline-fueled generator (G100). **a** Temperature and **b** voltage, current and power

Table 4.6 Emission values for the case which is 100% gasoline-fueled generator (G100)

CO_2 (vol.%)	CO (vol.%)	HC (ppm)	O_2 (vol.%)	NO_2 (ppm)	Lambda	AFR
8.593	9.974	241.891	0.391	37.239	0.711	10.461

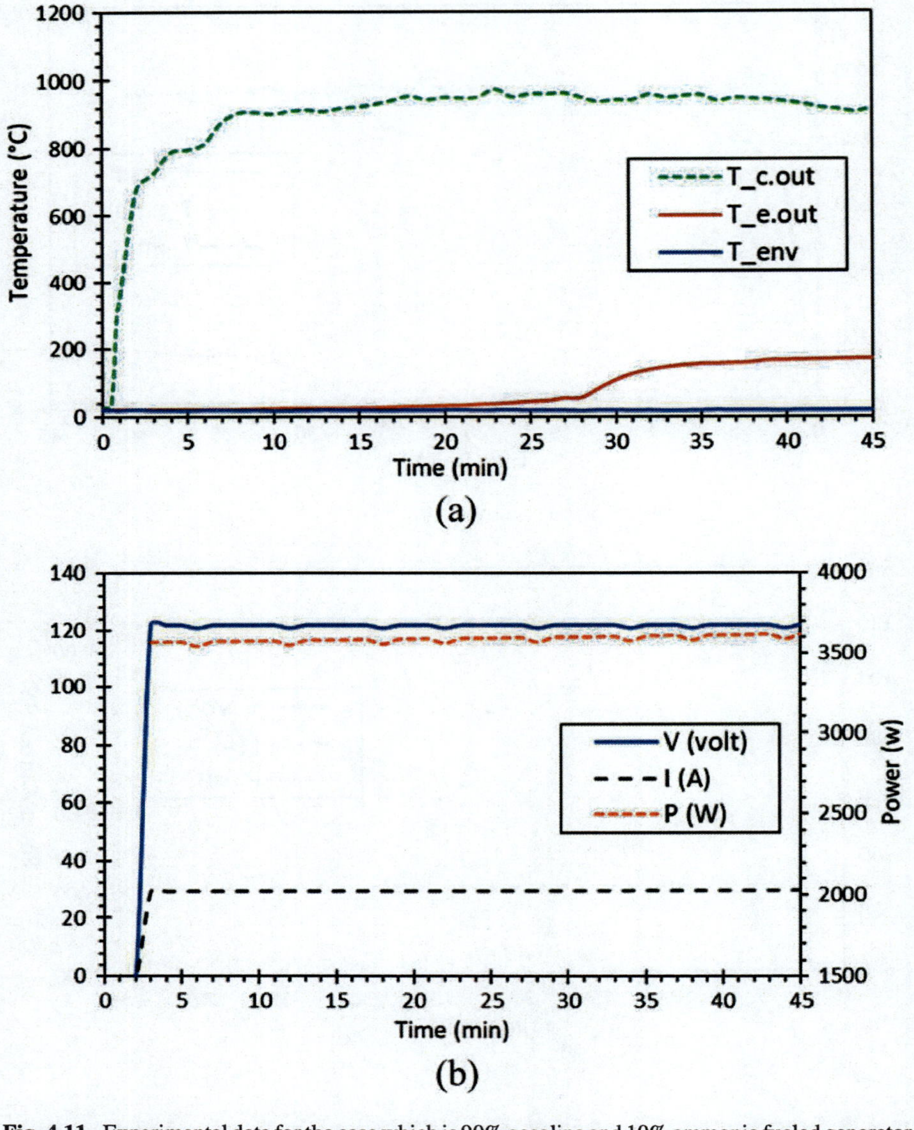

Fig. 4.11 Experimental data for the case which is 90% gasoline and 10% ammonia fueled generator (G90-A10). **a** Temperature and **b** voltage, current and power

Table 4.7 Emission values for the case which is 90% gasoline and 10% ammonia fueled generator (G90-A10)

CO_2 (vol.%)	CO (vol.%)	HC (ppm)	O_2 (vol.%)	NO_2 (ppm)	Lambda	AFR
8.336	9.774	232.216	0.587	74.478	0.711	10.461

80% Gasoline and 20% Ammonia

Figure 4.12 and Table 4.8 demonstrate the effect of 80% gasoline and 20% ammonia mixture on the generator performance and emission values, respectively. While there are no significant changes in temperature changes with time, the electricity current has reduced to 27.55 A. The generator has produced the power of 3306 W, which corresponds a PR = 0.90. The power output has reduced by 9.4% according to the case with 100% gasoline-powered one. In Table 4.8, CO_2 and CO are measured as 8.164 and 9.575 vol.%, respectively. Lambda and AFR values have not changes. HC is 220 ppm. NO_2 emission is observed to be 104.3 ppm. According to the gasoline-powered generator, while adding 20% ammonia has reduced CO_2 and CO emissions by 4.5 and 2.1%, respectively, it has increased NO_2 emission by 180%.

70% Gasoline and 30% Ammonia

The effect of 70% gasoline and 30% ammonia fuel blend on the temperatures, power output and emissions are shown in Fig. 4.13 and Table 4.9. The trends of temperatures are almost the same with the previous cases. The electricity current which is generated by the generator has reduced to 26.97 A. So, the power capacity of the generator is 3297 W. PR value is 0.88. So, the power output of the generator has reduced by 11.3%, parallel to 18 °C temperature reduction at the cylinder exit. 30% ammonia content in the fuel mixture has reduced CO_2 and CO emissions by 8 and 5.6%, respectively. On the other hand, NO_2 emission has increased up to 115 ppm.

60% Gasoline and 40% Ammonia

Figure 4.14 and Table 4.10 show the experimental data for the case with 60% gasoline and 40% ammonia mixture. The temperature changes at the cylinder exit and exhaust outlet is almost the same with the previous cases. There is about 20 °C temperature drop in the cylinder exit. 40% ammonia content in the fuel mixture has reduced the power output by 15.6%, since the electricity current has reduced to 25.6 A. PR is calculated to be 0.84. NO_2 values is increased to 243 ppm. In Table 4.10, CO_2 and CO emissions are observed to be 7.205 and 8.261 vol.%, respectively. There are no significant changes in HC, lambda, AFR and O_2 values.

50% Gasoline and 50% Ammonia

Figure 4.15 demonstrates the curves the effect of use of 50% gasoline and 50% ammonia fuel blend on the temperature and power values. The temperature at the cylinder exit is around 890–920 °C. The temperature changes at the exhaust outlet are the same with the previous cases. The electricity current is measured to be 23.88 A. The power output is 2866 W which corresponds to PR = 0.785. The CO_2 and CO values are observed to be 7.648 and 8.261 vol.%, respectively. G50-A50 fuel blend has reduced the CO_2 and CO emissions 3.5 and 8.7%, respectively. The NO_2 has increased up to 260 ppm (Table 4.11).

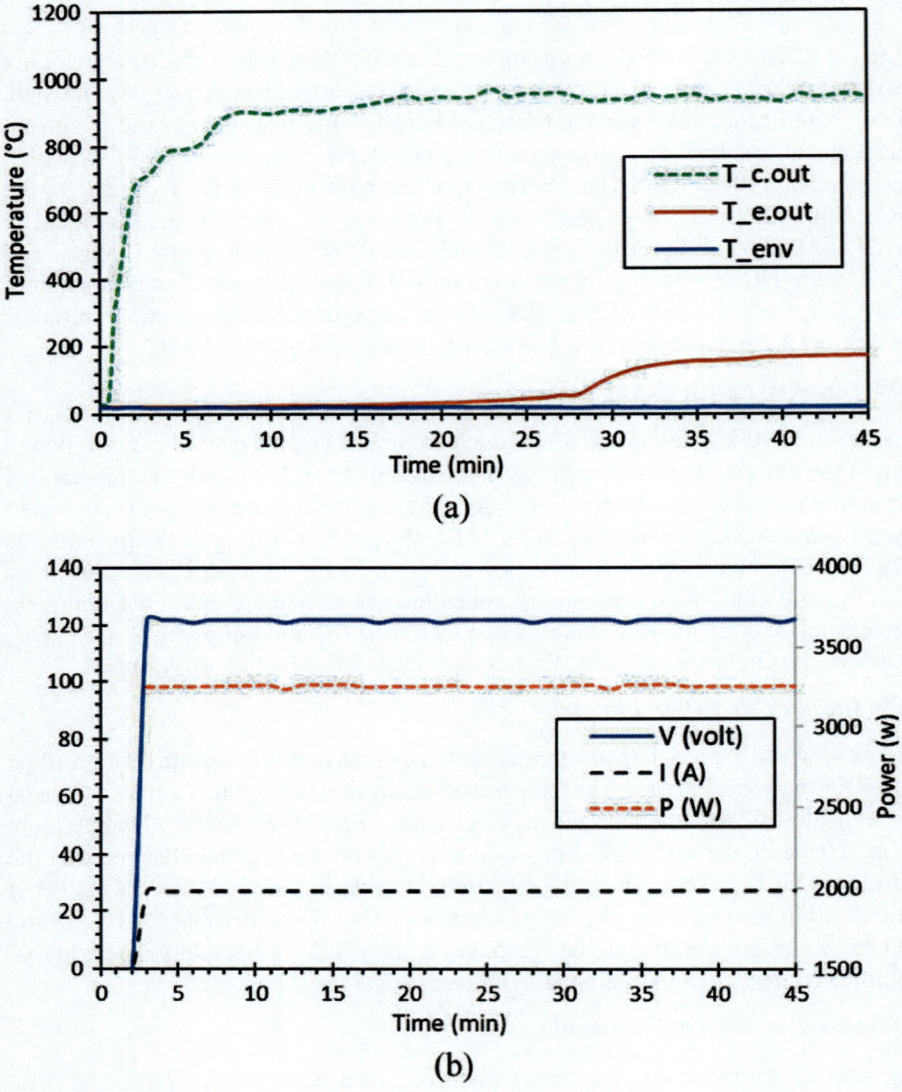

Fig. 4.12 Experimental data for the case which is 80% gasoline and 20% ammonia fueled generator (G80-A20). **a** Temperature and **b** voltage, current and power

Table 4.8 Emission values for the case which is 80% gasoline and 20% ammonia fueled generator (G80-A10)

CO_2 (vol.%)	CO (vol.%)	HC (ppm)	O_2 (vol.%)	NO_2 (ppm)	Lambda	AFR
8.164	9.575	220.121	0.732	104.270	0.711	10.461

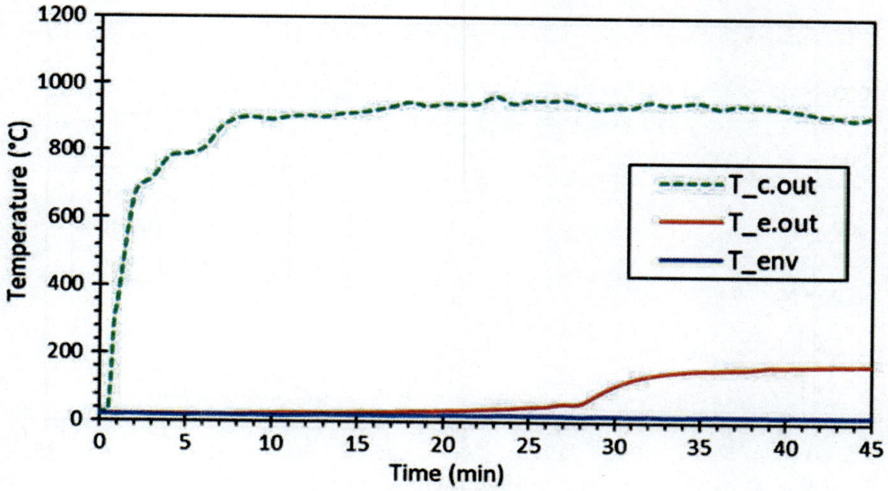

(a)

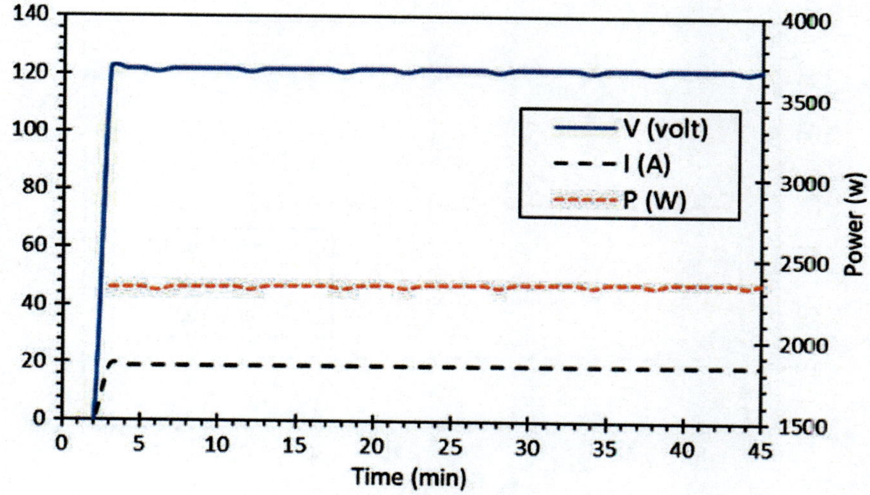

(b)

Fig. 4.13 Experimental data for the case which is 70% gasoline and 30% ammonia fueled generator (G80-A20). **a** Temperature and **b** voltage, current and power

Table 4.9 Emission values for the case which is 70% gasoline and 30% ammonia fueled generator (G90-A10)

CO_2 (vol.%)	CO (vol.%)	HC (ppm)	O_2 (vol.%)	NO_2 (ppm)	Lambda	AFR
7.906	8.905	215.283	0.747	115.441	0.711	10.461

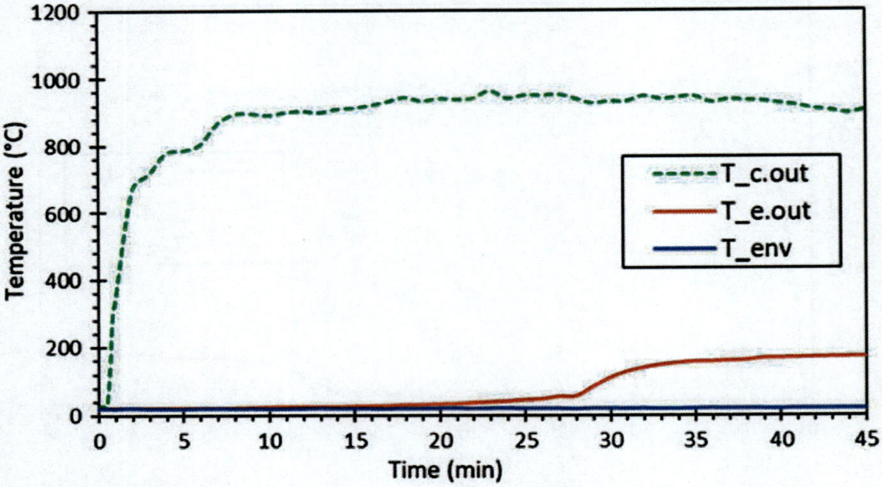

(a)

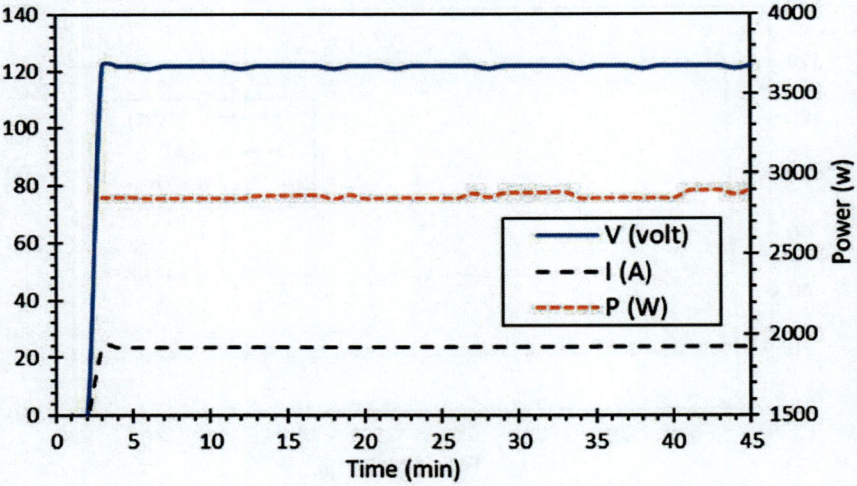

(b)

Fig. 4.14 Experimental data for the case which is G60-A40. **a** Temperature and **b** voltage, current and power

Table 4.10 Emission values for the case which is 60% gasoline and 40% ammonia fueled generator (G60-A40)

CO_2 (vol.%)	CO (vol.%)	HC (ppm)	O_2 (vol.%)	NO_2 (ppm)	Lambda	AFR
7.205	8.261	215.283	0.747	243.714	0.711	10.461

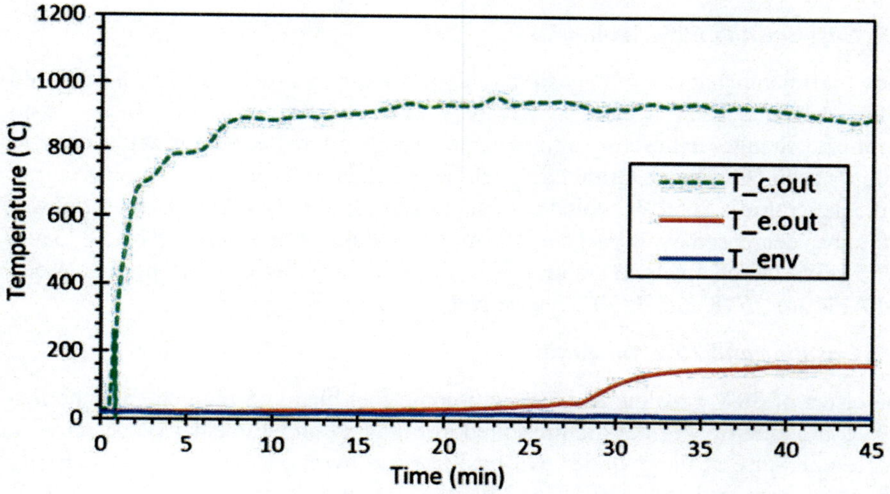

(a)

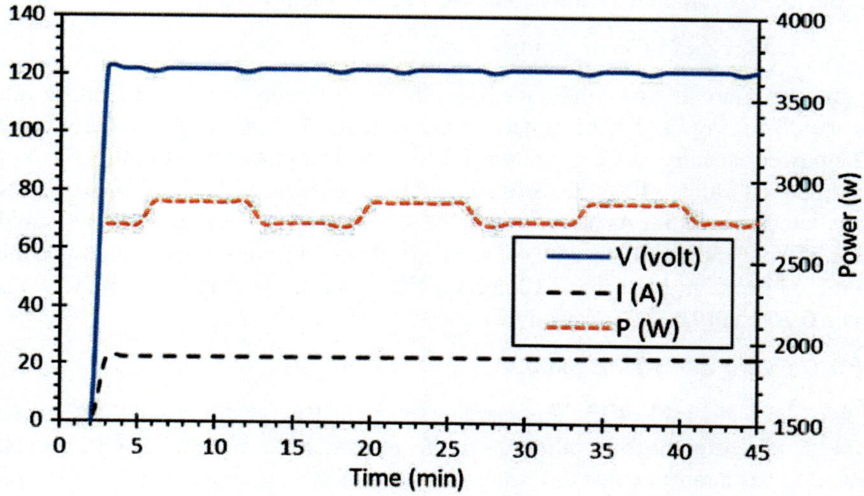

(b)

Fig. 4.15 Experimental data for the case which is G50-A50. **a** Temperature and **b** voltage, current and power

Table 4.11 Emission values for the case which is 50% gasoline and 50% ammonia fueled generator (G50-A50)

CO_2 (vol.%)	CO (vol.%)	HC (ppm)	O_2 (vol.%)	NO_2 (ppm)	Lambda	AFR
7.648	8.713	209.236	0.775	260.061	0.709	10.485

40% Gasoline and 60% Ammonia

The experimental data for 40% gasoline and 60% ammonia fuel mixture are exhibited in Fig. 4.16 and Table 4.12, respectively. Like in the previous cases, there are no significant changes in the temperature values. Temperature decreasing at the cylinder outlet is a 40 °C. The electricity current has reached to 21.9 A. The power output of the generator is 2650 W, which corresponds to PR = 0.726. The generator power output has decreased by 28%. The CO_2 and CO values are measured to be 6.875 and 7.756 vol.%, respectively. HC and NO_2 is observed to be 198 and 442 ppm. Lambda and AFR are 0.718 and 10.503, respectively.

30% Gasoline and 70% Ammonia

The effect of 30% gasoline and 70% ammonia fuel blend on the generator performance and emission values are presented in Fig. 4.17 and Table 4.13, respectively. The temperature at the cylinder exit has dropped by about 50 °C. The electricity current is measured to be 19.49 A. The power output is observed to be 2339 W. PR value is 0.641. 70% ammonia content in the fuel mixture has caused 6.273 and 6.702 vol.% CO_2 and CO emissions. NO_2 has reached 635 ppm.

20% Gasoline and 80% Ammonia

Similar to previous cases, there is no significant difference in the temperature values as seen from Fig. 4.18. The generator has produced 15.96 A of electricity current. The power capacity of the generator is 1916 W. The power output of the generator has reduced almost 48% due to 80% ammonia content in the fuel mixture. PR is calculated to be 0.52. As seen from Table 4.14, while CO_2 emission is measured to be 2.75 vol.%, the CO is 3.351 vol.% HC is about 141 ppm. 80% ammonia content in the fuel content has caused 1065 ppm NO_2 emission. Lambda and AFR is found to be 0.704 and 10.349, respectively.

10% Gasoline and 90% Ammonia

Figure 4.19 and Table 4.15 demonstrate the use of 10% gasoline and 90% ammonia fuel blend. Different from other cases, the generator has been run for 32 min. This situation has happened due to the auto-ignition of the ammonia. With the increasing ammonia rate in the fuel blend, the combustion performance has reduced. The temperature at the cylinder temperature has dramatically decreased after 25 min working period. As seen from Table 4.15, while CO_2 emission is measured to be 2.047 vol.%, the CO is 1.725 vol.% HC is about 79 ppm. 90% ammonia content in the fuel content has increased the NO_2 emission to 1065 ppm. Lambda and AFR is found to be 0.681 and 10.011, respectively.

100% Ammonia

Figure 4.20 and Table 4.16 show the use of pure gaseous ammonia as the fuel in the generator. As seen from Fig. 4.20, the generator has worked for 15 min. At the beginning of the test period, since the engine is warm due to the engine pre-heating,

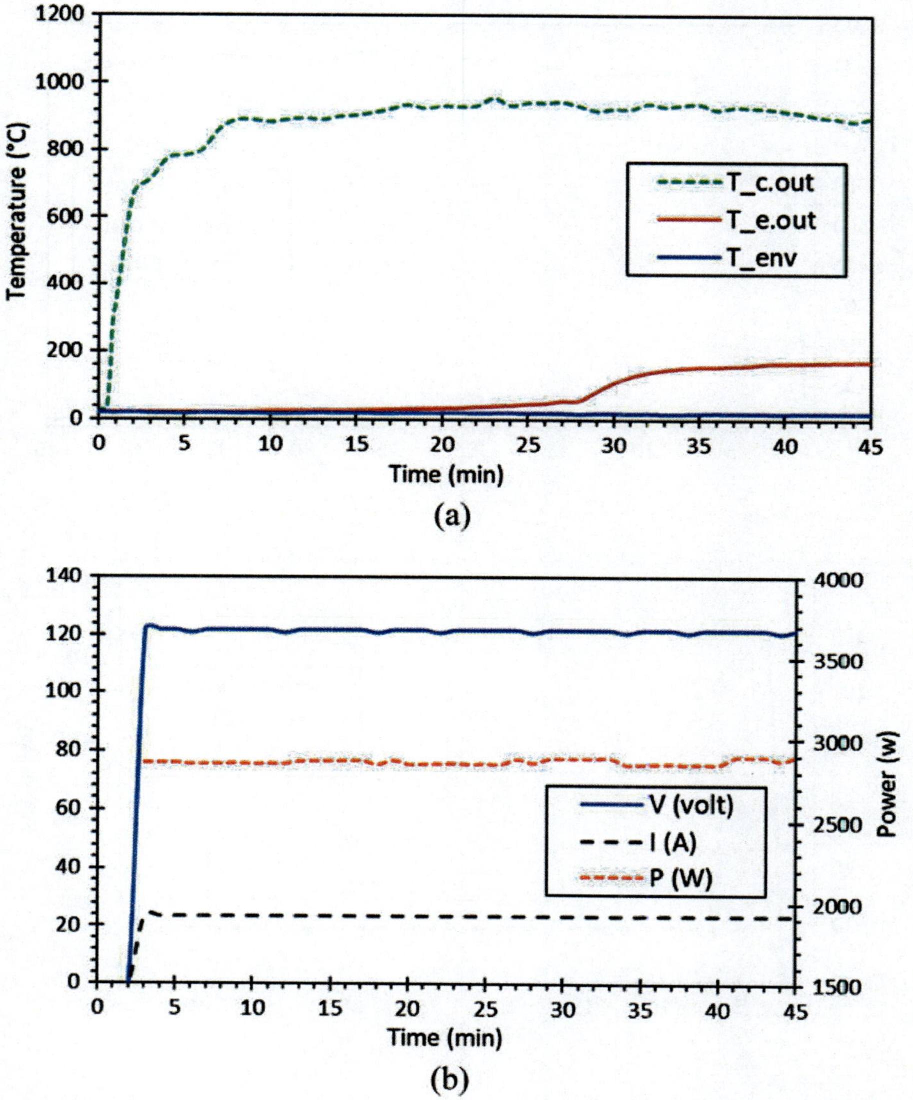

Fig. 4.16 Experimental data for the case which is G40-A60. **a** Temperature and **b** voltage, current and power

Table 4.12 Emission values for the case which is 40% gasoline and 60% ammonia fueled generator (G40-A60)

CO_2 (vol.%)	CO (vol.%)	HC (ppm)	O_2 (vol.%)	NO_2 (ppm)	Lambda	AFR
6.875	7.756	198.109	0.861	442.435	0.718	10.503

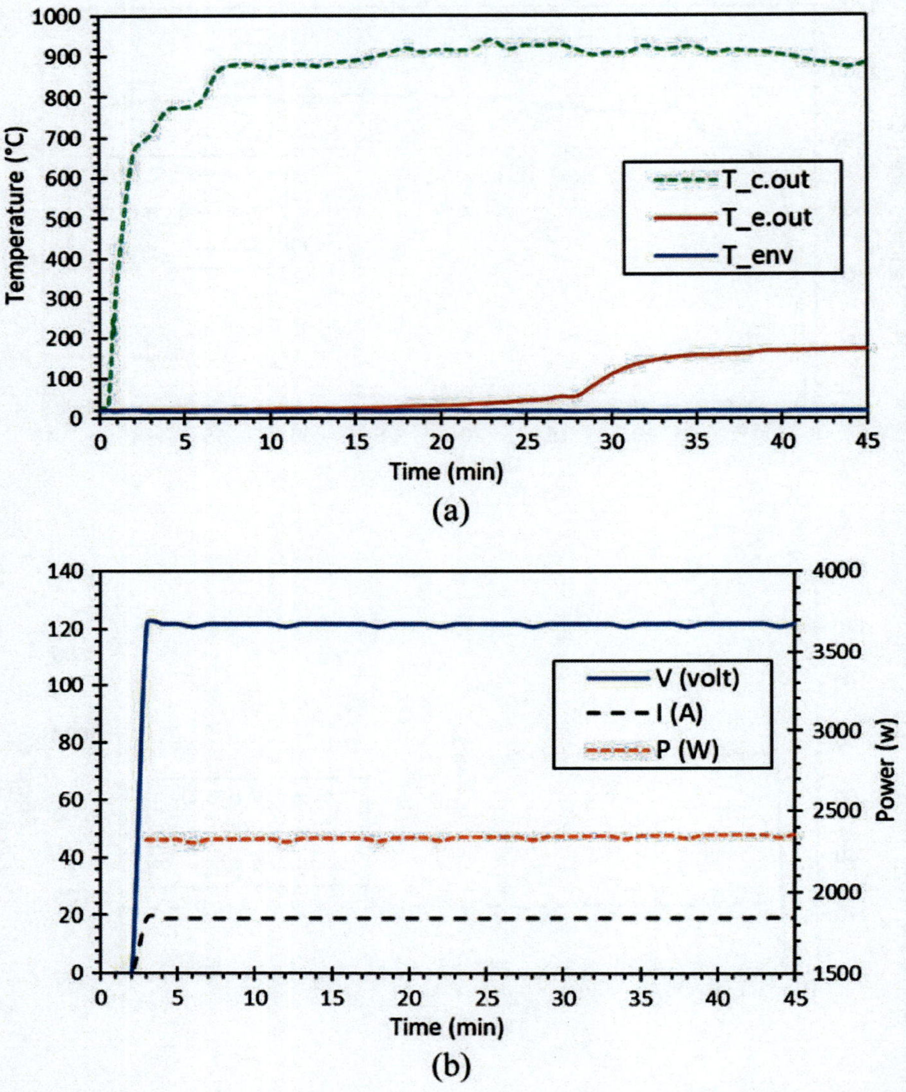

Fig. 4.17 Experimental data for the case which is G30-A70. **a** Temperature, **b** voltage, current and power, and **c** emission values

Table 4.13 Emission values for the case which is 30% gasoline and 70% ammonia fueled generator (G30-A70)

CO_2 (vol.%)	CO (vol.%)	HC (ppm)	O_2 (vol.%)	NO_2 (ppm)	Lambda	AFR
6.273	6.702	174.162	1.213	335.152	0.711	10.461

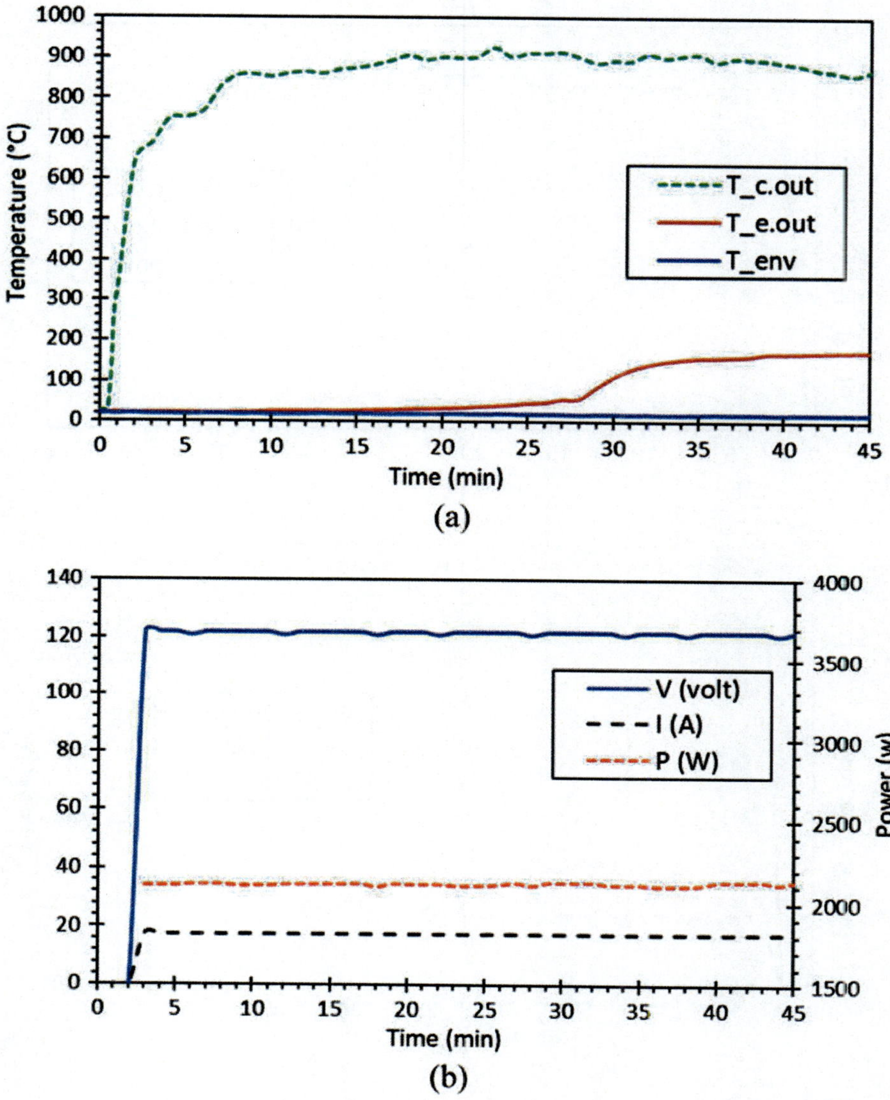

Fig. 4.18 Experimental data for the case which is G20-A80. **a** Temperature and **b** voltage, current and power

Table 4.14 Emission values for the case which is 20% gasoline and 80% ammonia fueled generator (G20-A80)

CO_2 (vol.%)	CO (vol.%)	HC (ppm)	O_2 (vol.%)	NO_2 (ppm)	Lambda	AFR
2.750	3.351	159.648	1.526	595.826	0.711	10.461

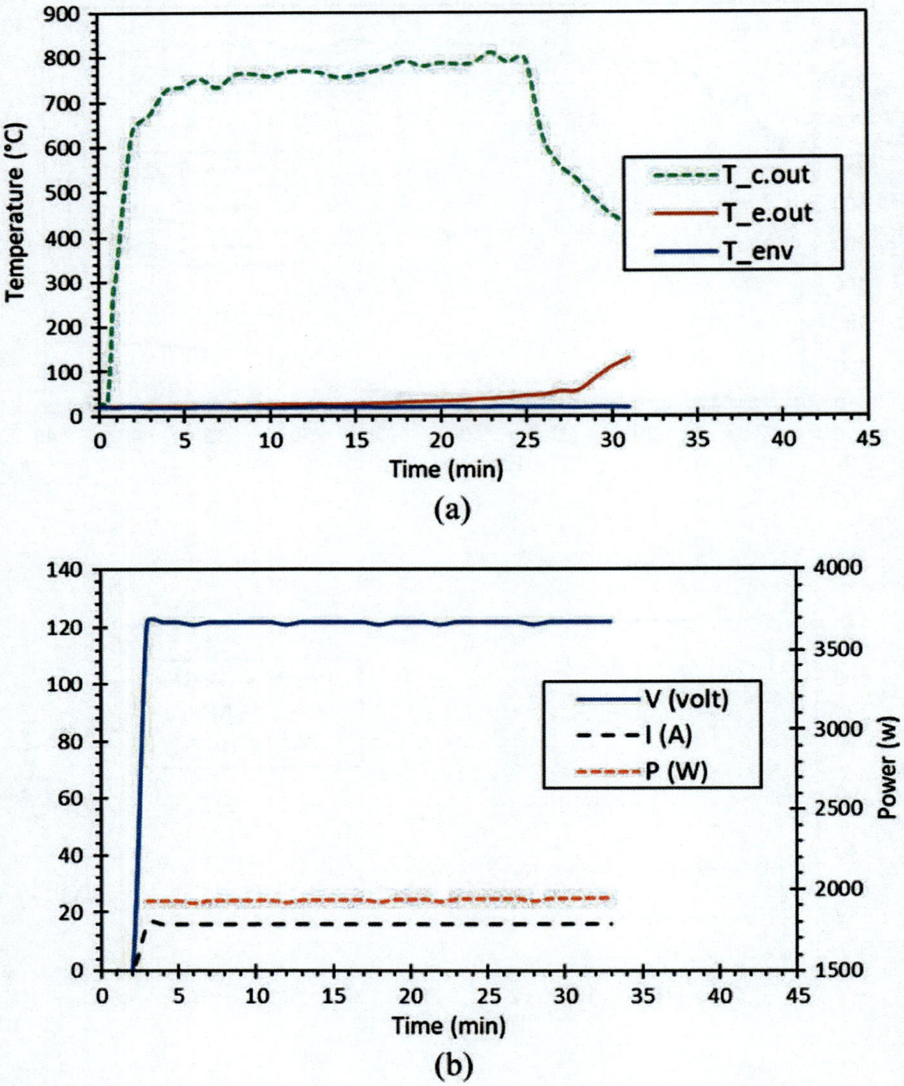

Fig. 4.19 Experimental data for the case which is G10-A90. **a** Temperature, **b** voltage, current and power, and **c** emission values

Table 4.15 Emission values for the case which is 10% gasoline and 90% ammonia fueled generator (G10-A90)

CO_2 (vol.%)	CO (vol.%)	HC (ppm)	O_2 (vol.%)	NO_2 (ppm)	Lambda	AFR
2.047	1.725	141.618	1.976	1065.750	0.704	10.349

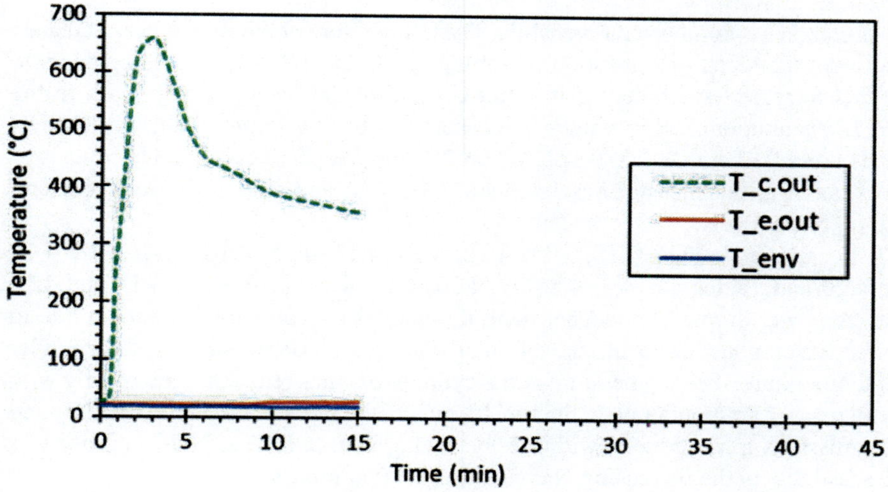

(a)

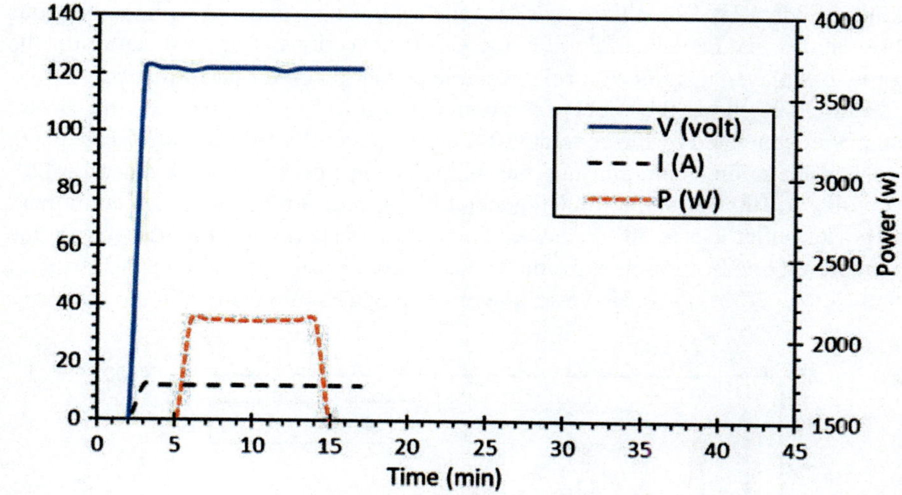

(b)

Fig. 4.20 Experimental data for the case which is A100. **a** Temperature, **b** voltage, current and power, and **c** emission values

Table 4.16 Emission values for the case which is 100% ammonia fueled generator (G10-A90)

CO_2 (vol.%)	CO (vol.%)	HC (ppm)	O_2 (vol.%)	NO_2 (ppm)	Lambda	AFR
0.908	0.830	79.029	3.413	1562.000	0.681	10.011

the ammonia has combusted for a while. The temperature of the engine has decreased due to the ineffective combustion of ammonia. 90 and 100% ammonia contents in the fuel mixture have failed as it is not completed the 45 min text period. In the case 100% ammonia, NO_2 values is measured to be 1562 ppm. CO_2, CO and HC values are 2.047 vol.%, 0.830 vol.%, and 79 ppm, respectively. The O_2 content of the exhaust gases is measured as 3.413 vol.% Lambda and AFR values are 0.681 and 10.011, respectively.

It is clear from Figs. 4.10, 4.11, 4.12, 4.13, 4.14, 4.15, 4.16, 4.17, 4.18, 4.19 and 4.20 and Tables 4.6, 4.7, 4.8, 4.9, 4.10, 4.11, 4.12, 4.13, 4.14, 4.15 and 4.16 that the power output has reduced with the increasing volume of ammonia due to ineffective combustion inside the cylinder. This can be observed from the cylinder exit temperature. The temperature at the cylinder exit has reduced dramatically with the increasing ammonia rate in the fuel blend. On the other hand, while the CO_2 and CO emissions have reduced with the increasing ammonia rate, NO_2 emission has increased due to the increasing N content of the fuel mixture.

For the cases of 90% and pure ammonia, the engine has stopped in 32 and 14 min. The ammonia has not combusted effectively as observed from temperature values at the cylinder exit. When the current results have been compared with the previous observations that have been done in the previous sections. The pre-heating for the engine has played a significant role in stable power generator performance.

Figure 4.21 shows the effect of ammonia volume in the fuel blend on the current and power generated by the generator. When stable combustion conditions are taken into consideration, ammonia use has reduced the power output by nearly 42% according to 100% gasoline-fueled generator. For pure ammonia (100% ammonia), the power outlet loss is 50%. As seen from these observations, the use of gasoline in ammonia combustion as a promoter has been seemed effective in the terms of combustion performance. However, the increasing gasoline volume in the blend has

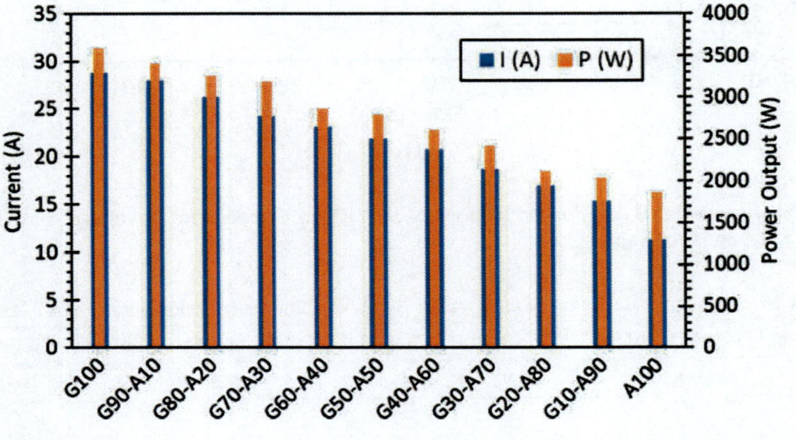

Fig. 4.21 The current and power output values versus ammonia rate in the fuel blend

increased the CO_2 and CO emission, too. It has been concluded in these experiments that ammonia is able to be used in the gasoline engine when it is provided in certain conditions. However, in order to enhance the engine performance, some actions should be taken place.

4.5 Experimental Works with Pure Ammonia and Air Preheating

One of the options to be used for enhancing the ammonia combustion is to perform preheating for fuel mixture before entering the engine cylinder. For this propose, a hot air blower is included in the system to increase the temperature air intake. Thus, it is aimed to increase the temperature of the fuel and air mixture. Figure 4.22 demonstrates the photography of the experimental setup. The schematic of the experimental setup is also shown in Fig. 4.23. The hot air blower has a temperature controller on it. The blower is able to produce hot air an adjusted temperature. Air temperature has been studied parametrically. The effects of the air temperatures, which are 80 °C, 100 °C, 120 °C, 140 °C, 160 °C and 180 °C on the generator performance and emission values, have been investigated in the present section.

The following procedures are followed to perform the recent experiments accordingly:

1. The engine is warmed up for 5 min with gasoline.
2. The flowrate of the ammonia is set.

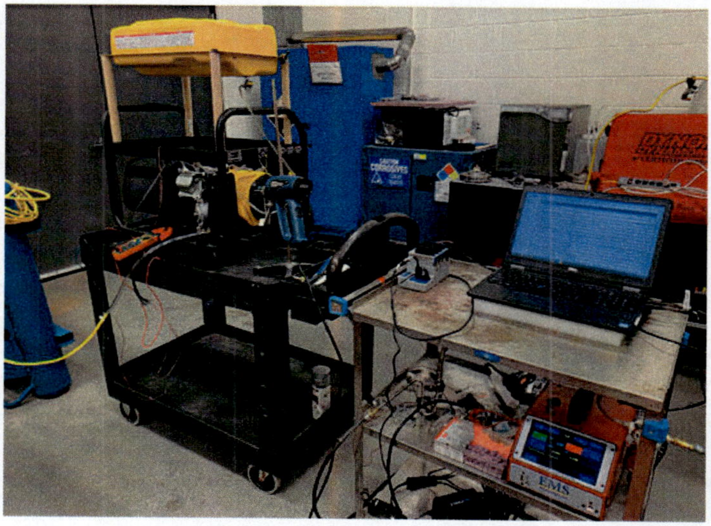

Fig. 4.22 A photograph of the final experimental setup

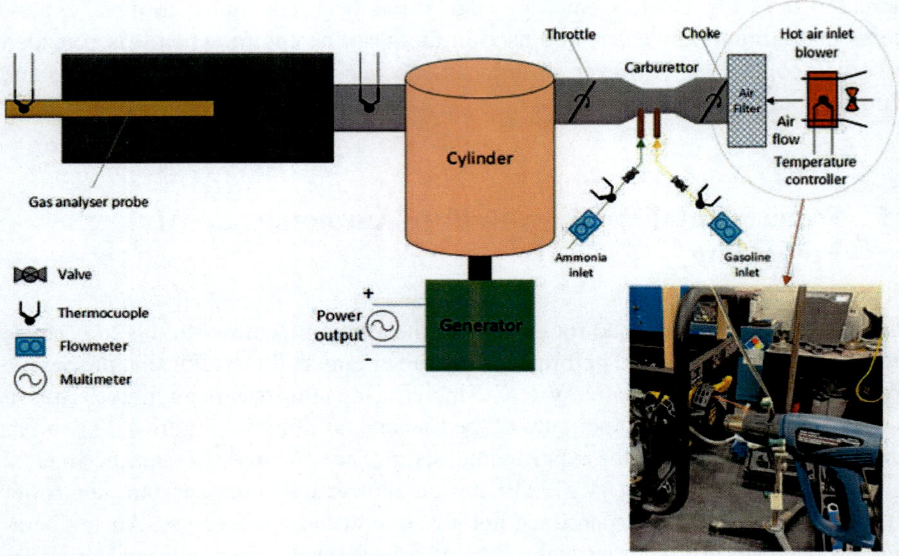

Fig. 4.23 A schematic view of the experimental setup with the additional hot air blower option

3. The temperature of blower is set.
4. A quick check has been done for system stability and safety.
5. The experimental run has commenced, and the data is collected for analysis and assessment.
6. Each experimental run is conducted for 45 min.
7. Each experiment has been repeated two times for checking data reliability.

Results of Experiments with Pure Ammonia and Air Preheating

It should be noted that the ambient temperature in the present section is lower than in the experiments in the previous section. Therefore, first, it has been started the experiment with a gasoline-fueled engine to observe the effect of a lower ambient temperature on the engine performance. Figure 4.24 demonstrates the experimental data for the case fueled by 100% gasoline. As seen in Fig. 4.24, the temperature at the cylinder outlet is around 780 °C, which is almost 150 °C lower than the case in the previous section (Fig. 4.10 and Table 4.6). The temperature drops at the cylinder outlet have occurred due to the ambient temperature. The environmental temperature is almost 10–12 °C lower than the one in the previous section. The gasoline consumption rate is almost 9% higher than the cases done in the previous experimental works with 100% gasoline. As seen from the results graphed in these graphs, the gasoline-fueled generator and the cold weather conditions cause higher fuel consumption.

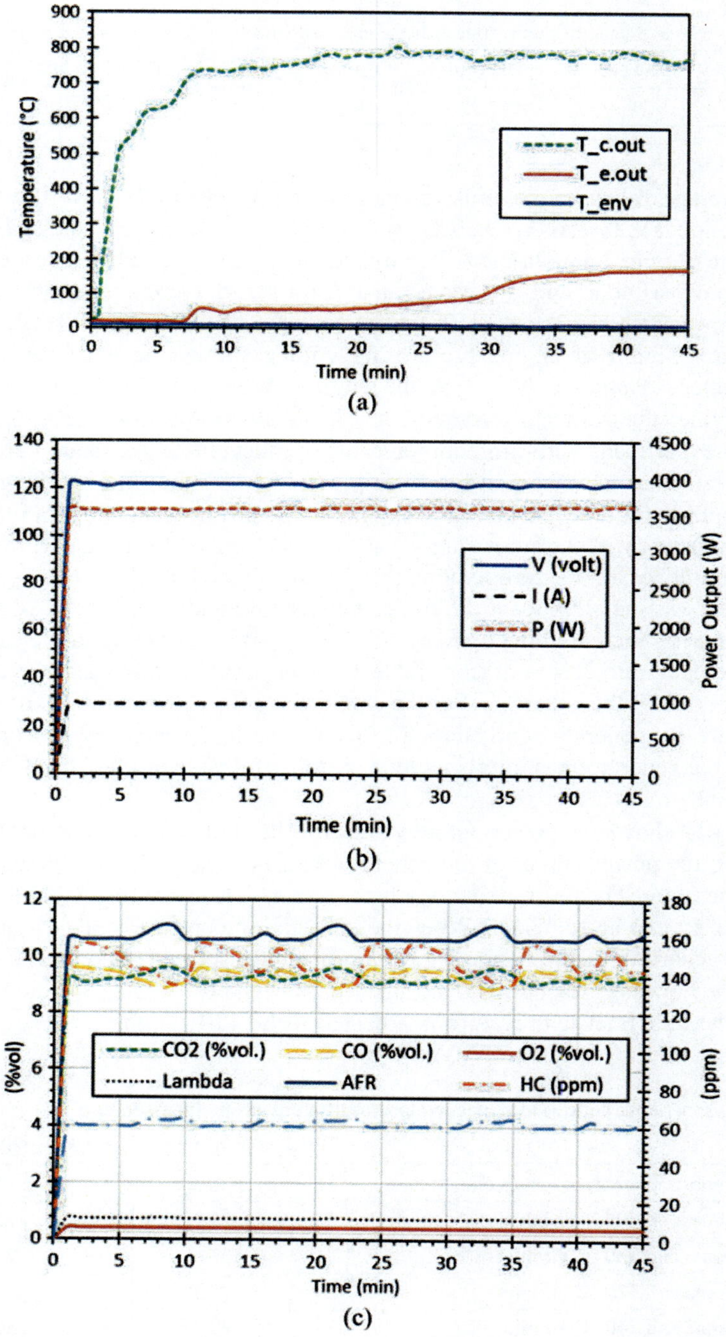

Fig. 4.24 Experimental data for 100% gasoline fueled generator. **a** Temperature, **b** voltage, current and power, and **c** emission values

Table 4.17 The average emission values for the case with 100% gasoline powered generator

CO_2 (vol.%)	CO (vol.%)	HC (ppm)	O_2 (vol.%)	NO_2 (ppm)	Lambda	AFR
9.28	9.29	147.27	0.40	61.07	0.74	10.69

The average values of the emission values are given in Table 4.17. The average CO_2 emission is 9.28%, CO is 9.29%, O_2 is 0.40%, respectively. HC and NO_2 values are 147 and 61 ppm. Lambda and AFR are observed to be 0.74 and 10.69, respectively.

A set of experiments with pre-heated air are performed. The temperature controller of the blower is set to be 80, 100 and 120 °C; 140, 160 and 180 °C, respectively. After 5 min of engine pre-heating with gasoline, the gasoline flow has been stopped. Ammonia is fed to the engine. When pure ammonia was sent to the engine, the generator worked for 15 min in the previous section. In this term, the experiment with ammonia and environmental air has been performed. The generator has now worked for 13 min. Table 4.18 shows the working period of the generator for different air inlet temperatures. The operating time for the generator has increased with the increasing air inlet temperature. When the air is provided at the temperature of 80 °C, the operating time has increased by approximately 10 min. While there was no significant difference between 80 and 100, 120 °C of air inlet temperature has increased the operating time by 10 min according to the case with ambient temperature. It is clear from Table 4.18 that air pre-heating has increased the working period of the generator. For the cases with 140, 120 and 180 °C of air inlet temperature, the generator worked for 22, 24 and 26 min, respectively. The operating period of the generator has improved by 4–5 min, against increasing each 20 °C air temperature.

Table 4.19 shows the power output values for the studied cases. Parallel to operating time, the power output of the generator has increased with the increasing air inlet temperature. The power reduction for the case with environmental temperature has ranged from 55.8 to 57.1%. When the inlet air temperature is 80 °C, the power reduction according to the standard generator which gasoline-fueled, is around 54.4 and 56.4%. For the case of 100 °C air inlet, it has been observed as 54.2 and 53.4%. Lastly, when air inlet temperature is adjusted to be 120 °C, the power reduction

Table 4.18 Operating periods of the generator for different air inlet temperature

Case	Test 1	Test 2	Test 3	Test 4
Pure ammonia and environmental temperature	14	13	13	11
Pure ammonia and 80 °C air inlet	23	21	21	22
Pure ammonia and 100 °C air inlet	24	25	24	24
Pure ammonia and 120 °C air inlet	26	25	26	27
Pure ammonia and 140 °C air inlet	30	29	29	30
Pure ammonia and 160 °C air inlet	34	36	35	36
Pure ammonia and 180 °C air inlet	39	38	41	40

Table 4.19 Power outputs of the generator for different air inlet temperatures

Case	Test 1	Test 2	Test 3	Test 4
Pure ammonia and environmental temperature	1565	1613	1595	1597
Pure ammonia and 80 °C air inlet	1591	1589	1656	1665
Pure ammonia and 100 °C air inlet	1673	1688	1702	1699
Pure ammonia and 120 °C air inlet	1856	1878	1891	1906
Pure ammonia and 140 °C air inlet	1913	1948	1936	1925
Pure ammonia and 160 °C air inlet	1987	1985	1994	1999
Pure ammonia and 180 °C air inlet	2012	2004	2009	2001

has dropped to 47.8 and 49.2%. As seen from Table 4.19, the power losses in the generator have been reduced by providing pre-heated air.

Figure 4.25 demonstrates the use of pure ammonia and 80 °C air inlet. The temperature at the cylinder exit has increased up to 470 °C and dropped suddenly. NO_2 emission is around 1780 ppm. Figure 4.26 shows the effect of 100 °C air inlet temperature on the generator. The temperature at the cylinder outlet is about 450 °C, which 20 °C lower than the case with 80 °C air inlet temperature. The NO_2 emissions result in approximately 1780 ppm. Figure 4.27 demonstrates the experimental data for the case with 120 °C air inlet temperature. The emissions values are almost same with the cases with 80 and 100 °C air inlet temperature. The enhancement in the power outlet due to pre-heated air is seen in current values.

The experimental data for the cases with pre-heated air are shown in the figures inclusively in Figs. 4.25, 4.26, 4.27, 4.28, 4.29 and 4.30 respectively. As seen from these figures, the there is no significant changes in the emission values against to changing air inlet temperature. The experimental data for the cases with pre-heated air at the temperature of 140, 160 and 180 °C are shown in the figures inclusively, Figs. 4.28, 4.29 and 4.30, respectively. The average emission values for the studied cases are given in Tables 4.20, 4.21 and 4.22, respectively. As seen from Fig. 4.28, the 140 °C air preheating has increased the combustion temperature almost 25 °C according to the previous case which has 120 °C air preheating. We have reached this observation from the exhaust outlet temperature at the cylinder exit. The average emission values are given in Table 4.20.

Figure 4.29 and Table 4.21 shows the experimental data for the case with 160 °C air inlet temperature. The temperature at the cylinder exit is almost 20 °C higher than the previous case with 140 °C air inlet temperature. When the emission values are compared for the cases with 140 °C and 160 °C, there is no significant change in CO_2, CO, HC, O_2, and lambda values, NO_2 values has increased from 1290 to 1345 ppm with the increase air inlet temperature. Also, AFR values has increased with the increasing air inlet temperature.

Figure 4.30 and Table 4.22 demonstrates the experimental data for the case which has 160 °C air inlet temperature. The cylinder exit temperature has increased 15 °C according to the case with 140 °C air inlet temperature. Similar to other cases, while

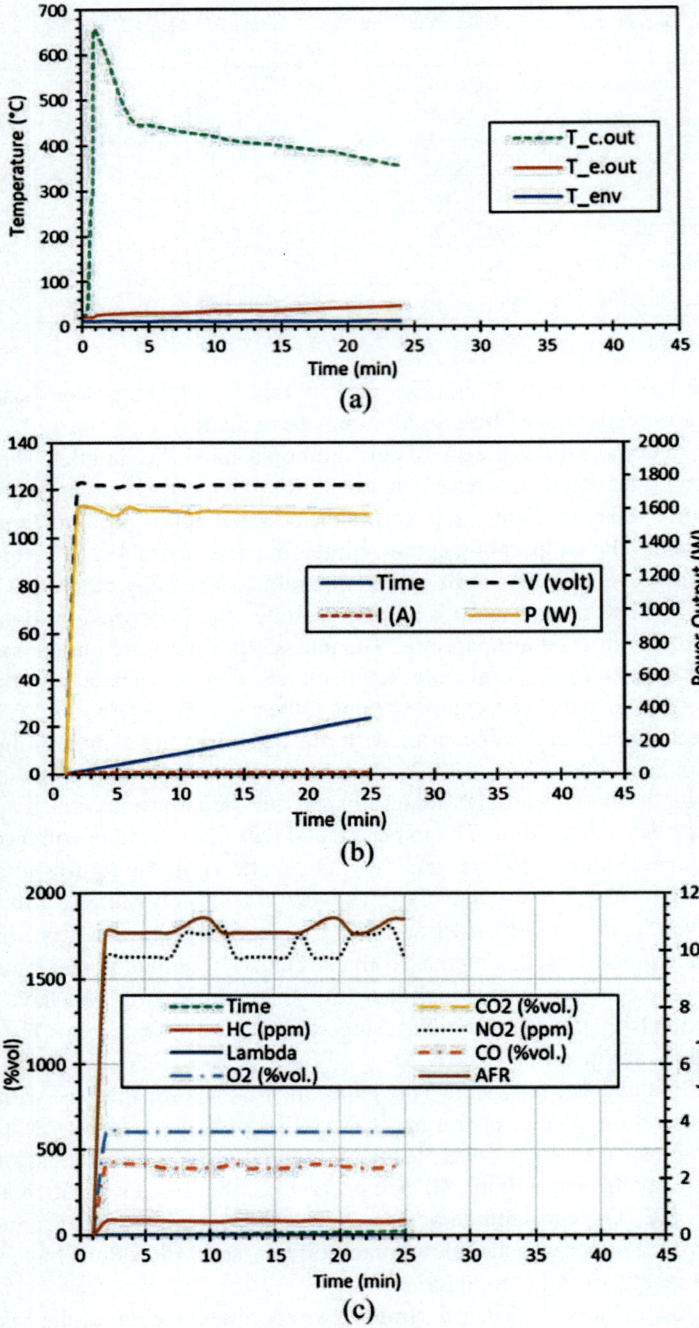

Fig. 4.25 Experimental data for pure ammonia and 80 °C air inlet temperature generator. **a** Temperature, **b** voltage, current and power, and **c** emission values

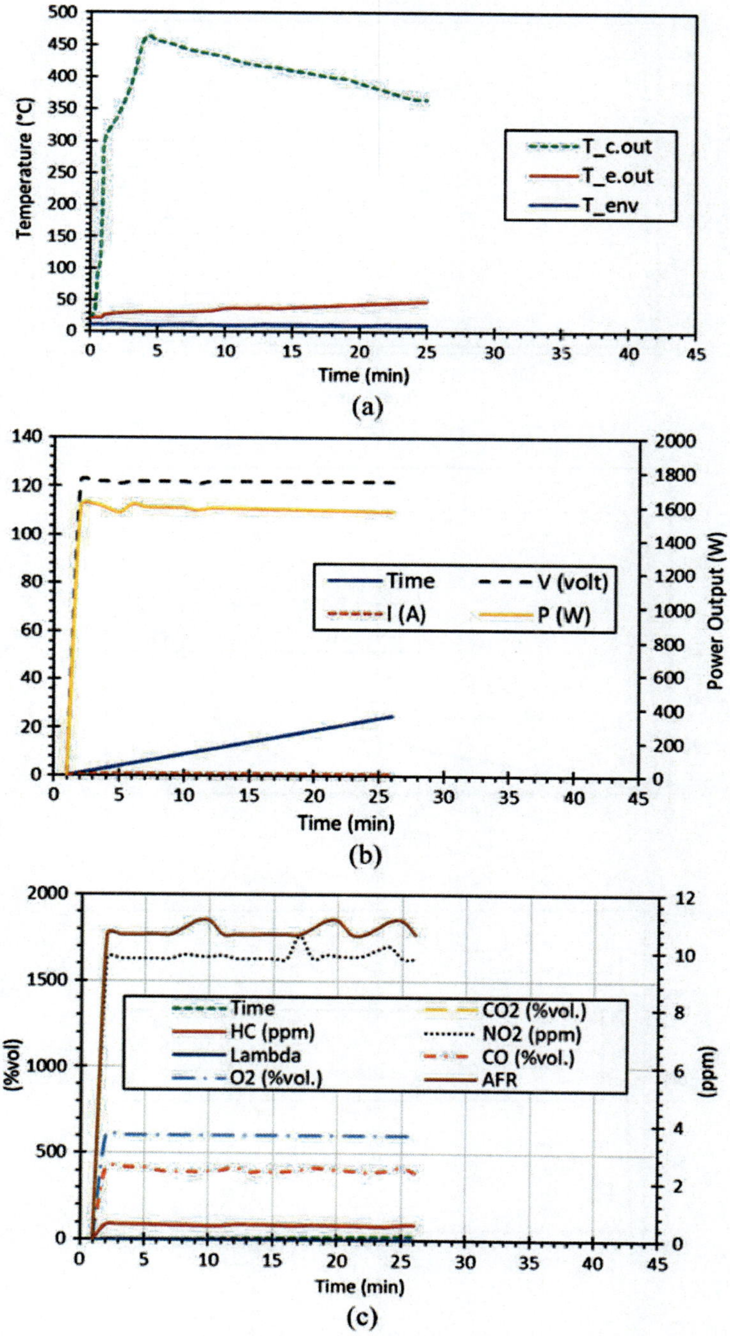

Fig. 4.26 Experimental data for pure ammonia and 100 °C air inlet temperature generator. **a** Temperature, **b** voltage, current and power, and **c** emission values

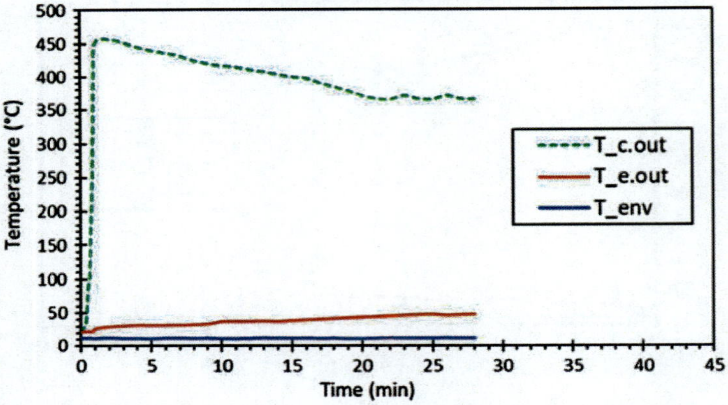

(a)

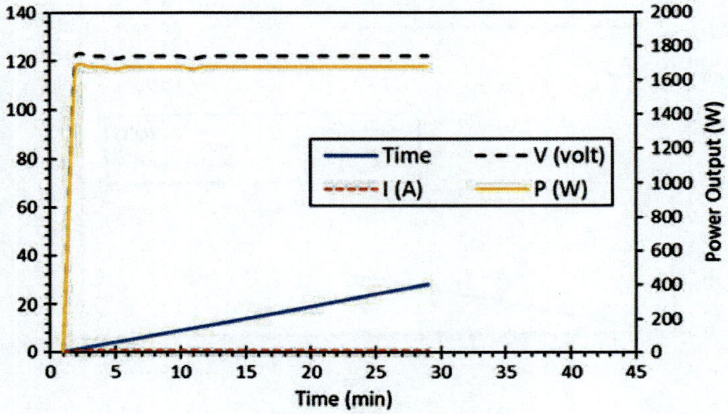

(b)

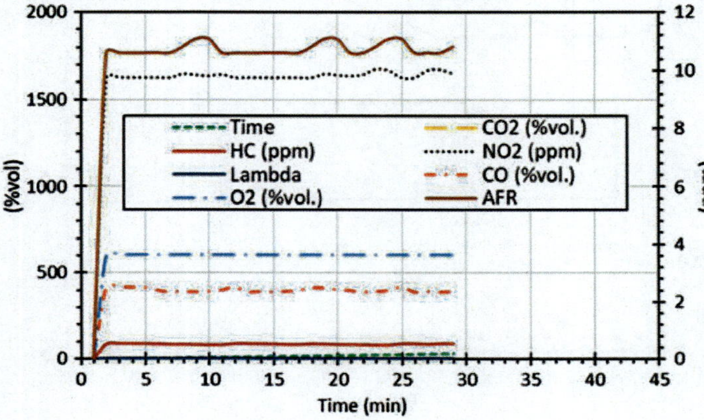

(c)

Fig. 4.27 Experimental data for pure ammonia and 120 °C air inlet temperature generator. **a** Temperature, **b** voltage, current and power, and **c** emission values

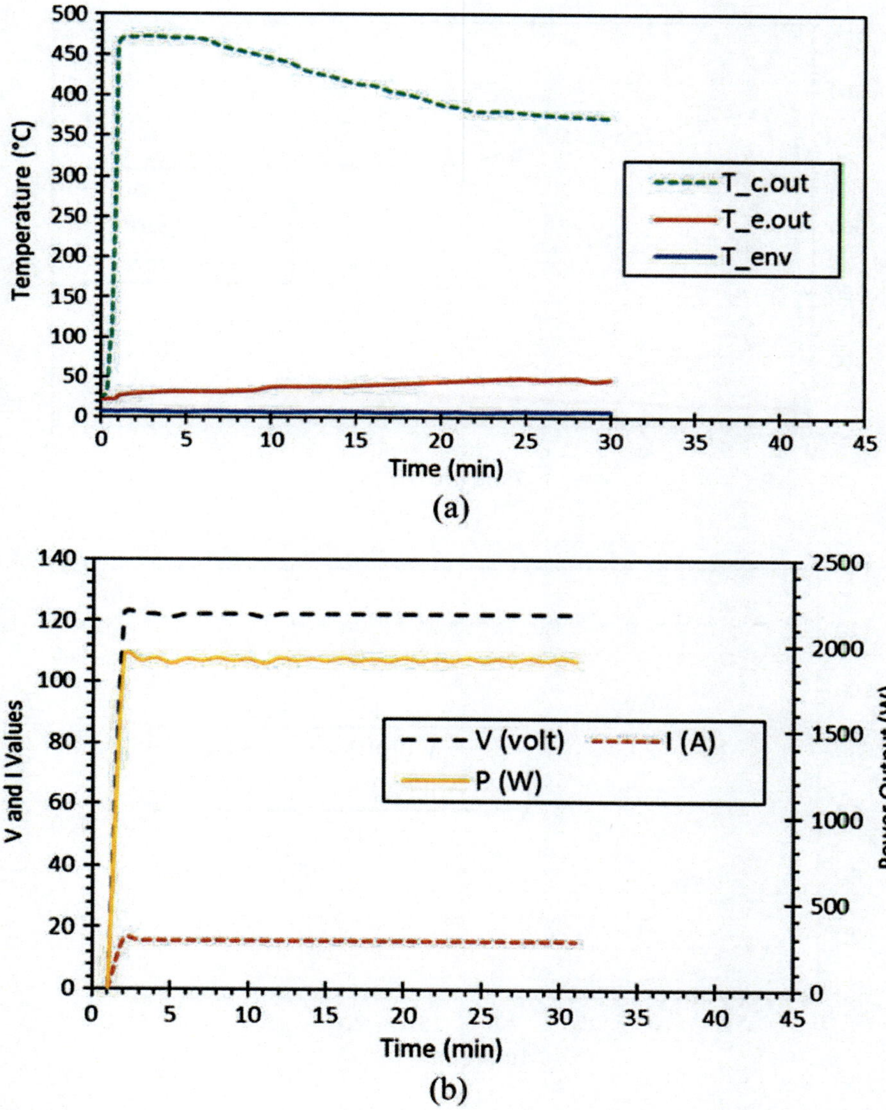

Fig. 4.28 Experimental data for pure ammonia and 140 °C air inlet temperature generator. **a** Temperature and **b** voltage, current and power

there is no significant change in CO_2, CO, HC, O_2, and lambda values, the NO_2 and AFR have increased to 1510 and 17.21 ppm, respectively.

Figure 4.31 demonstrates the comparisons of the studied cases in the terms of power output and operating time. The power output and operating time were 1620 W and 15 min for the experiment of pure ammonia which was done in the previous section where the environmental temperature is 20 °C. Due to lower environmental

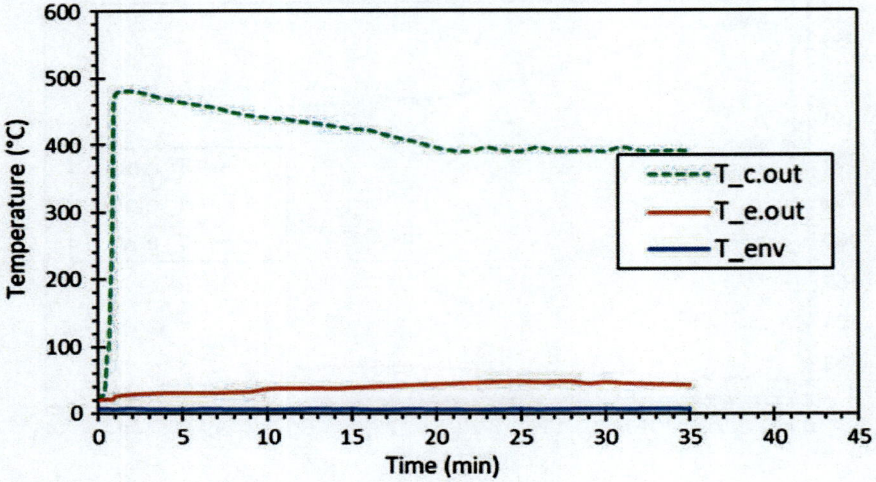

(a)

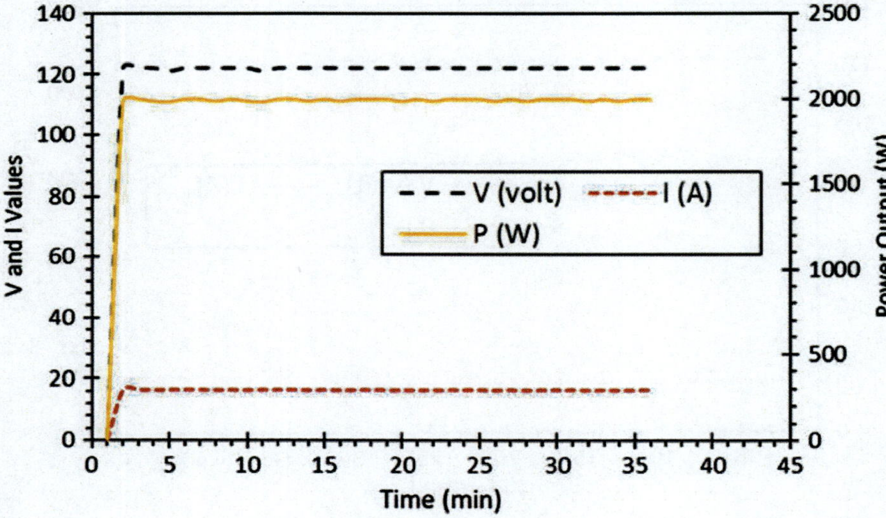

(b)

Fig. 4.29 Experimental data for pure ammonia and 160 °C air inlet temperature generator. **a** Temperature and **b** voltage, current and power

temperature, the power output and operating time have decreased to 1592 W and 12 min, respectively. The effect of pre-heated air is clearly seen in Fig. 4.31. Both power output and operating time have increased with air inlet temperature. Although

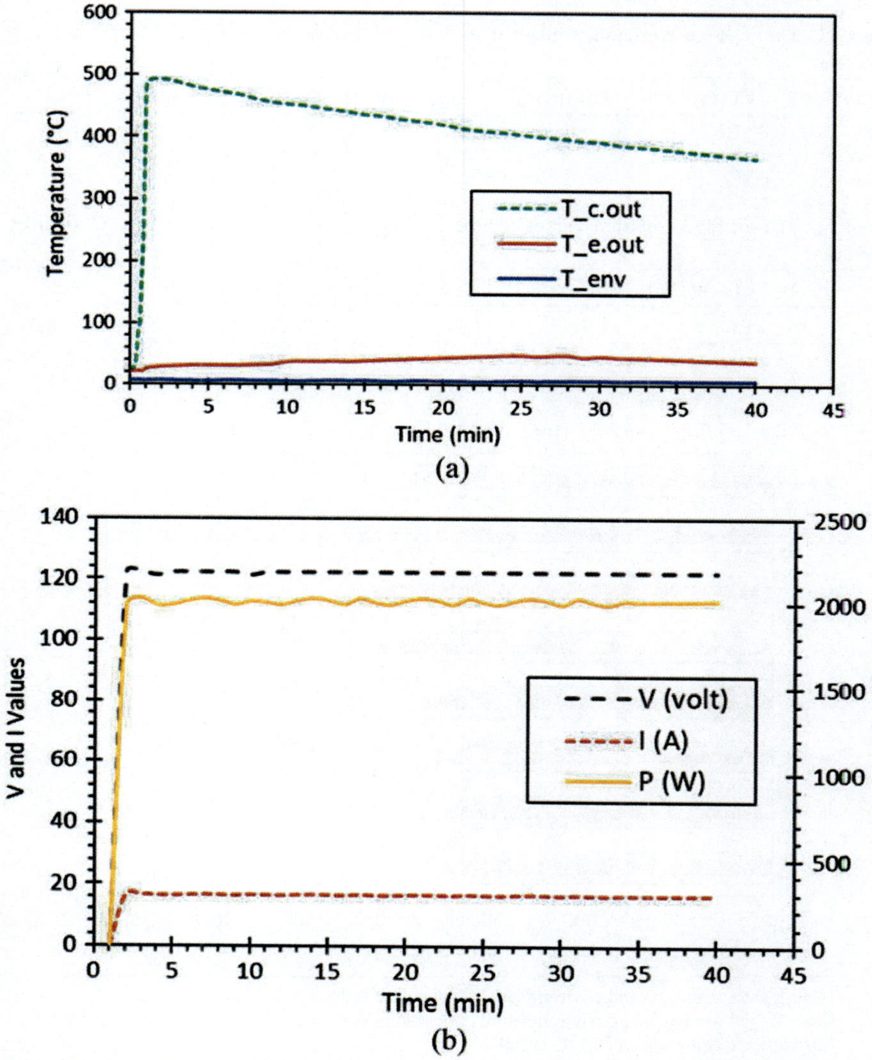

Fig. 4.30 Experimental data for pure ammonia and 180 °C air inlet temperature generator. **a** Temperature and **b** voltage, current and power

Table 4.20 The average emission values for the case with pure ammonia and 140 °C air inlet temperature

CO_2 (vol.%)	CO (vol.%)	HC (ppm)	O_2 (vol.%)	NO_2 (ppm)	Lambda	AFR
1.28	1.29	2.27	0.60	1290.07	0.74	15.69

Table 4.21 The average emission values for the case with pure ammonia and 160 °C air inlet temperature

CO_2 (vol.%)	CO (vol.%)	HC (ppm)	O_2 (vol.%)	NO_2 (ppm)	Lambda	AFR
1.35	1.49	1.27	0.9	1345.91	0.79	16.1

Table 4.22 The average emission values for the case with pure ammonia and 180 °C air inlet temperature

CO_2 (vol.%)	CO (vol.%)	HC (ppm)	O_2 (vol.%)	NO_2 (ppm)	Lambda	AFR
1.72	1.68	1.91	0.80	1510.07	0.79	17.21

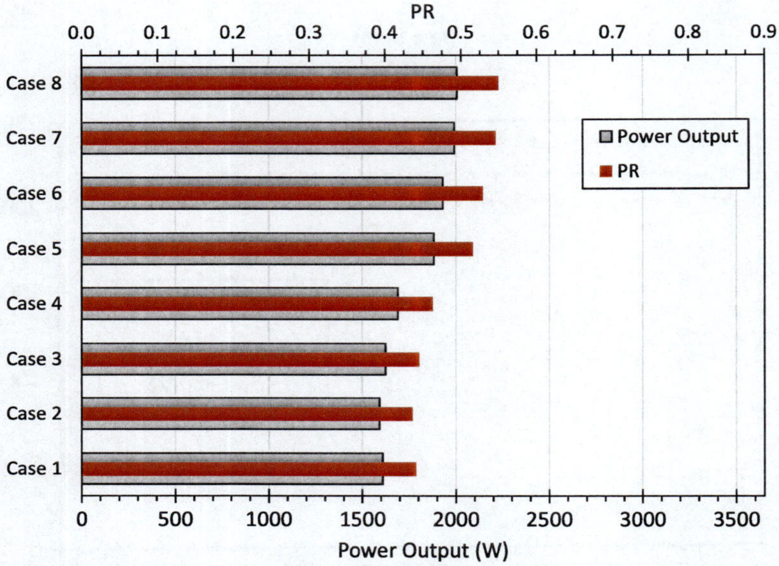

Case 1: Pure ammonia and environmental temperature (20°C)
Case 2: Pure ammonia and environmental temperature (12°C)
Case 3: Pure ammonia and 80 °C air inlet
Case 4: Pure ammonia and 100 °C air inlet
Case 5: Pure ammonia and 120 °C air inlet
Case 6: Pure ammonia and 140 °C air inlet
Case 7: Pure ammonia and 160 °C air inlet
Case 8: Pure ammonia and 180 °C air inlet

Fig. 4.31 The comparison of the studied cases in terms of the power output and PR

the environmental temperature has decreased, the generator performance has been improved with preheated air. 120 °C air inlet has increased power output up to 16% according to the case with 20 °C of the environmental temperature. However, there is no significant change in the emission values.

4.6 Experimental Works with Ammonia and Air Preheaters

After the test with air preheating, the effect of ammonia preheating on the generator performance has been researched experimentally. For this purpose, the apparatus shown in Fig. 4.32 has been developed and tested. Fibre heating tapes are used for providing the heat input for ammonia heating. The view of the fibre tape is shown in Fig. 4.32. The heat flux is controlled by the length of the tape. A 50, 75 and 100 cm of length have been tested, respectively. In order to reduce heat loss from the ammonia preheater, it is covered with a rubber insulation material which is 1.5 cm thickness, as shown in Fig. 4.32. Before each tape length adjustment, the insulation has been removed and then applied back. The fibre tape works with 110 V and 2.4 A. The tape produces 150 W of the heat per 30 cm. Namely, 250, 375 and 500 W of the heating power has been applied, respectively. Air preheating, which is performed in previous experiments, is also performed simultaneously.

Figure 4.33 illustrates the schematic view of the experimental setup with the hot air blower and ammonia-preheater options.

The following procedures have been followed to perform the recent experiments accordingly:

1. The engine has been warmed up for 10 min with gasoline.
2. The gas analyzer has been reset.
3. The flowrates of the ammonia have been set.
4. The temperatures of blower have been set.
5. Blower has been operated.

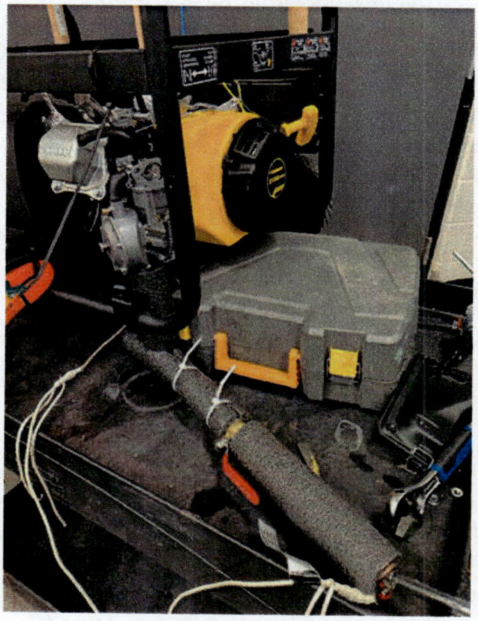

Fig. 4.32 A photo of the ammonia preheating apparatus

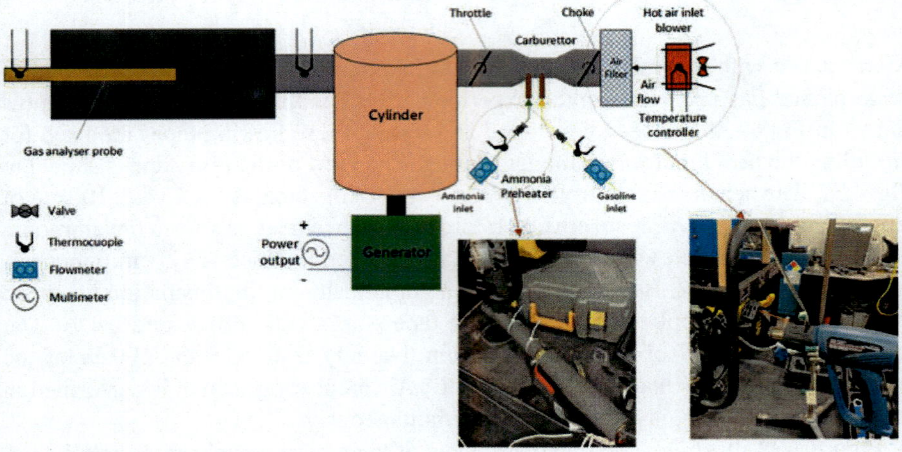

Fig. 4.33 Schematic views of the experimental setup with the hot ait blower and ammonia-preheater options

6. A quick check has been done for system stability and safety.
7. The experimental run has commenced, and the data have been collected for analysis and assessment.
8. Each experimental run has been conducted for 45 min.
9. Each experiment has been repeated at least two times for checking data reliability.

Results of Experiments with Ammonia and Air Preheaters

All experimental data have been illustrated in three divisions, which are temperature and electricity versus time, and the average emission values. We have started the experiments with a gasoline-fueled engine in order to observe the effect of environmental temperature on the engine performance. Figure 4.34 and Table 4.17 show the experimental data for the case powered by 100% gasoline with 180 °C air-preheating. As seen from Fig. 4.34, the temperature at the cylinder outlet is approximately 850 °C. The environmental temperature is around – 2 °C. The temperature at the gas analyzer probe has reached up to 81 °C, which is 50 °C lower than previous cases. The generator has provided the stable power output. Table 4.17 demonstrates the emission values. The average CO_2 emission is 10.08%, CO is 8.92%, O_2 is 0.35%, respectively. HC and NO_2 values are 157 and 82 ppm. Lambda and AFR are observed to be 0.72 and 11.53, respectively.

A set of experiments with preheated ammonia and preheated air have been performed. The preheated air temperature has been set as 180 °C. The fibre heating tape has been used for heating up the ammonia. The length of the heating tape has been studied parametrically. A 50, 75 and 100 cm length of the heating tape have been tested. Each experiment has been performed at least four times for data reliability.

The fibre tape used for ammonia preheating has been applied as 50, 75 and 100 cm, which corresponds to 250, 375 and 500 W of heating power, respectively. Table 4.24

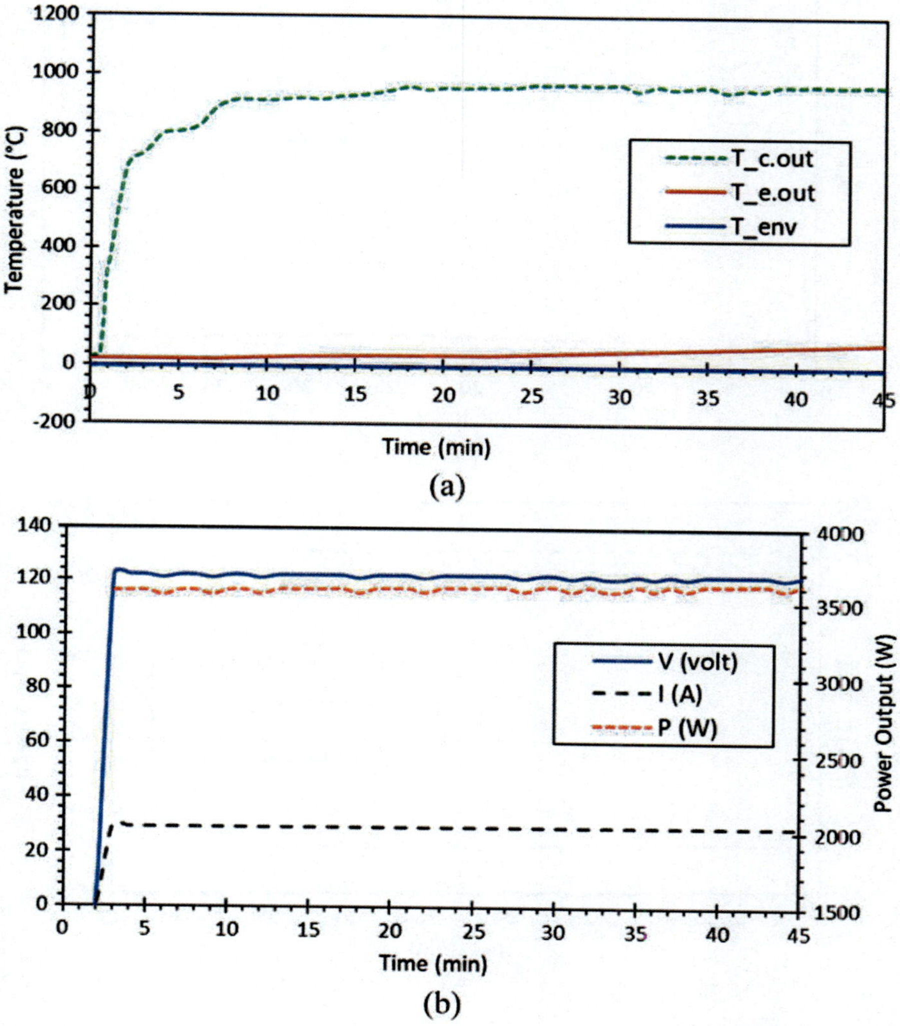

Fig. 4.34 Experimental data for gasoline fueled generator with 180 °C air-preheating. **a** Temperature, **b** voltage, current and power values

demonstrates the operating periods of the generator for the different cases studied in previous and current terms. As seen from Table 4.24, the ammonia preheating has not affected the operating period for the case of 50 cm of fibre tape length. For the case of 75 cm tape length, the operating period has extended up to 41 min. The generator has run for 45 min with a 100 cm tape length. The generator has stopped working at the 45th minute.

The experimental data for the cases with 50, 75 and 100 cm of the length of the heater tape are shown in the graphs inclusively, Figs. 4.30, 4.31, 4.32, 4.33, 4.34

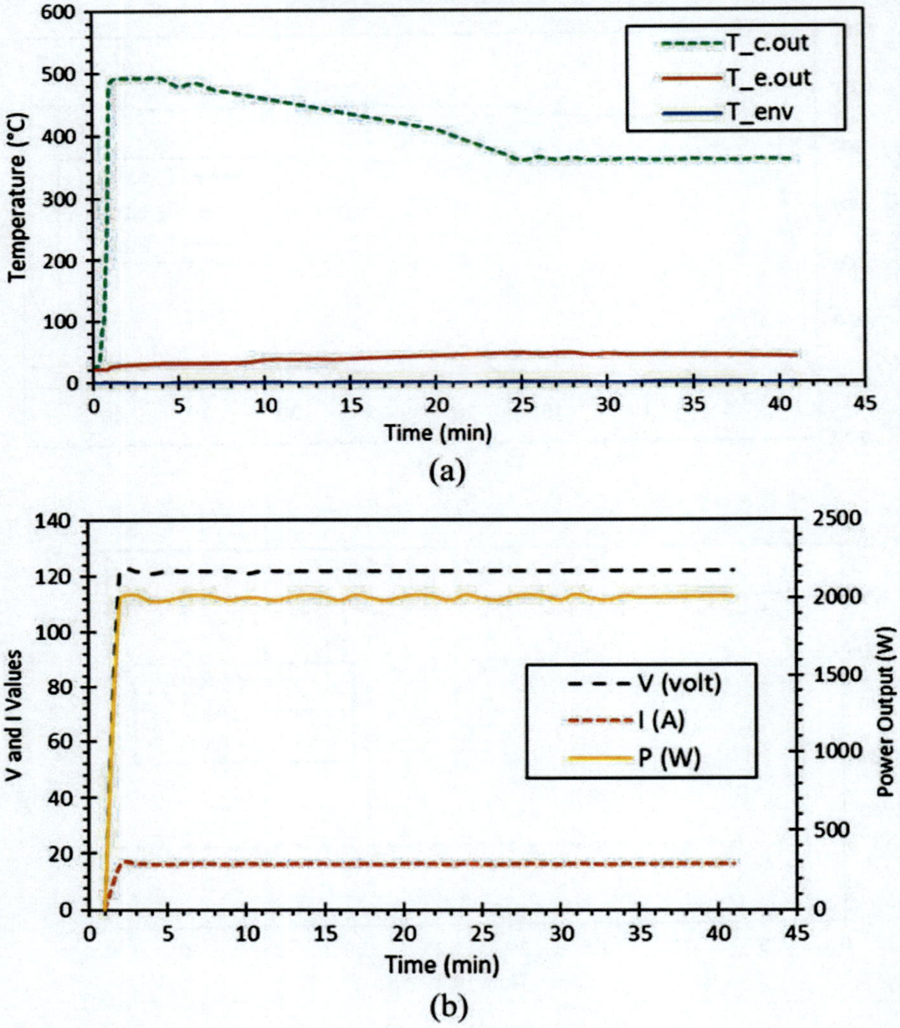

Fig. 4.35 Experimental data for ammonia-fueled generator with 180 °C air-preheating and 50 cm heater tape. **a** Temperature changes and **b** power values

Table 4.23 The average emission values for the case with 100% gasoline powered generator

CO_2 (vol.%)	CO (vol.%)	HC (ppm)	O_2 (vol.%)	NO_2 (ppm)	Lambda	AFR
10.08	8.92	157.47	0.35	81.68	0.72	11.53

and 4.35, respectively. The average emission values for the studied cases are given in Tables 4.6, 4.7, 4.8, 4.9, 4.10, 4.11, 4.12, 4.13, 4.14, 4.15, 4.16, 4.17, 4.18, 4.19, 4.20, 4.21, 4.22, 4.23, 4.24 and 4.25, respectively. For each experiment configuration, two datasets have been illustrated.

Table 4.24 Operating periods of the generator for different air inlet temperature and ammonia preheater length conditions

Case	Test 1	Test 2	Test 3	Test 4
Pure ammonia and 180 °C air inlet 50 cm length of heating tape	39	40	40	39
Pure ammonia and 180 °C air inlet 75 cm length of heating tape	40	41	41	41
Pure ammonia and 180 °C air inlet 100 cm length of heating tape	43	42	44	45

Table 4.25 The average emission values for ammonia-fueled generator with 180 °C air-preheating and a 55 cm heater tape

CO_2 (vol.%)	CO (vol.%)	HC (ppm)	O_2 (vol.%)	NO_2 (ppm)	Lambda	AFR
1.22	1.29	2.03	0.6	1375	0.79	17

Figure 4.35 and Table 4.25 show the experimental data for the case with 180 °C air inlet temperature and 50 cm the length of the heating tape. In Fig. 4.35, the temperature, voltage, current and power changes are shown for test 1 and test 2. The voltage, current and power output are the same for both cases. On the other hand, there is a difference in the temperature change. While the maximum temperature is the same for both cases, the trend of the temperature decrease due to ammonia combustion is different. The surrounding temperatures is 0 and 5 °C for test 1 and test 2, respectively. The temperature at the gas analyzer probe is almost the same for both cases.

As seen from Table 4.25, there is no significant change in the emission values. For both cases, all values are almost equal. The CO_2 values are 1.22 and 1.37 vol.%, respectively, for the studied two cases. The CO emissions are 1.29 and 1.19. There is a remarkable in the value of NO_2 comparing the other emission values measured. It is 1375 in test 1 and 1305 for test 2.

Figure 4.36 and Table 4.26 show the experimental data for the case with 180 °C air inlet temperature and a 75 cm heater tape. Firstly, the temperature at the cylinder exit is almost 20 °C higher than the previous case with 50 cm heater tape. There is no significant case in the voltage, current and power output values. The power output is stable. On the other hand, when the temperature changes are taken into consideration, there is a remarkable change in the temperature at the cylinder exit. In the test 1, the cylinder outlet temperature has reduced suddenly after 25 min, despite there is no change in power output. The temperature changes on the gas analyzer probe is almost the same with the previous case with 50 cm heating tape (Fig. 4.35).

Table 4.26 illustrates the emission values of the case with 75 cm heating tape for test 1 and 2. There are a small change between test 1 and test 2. The CO_2 values are observed to be 1.28 and 1.03 vol.% for test 1 and test 2. The NO_2 emissions are measured to be 1386 and 1430 ppm. The respective AFR values are 17 and 16.5.

Figure 4.36 and Table 4.26 show the experimental observations with 180 °C air inlet temperature and 100 cm heating tape. A 100 cm heating tape provided a 500 W of the heating power. The temperature at the cylinder exit is almost the same with the

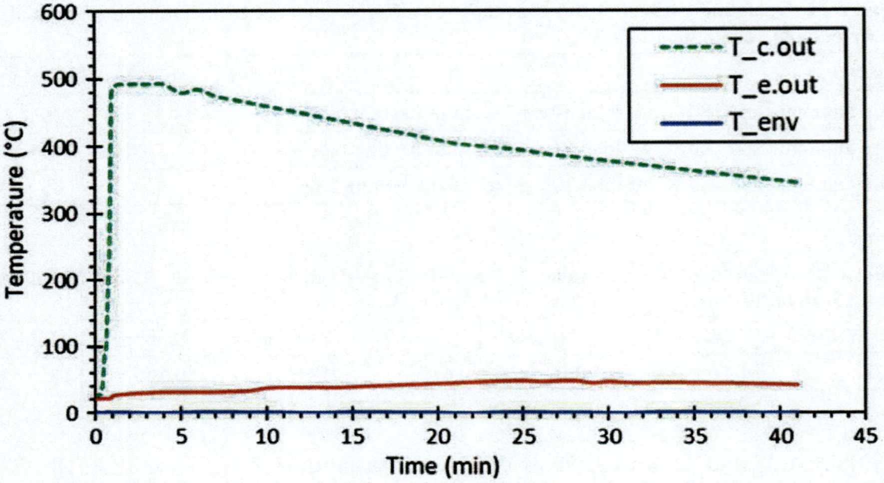

(a)

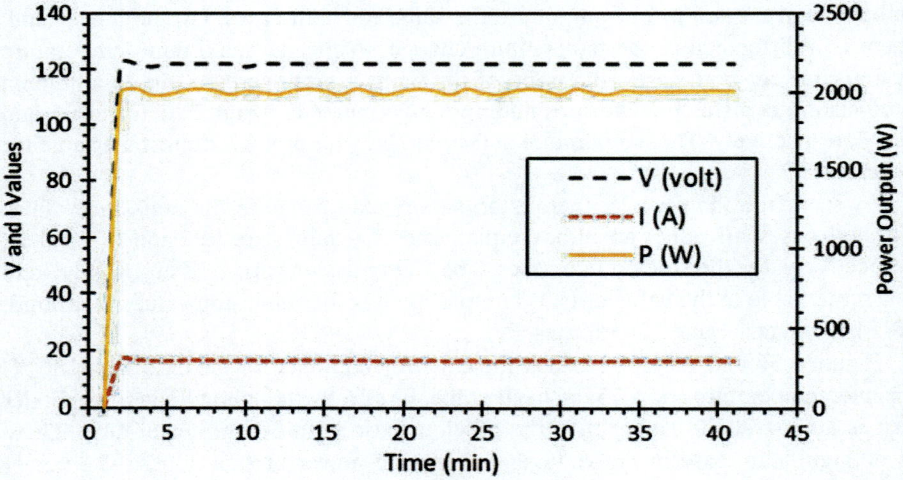

(b)

Fig. 4.36 Experimental data for ammonia-fueled generator with 180 °C air-preheating and 75 cm heater tape. **a** Temperature changes and **b** power values

Table 4.26 The average emission values for ammonia-fueled generator with 180 °C air-preheating and 75 cm heater tape

CO_2 (vol.%)	CO (vol.%)	HC (ppm)	O_2 (vol.%)	NO_2 (ppm)	Lambda	AFR
1.28	1.32	2.65	0.9	1386	0.79	17

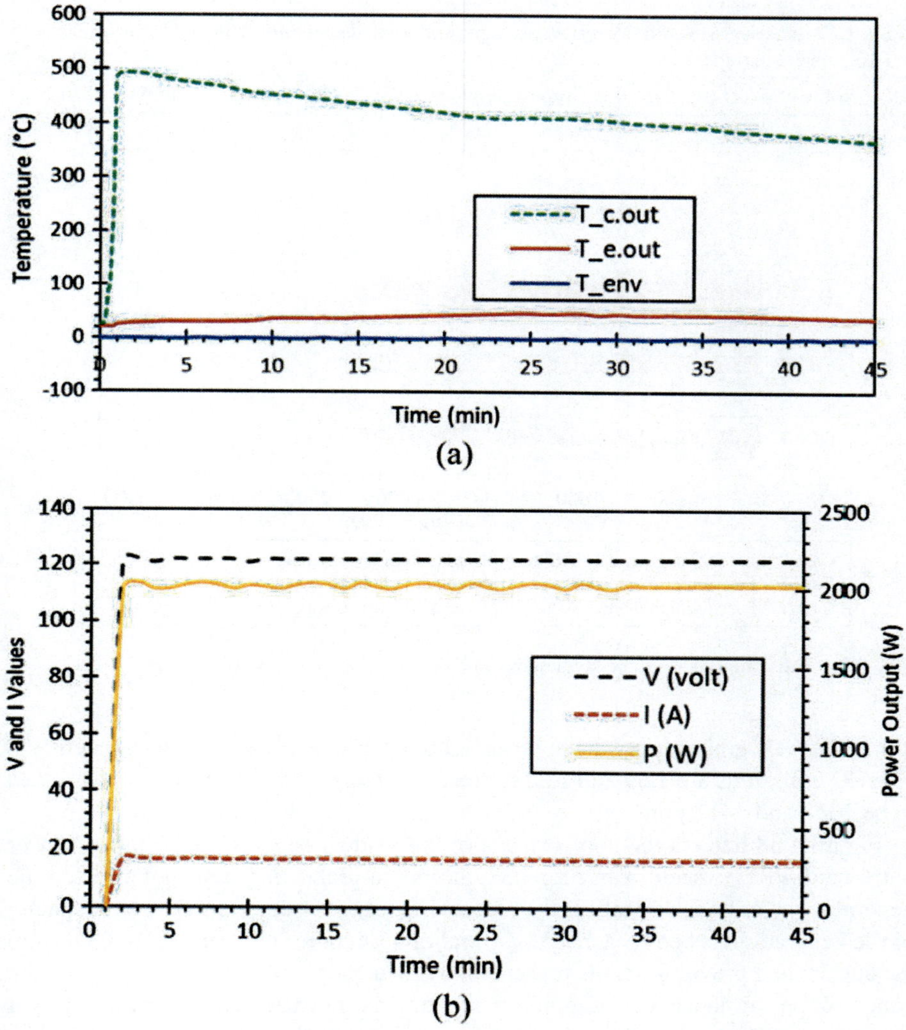

Fig. 4.37 Experimental data for ammonia-fueled generator with 180 °C air-preheating and 100 cm heater tape. **a** Temperature changes and **b** power values

cases with 75 and 50 cm the length of heating tape. The only remarkable variation is seen in the temperature at the cylinder exit. Similar to the previous cases studied in the present term, the voltage, current and power output values are stable. The ambient temperature is about 0 and 5 °C for test 1 and 2, respectively. The temperature taken from gas analyser probe is almost the same with the previous cases.

Table 4.27 gives the emission values measured for the case with a 100 cm the length of the heating tape. For the test 1 and test 2, there is no significant change in the emission values for the current case and previous cases studied in the present

Table 4.27 The average emission values for ammonia-fueled generator with 180 °C air-preheating and 100 cm heater tape

CO_2 (vol.%)	CO (vol.%)	HC (ppm)	O_2 (vol.%)	NO_2 (ppm)	Lambda	AFR
1.31	1.39	2.45	0.7	1368	0.81	17

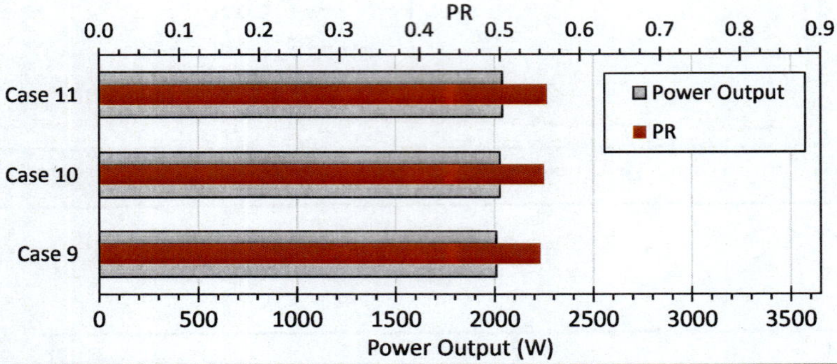

Case 9: Pure ammonia and 180°C air inlet and 50 cm heating tape
Case 10: Pure ammonia and 180°C air inlet and 75 cm heating tape
Case 11: Pure ammonia and 180°C air inlet and 100 cm heating tape

Fig. 4.38 The comparison of the studied cases in terms of the power output and PR

term. The CO_2 emission has been observed to be 1.31 and 1.24 vol.%, respectively. The CO emissions are 1.39 and 1.32 for test 1 and test 2. NO_2 values have been seen to be 1368 and 1417 ppm.

Figure 4.38 depicts the comparisons of the studied cases in the terms of power output and PR. As seen from Fig. 4.38, ammonia-preheating have not affected the generator performance significantly. Approximately 56% of nominal power capacity of the generator has been reached at the end of 45 minutes test period. As seen from the current and previous results presented, while the air preheating has enhanced the generator performance, the ammonia-preheating has not affected the performance of the generator. The main reason for this is the mixture ratio of ammonia and air. The air temperature is more dominant than ammonia temperature.

4.7 Experimental Works with Propane and Ammonia Blends

Propane is one of the commonly used fuels in the internal combustion engines. In this section, the effect of propane blend with ammonia is studied experimentally. For this purpose, a propane tank is included in the system. In order to provide propane and ammonia to the engine at the same time, T-connection has been used. The mixture ratios have been set by pressure adjustment and volumetric flowrates. Figure 4.39

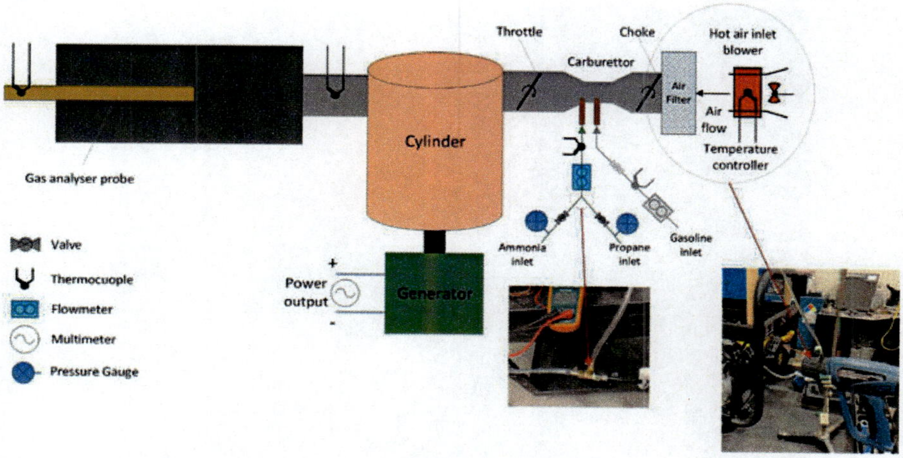

Fig. 4.39 The schematic view of the experimental setup

depicts the schematic view of the experimental setup. As seen from Fig. 4.39, the air preheating has been continued to use during the experiments.

Figure 4.40 shown the photography of the experimental setup. In this term, we have worked on the ammonia/propane blends. In this regard, 100% propane (P100), 80% propane and 20% ammonia (P80-A20), 60% propane and 40% ammonia (P60-A40), 50% propane and 50% ammonia (P50-A50), 40% propane and 60% ammonia (P40-A60) and 20% propane and 80% ammonia (P80-A20) have been studied experimentally.

In the present section, the engine preheating with gasoline for 5 min, which is performed just before the experiments, has not been performed, since propane is able to combust easily according to the ammonia. Engine preheating has been performed with propane. The following procedures have been followed to perform the recent experiments accordingly:

1. The engine has been warmed up for 10 min with propane.
2. The gas analyzer has been reset.
3. The flowrate and pressure of the ammonia have been set.
4. The pressure of the propane has been set.
5. The temperatures of blower have been set.
6. Blower has been operated.
7. A quick check has been done for system stability and safety.
8. The experimental run has commenced, and the data have been collected for analysis and assessment.
9. Each experimental run has been conducted for 45 min.

Fig. 4.40 A photo of the final experimental setup

Results of Experiments with Propane and Ammonia Blends

All experimental data have been illustrated in three divisions, which are temperature and electricity versus time, and the average emission values. We have started the experiments with a gasoline-fueled engine in order to observe the effect of environmental temperature on the engine performance. First, the operating time of the generator is given in Table 4.28. The operating times for all cases studied in the previous terms are also shown in Table 4.28 in order to compare the result clearly. As seen from Table 4.28, the generator has completed 45 min test period up to 50% ammonia content. The engine has stopped after 60% ammonia content in the fuel mixture. The case with propane/ammonia blends are more unsteady comparing to the case with gasoline and gasoline/ammonia blends. When the operating period is taken into consideration, the propane seems a significant promoter for ammonia combustion.

Table 4.28 Operating periods of the cases studied

Case	Test 1	Test 2	Test 3	Test 4
100% Propane	45	45	45	45
80% Propane–20% Ammonia	45	45	45	45
60% Propane–40% Ammonia	45	45	45	45
50% Propane–50% Ammonia	45	45	45	45
40% Propane–60% Ammonia	39	35	37	40
20% Propane–80% Ammonia	30	28	35	33

Figure 4.41 and Table 4.29 show the experimental data for the case powered by 100% propane with 180 °C air-preheating. As seen from Fig. 4.34, the temperature at the cylinder outlet is approximately 900 °C. The environmental temperature is around 5 °C. The temperature at the gas analyzer probe has reached up to 90 °C. The generator has provided the stable power output. Table 4.29 demonstrates the emission values for 100% propane-fueled generator. The average CO_2 emission is 8.60%, CO is 10.43%, O_2 is 0.4%, respectively. HC and NO_2 values are 283 and 41 ppm. Lambda and AFR are observed to be 0.72 and 11.58, respectively. The generator has generated 2920 kW of power, which is PR = 0.80. On other words, the generator has reached about 80% of its ideal capacity with gasoline. Propane has reduced the emission values according to the gasoline-powered one.

A set of experiments for propane/ammonia blends have been conducted with 180 °C air-preheating. The experimental results for 80% propane and 20% ammonia blend are given in Fig. 4.42 and Table 4.30, respectively. As seen from Fig. 4.42, the temperature at the cylinder exit is almost 40 °C lower than the case with 100% propane. 20% ammonia content has reduced the electricity current from 24.5 to 22.3 A. The reduction corresponds a 0.75 of PR value. As seen from Table 4.30, the CO_2 and CO emissions have tended to reduce with the ammonia content. The CO_2 is observed to be 7.98 vol.% and CO is measured to be 9.80. HC values is 267 ppm which is 16 ppm lower then 100% propane combustion. The NO_2 emissions has increased from 41 to 215 ppm with the ammonia combustion.

Figure 4.43 and Table 4.31 show the experimental data for the case 60% propane–40% ammonia fuel blend and 180 °C air inlet temperature. The temperature at the cylinder exit is 55 °C lower than the case with 100% propane. There is no significant change in the temperature obtained from the gas analyzer probe. As seen from Table 4.31, the CO_2 emission is 7.98 vol.% and CO emission is 9.80 vol.% HC and NO_2 values are measured to be 267 and 215 ppm, respectively. There is no change in the lambda values. AFR is observed to be 12. The electricity current is measured to be 20.8 A which corresponds to 2533 kW of power. PR is calculated as 0.70.

Figure 4.44 and Table 4.32 show the experimental observations for 50% propane and 50% ammonia fuel blends and 180 °C air inlet temperature. The temperature at the cylinder exit is around 880 °C. There is no significant change in the temperatures from exhaust probe and ambient. The generator has produced 19.2 A electricity current. So, the power provided by generator is 2291 kW which corresponds the 0.65 of PR. Table 4.32 gives the emission values measured for the case with 50% propane and 50% ammonia fuel blends and 180 °C air inlet temperature. The CO_2 and CO emissions have been observed to be 6.43 and 8.42 vol.%, respectively. NO_2 values are 417 ppm. Lambda and AFR values are measured to be 0.83 and 13, respectively.

Figure 4.45 and Table 4.33 show the experimental observations for the case with 40% propane and 60% ammonia fuel blend and 180 °C air preheating. It is clear from Fig. 4.45 that the temperature at the cylinder exit has reduced due to lower combustion efficiency. The engine has run for 35 min and then stopped. Before engine has stopped, the temperature at the cylinder exit has reduced. While CO_2 and

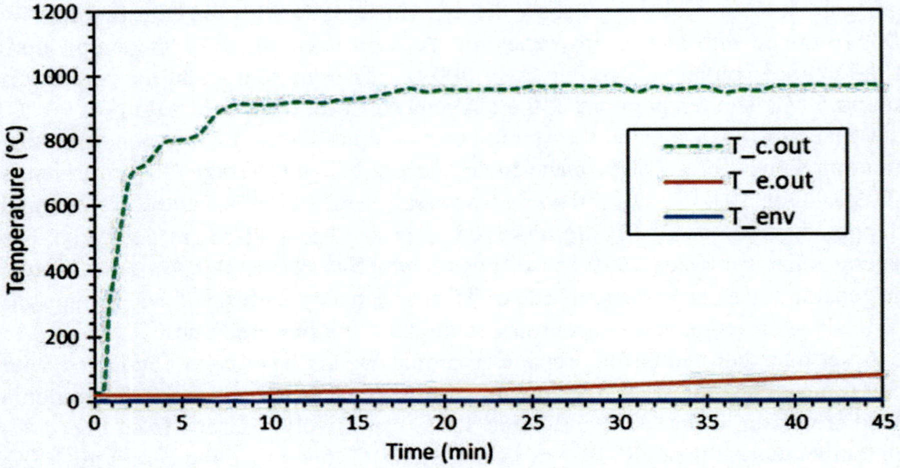

(a)

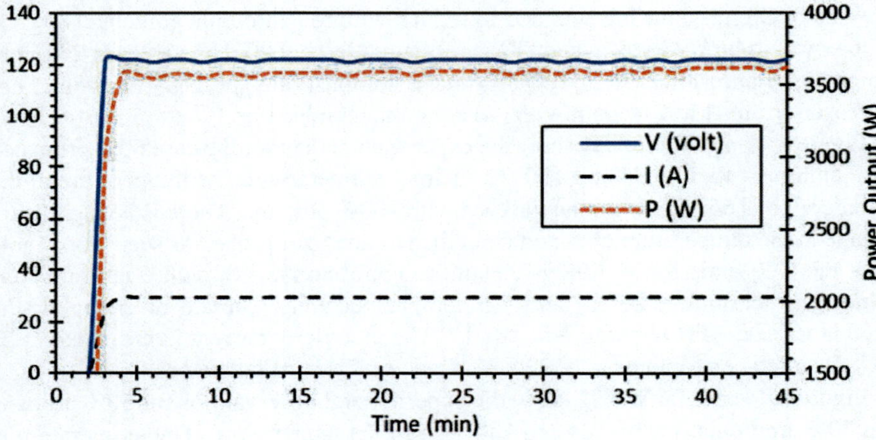

(b)

Fig. 4.41 Experimental data for propane fueled generator with 180 °C air-preheating. **a** Temperature, **b** voltage, current and power values

Table 4.29 The average emission values for the case with 100% propane powered generator

CO_2 (vol.%)	CO (vol.%)	HC (ppm)	O_2 (vol.%)	NO_2 (ppm)	Lambda	AFR
8.60	10.43	283	0.4	41	0.72	11.58

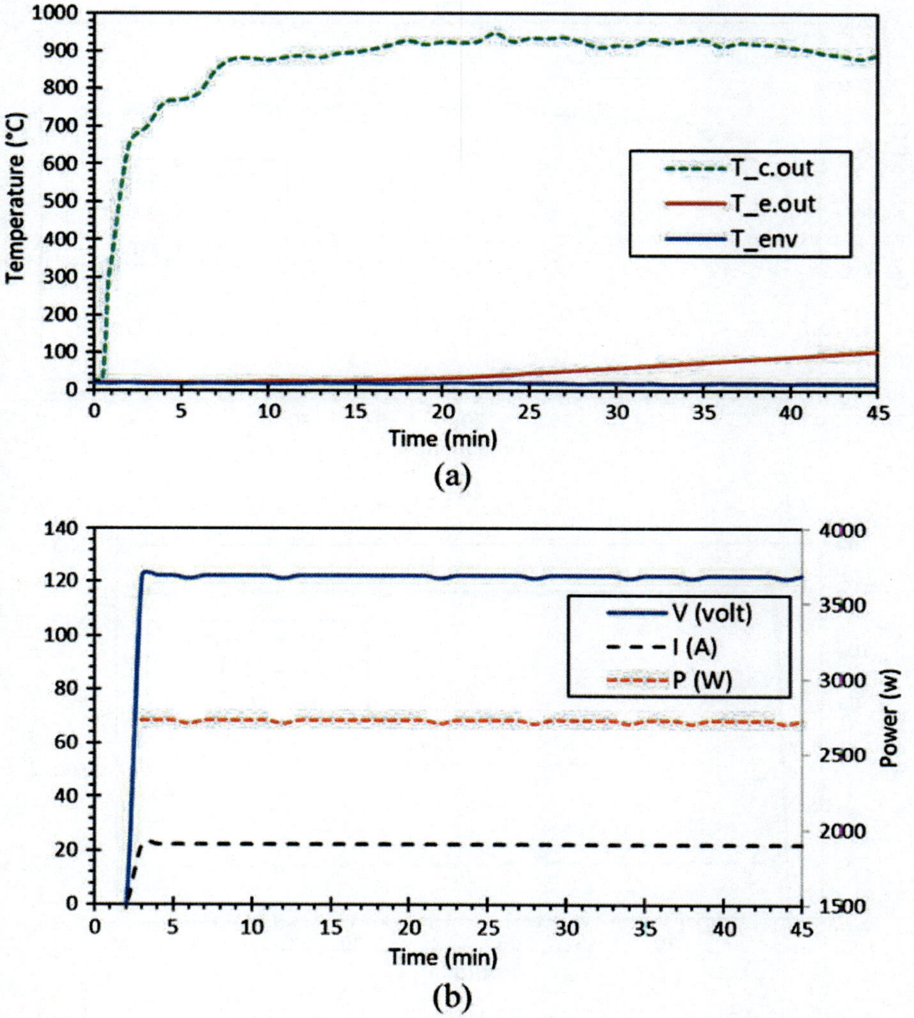

Fig. 4.42 Experimental data for 80% propane and 20% ammonia blend-fueled generator with 180 °C air-preheating. **a** Temperature changes and **b** power values

Table 4.30 The average emission values for 80% propane and 20% ammonia fueled generator

CO_2 (vol.%)	CO (vol.%)	HC (ppm)	O_2 (vol.%)	NO_2 (ppm)	Lambda	AFR
7.98	9.80	267	0.8	215	0.79	12

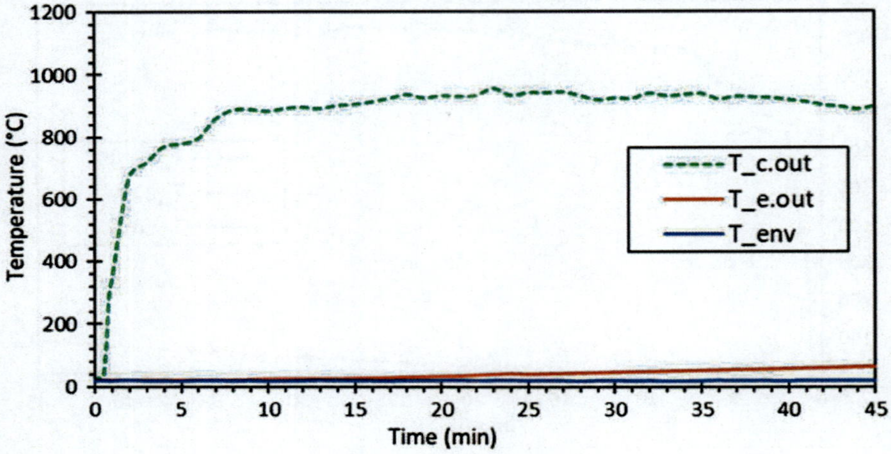

(a)

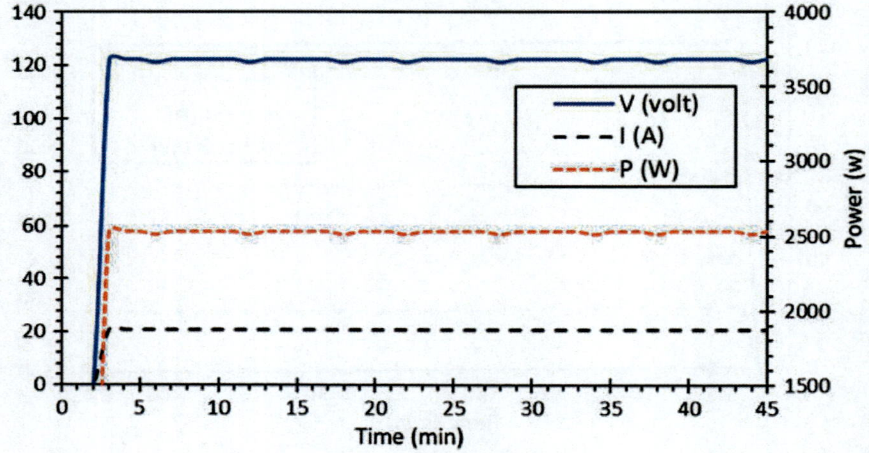

(b)

Fig. 4.43 Experimental data for 60% propane and 40% ammonia blend-fueled generator with 180 °C air-preheating. **a** Temperature changes and **b** power values

Table 4.31 The average emission values for 60% propane and 40% ammonia fueled generator

CO_2 (vol.%)	CO (vol.%)	HC (ppm)	O_2 (vol.%)	NO_2 (ppm)	Lambda	AFR
7.17	9.21	251	0.85	324	0.79	12

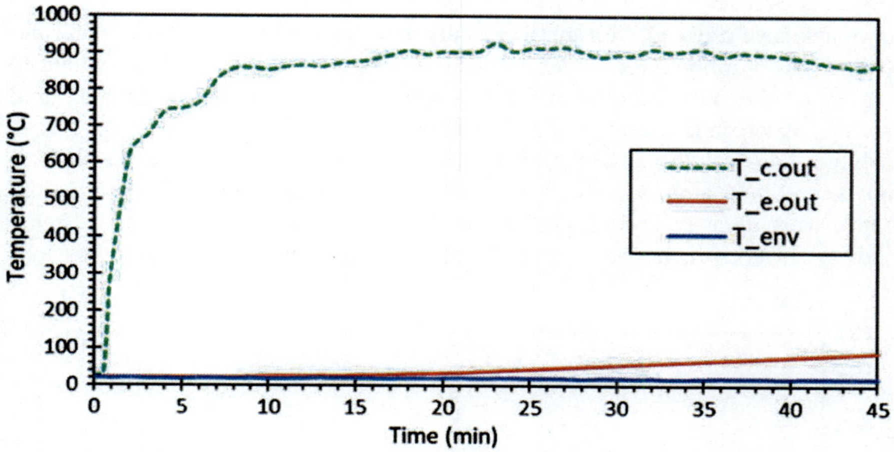

(a)

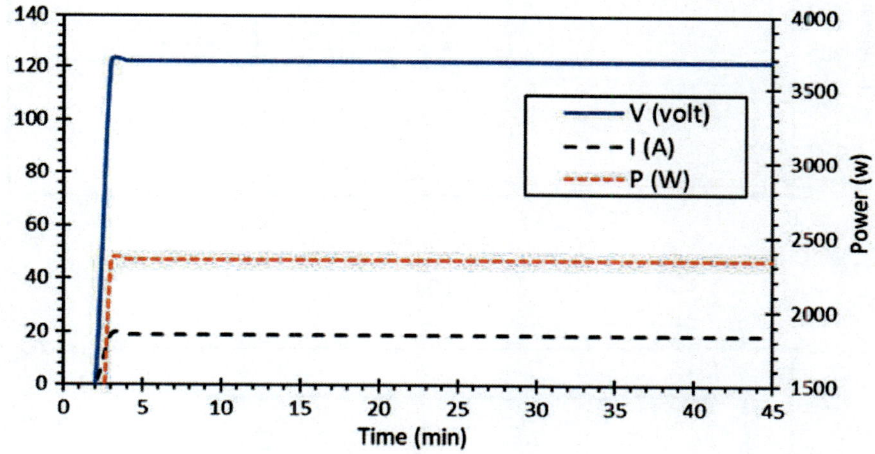

(b)

Fig. 4.44 Experimental data for 50% propane and 50% ammonia blend-fueled generator with 180 °C air-preheating. **a** Temperature changes and **b** power values

Table 4.32 The average emission values for 50% propane and 50% ammonia fueled generator

CO_2 (vol.%)	CO (vol.%)	HC (ppm)	O_2 (vol.%)	NO_2 (ppm)	Lambda	AFR
6.43	8.42	245	1	417	0.83	13

CO values have reduced with the increasing ammonia content, NO_2 emission has increased to 647 ppm. AFR values has also increased to 14. HC values has reduced to 187 ppm. The during the experiments, the electricity current has been measured to be 17.1 A which corresponds a 2060 kW of power.

Figure 4.46 and Table 4.34 give the experimental data for the case with 20% propane and 80% ammonia fuel blend and 180 °C air preheating. The engine has stopped working after 31 min. During the working time, it has generated a 15.9 A electivity current which means 1945 kW of the power. The CO_2 and CO emissions

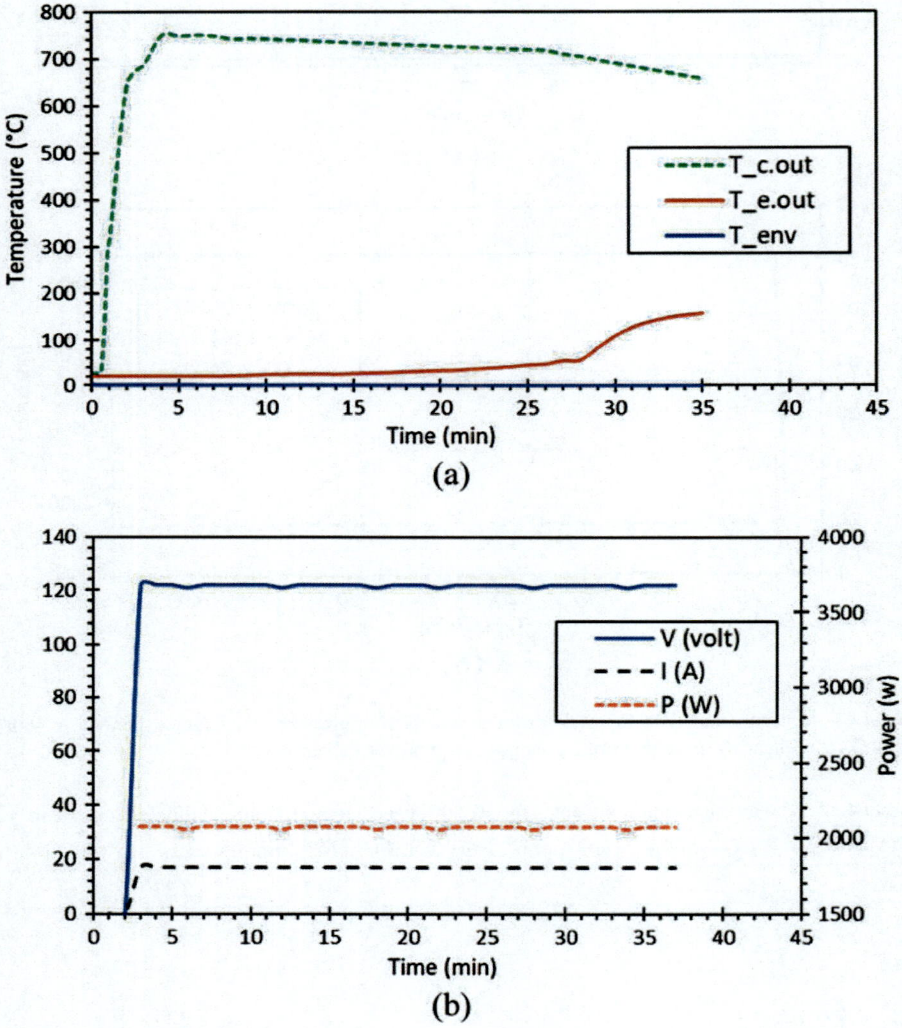

Fig. 4.45 Experimental data for 40% propane and 60% ammonia blend-fueled generator with 180 °C air-preheating. **a** Temperature changes and **b** power values

Table 4.33 The average emission values for 40% propane and 60% ammonia fueled generator

CO_2 (vol.%)	CO (vol.%)	HC (ppm)	O_2 (vol.%)	NO_2 (ppm)	Lambda	AFR
5.03	6.42	187	1.2	647	0.83	14

are measured to be 3.47 and 4.13 vol.%, respectively. NO_2 values has increased up to 580 ppm. HC has reduced to 98 ppm.

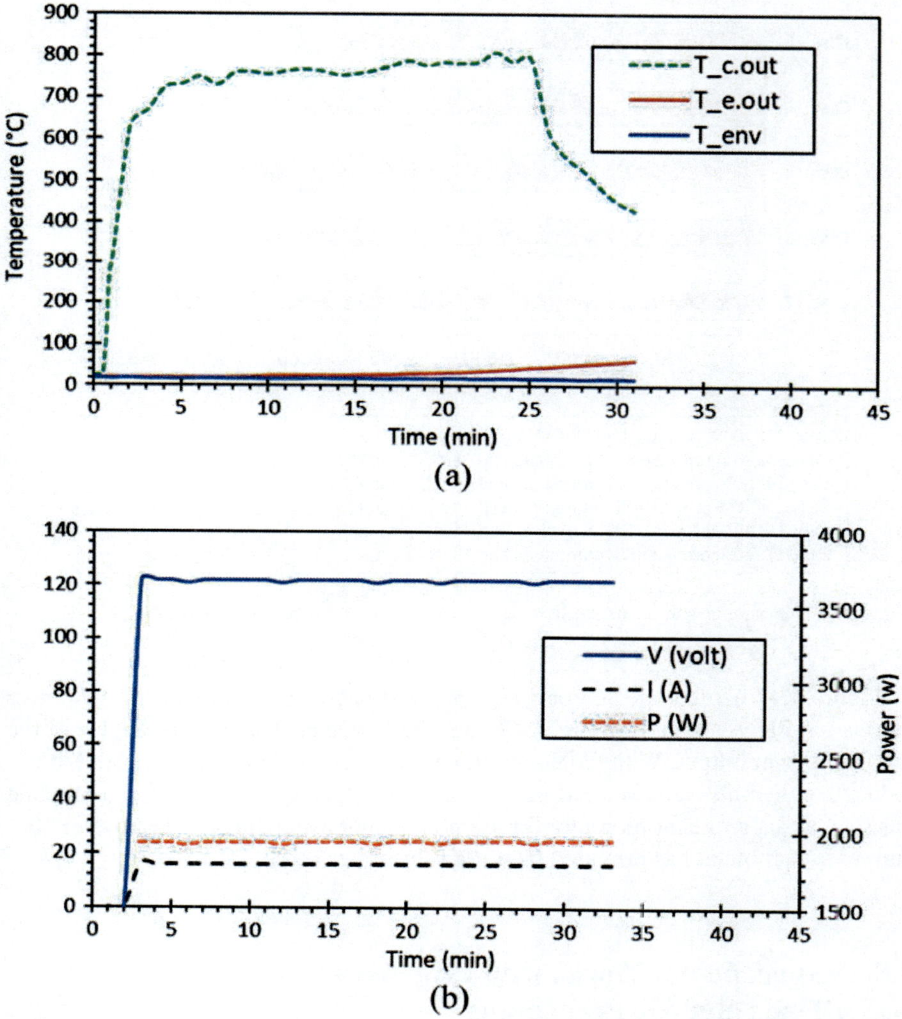

Fig. 4.46 Experimental data for 20% propane and 80% ammonia blend-fueled generator with 180 °C air-preheating. **a** Temperature changes and **b** power values

Table 4.34 The average emission values for 20% propane and 80% ammonia fueled generator

CO_2 (vol.%)	CO (vol.%)	HC (ppm)	O_2 (vol.%)	NO_2 (ppm)	Lambda	AFR
3.47	4.13	98	1.1	850	0.85	15

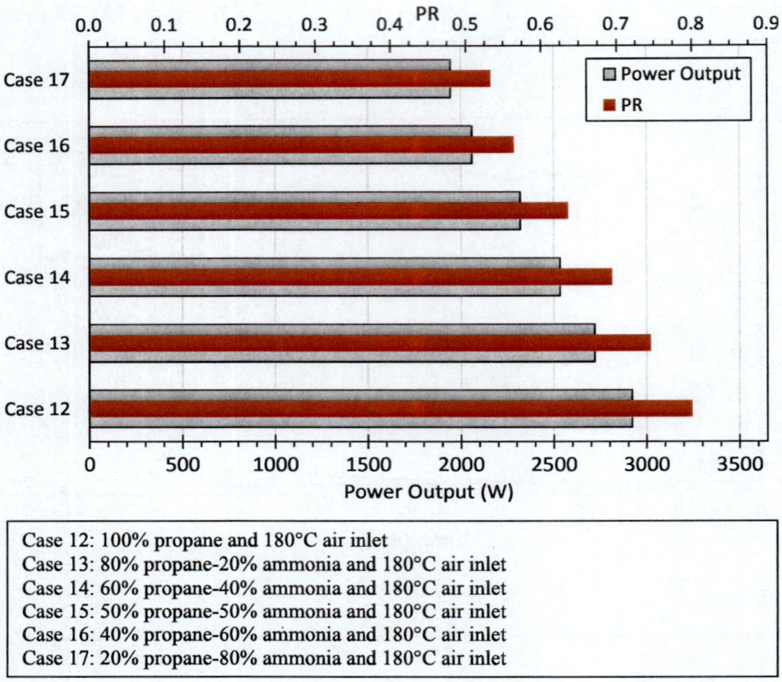

Fig. 4.47 The comparison of the studied cases in terms of the power output and PR

Figure 4.47 depicts the comparisons of the studied cases in the terms of power output and PR. As seen from Fig. 4.47, the 100% propane has reached 81.1% of the nominal power output. With the increasing ammonia content in the fuel blend, it has reduced. When all cases studied given in the Fig. 4.47 are compared, the propane has seen as a significant promoter for ammonia combustion. 50% propane and 50% ammonia fuel blend has provided 0.64 PR values.

4.8 Experimental Works with Propane and Ammonia Blends and Air Preheating

In the previous section, the only propane and ammonia blends are experimented with air intake at the ambient conditions. As propane seems a significant promoter for the ammonia combustion, it has been tried to enhance the performance of the

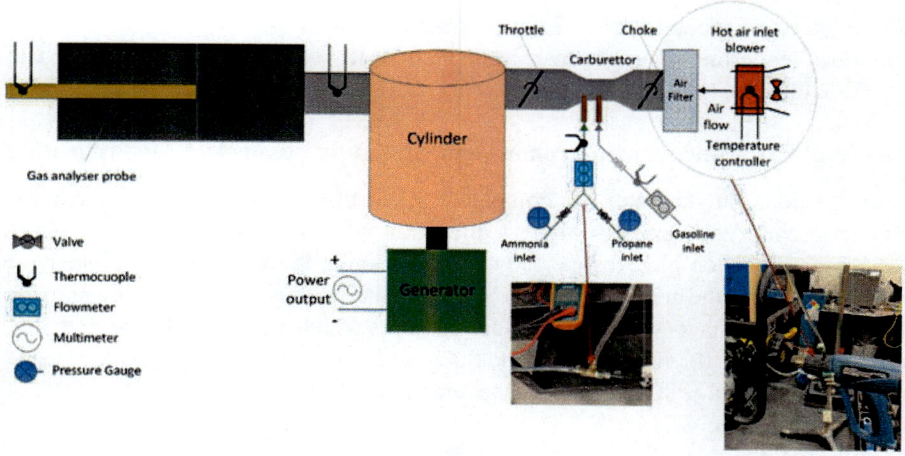

Fig. 4.48 The schematic view of the experimental setup

propane/ammonia blend fueled generator by integrating the air preheating. The same experimental setup which is presented in the previous section has been used. The schematic of the experimental setup is demonstrated in Fig. 4.48. In the present term, we have changed the air inlet temperature via a temperature controller on the hot air blower. A 180, 210, and 240 °C of temperatures have been studied experimentally. In the present term, we have included a thermal camera in the experiments in order to investigate the temperature distribution in generator surfaces.

In this section, we have worked on the ammonia/propane blends with a higher air inlet temperature. In this regard, we have capability to perform the experiments for the cases with 100% propane (P100), 80% propane and 20% ammonia (P80-A20), 60% propane and 40% ammonia (P60-A40), 50% propane and 50% ammonia (P50-A50), 40% propane and 60% ammonia (P40-A60) and 20% propane and 80% ammonia (P80-A20).

In this term, we haven't used gasoline preheating since propane can easily combust according to the ammonia. Engine preheating has been performed with propane. The following procedures have been followed to perform the recent experiments accordingly:

1. The engine has been warmed up for 10 min with propane.
2. The gas analyzer has been reset.
3. The flowrate and pressure of the ammonia have been set.
4. The pressure of the propane has been set.
5. The temperatures of blower have been set.
6. Blower has been operated.
7. A quick check has been done for system stability and safety.
8. The experimental run has commenced, and the data have been collected for analysis and assessment.

9. Each experimental run has been conducted for 45 min.
10. Each experiment has been repeated at least two times for checking data reliability.

Results of Experiments with Propane and Ammonia Blends and Air Preheating

In the section, the generator has completed the test period up to 50% propane and 50% ammonia fuel blends. A higher concentration of the ammonia in the mixture has caused the incomplete test period which is 45 min. In the present term, we have aimed to enhance the generator performance and increase the ammonia content in the fuel mixture. We have work with the fuel mixtures which are 50% propane and 50% ammonia (P50-A50), 40% propane and 60% ammonia (P40-A60) and 20% propane and 80% ammonia (P80-A20). We have tested the effect of 180, 210 and 240 °C of air inlet temperatures.

First, the operating time of the generator is given in Table 4.35. As seen from Table 4.35, the generator has completed 45 min test period up to 50% propane and 50% ammonia mixture. The engine has not completed the 45 min test period after 60% ammonia content in the fuel mixture. It is clear from Table 4.35 that a higher air inlet temperature has enhanced the operating period of the generator. There is a significant improvement in the operating period with 210 and 240 °C. It has been observed from operating periods that propane is a significant promoter for ammonia combustion with the air heater.

Figure 4.49 and Table 4.36 show the experimental data for the cases powered by 50% propane–50% ammonia and 180 °C air preheating. It is clear from Fig. 4.49 that the temperature at the cylinder outlet is approximately 880 °C. The environmental temperature is around 4 °C. The temperature at the gas analyzer probe has reached up to 80 °C. The generator has provided the stable power output. The generator has produced 19.2 A electricity current. So, the power provided by generator is 2291 kW which corresponds the 0.64 of PR. On other words, the generator has reached about 65% of its ideal capacity with gasoline.

On the other hand, propane has reduced the emission values according to the gasoline-powered one. Table 4.36 gives the emission values measured for the case

Table 4.35 Operating periods of the cases studied

Case	Test 1	Test 2	Test 3	Test 4
50% Propane–50% Ammonia and 180 °C air inlet	45	45	45	45
50% Propane–50% Ammonia and 210 °C air inlet	45	45	45	45
50% Propane–50% Ammonia and 240 °C air inlet	45	45	45	45
40% Propane–60% Ammonia and 180 °C air inlet	39	35	37	40
40% Propane–60% Ammonia and 210 °C air inlet	43	45	44	43
40% Propane–60% Ammonia and 240 °C air inlet	45	45	45	45
20% Propane–80% Ammonia and 180 °C air inlet	30	28	35	33
20% Propane–80% Ammonia and 210 °C air inlet	42	43	43	44
20% Propane–80% Ammonia and 240 °C air inlet	45	45	45	45

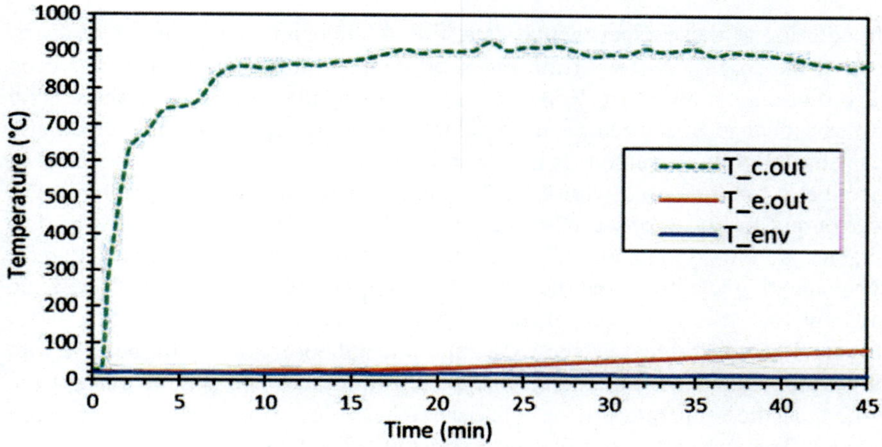

(a)

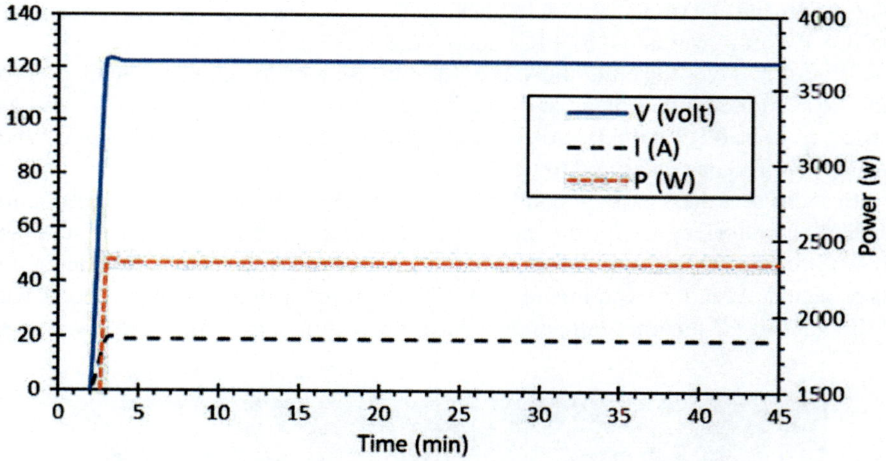

(b)

Fig. 4.49 Experimental data for 50% propane and 50% ammonia blend-fueled generator with 180 °C air-preheating. **a** Temperature changes and **b** power values

Table 4.36 The average emission values for 50% propane and 50% ammonia fueled generator

CO_2 (vol.%)	CO (vol.%)	HC (ppm)	O_2 (vol.%)	NO_2 (ppm)	Lambda	AFR
6.46	8.37	251	0.7	434	0.81	14

with 50% propane and 50% ammonia fuel blends and 180 °C air inlet temperature. The CO_2 and CO emissions have been observed to be 6.43 and 8.42 vol.%, respectively. NO_2 values are 417 ppm. Lambda and AFR values are measured to be 0.83 and 13, respectively.

In addition to those experimental data, Fig. 4.50 depicts the temperature distribution of the engine. As seen from Fig. 4.50, the maximum temperature has been observed at the cylinder exit. With the increasing air inlet temperature, there is no significant change in the temperature values shown in Fig. 4.49a. The electricity current has slightly increased. It has been observed to be 19.2, 19.7 and 20.1 A, respectively, for the cases with 180, 210 and 240 °C air inlet temperature. The power output has been observed to be 2342, 2403 and 2452 W, respectively. The PR values are 0.64, 0.65 and 0.67, respectively. As seen from those values, 240 °C of air preheating has improved the power output approximately by 3%. There is no significant change in emission values observed with gas analyzer (Fig. 4.51).

Figure 4.54 and Table 4.37 show the experimental observations for the case with 40% propane and 60% ammonia fuel blend and 180 °C air preheating. It is clear from Fig. 4.54 that the temperature at the cylinder exit has reduced due to lower combustion efficiency. The engine has run for approximately 37 min and then stopped. Before engine has stopped, the temperature at the cylinder exit has reduced. As seen in Fig. 4.54b, the electricity current has been measured to be 17.2 A which corresponds a 2098 kW of power and 0.57 PR value.

Table 4.37 gives the emission values for the case with the case with 40% propane and 60% ammonia fuel blend and 180 °C air preheating. CO_2 and CO values have observed to be 4.95 and 6.02 vol.%, respectively. NO_2 emission is 678 ppm. While AFR is 14, HC has measured to be 178 ppm.

In addition to these experimental data, Fig. 4.52 shows the temperature distribution of the generator engine for the case with 50% propane–50% ammonia blend and 180 °C air inlet temperature. The maximum temperature is seen in the cylinder outlet, as expected. After the experiments with 180 °C air preheating, the experiments with 210 and 240 °C air inlet temperature have been conducted. Although there is no

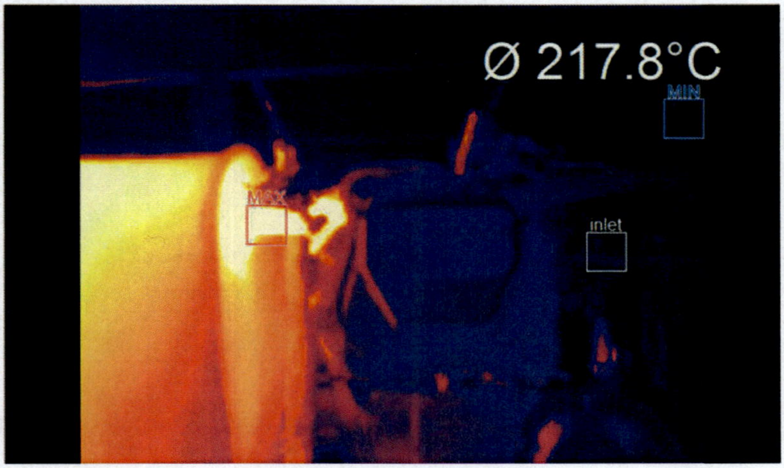

Fig. 4.50 The temperature distribution of the generator engine for the case with 50% propane–50% ammonia blend and 180 °C air inlet temperature

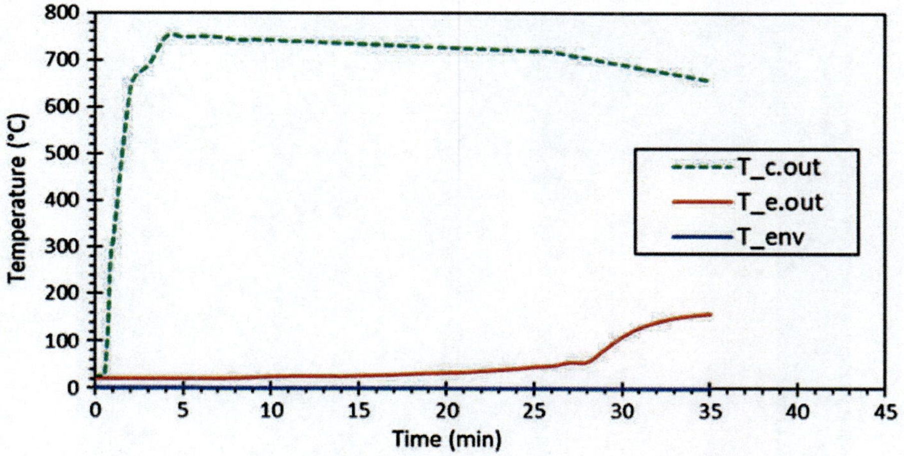

(a)

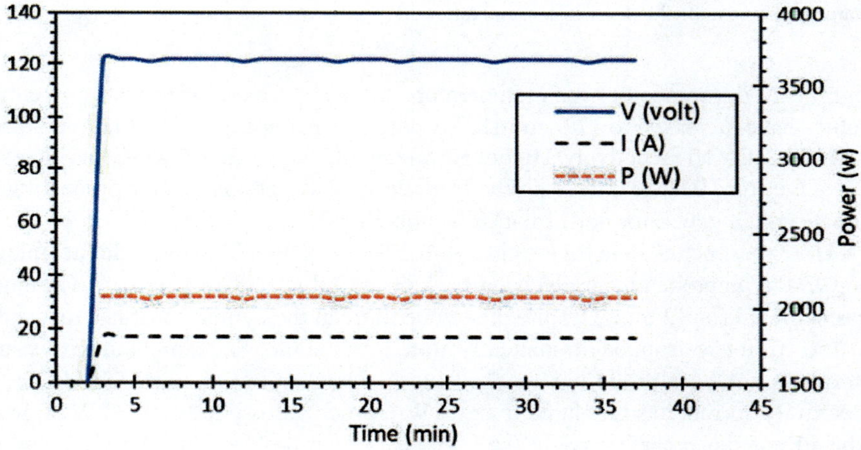

(b)

Fig. 4.51 Experimental data for 40% propane and 60% ammonia blend-fueled generator with 180 °C air-preheating. **a** Temperature changes and **b** power values

Table 4.37 The average emission values for 40% propane and 60% ammonia fueled generator

CO_2 (vol.%)	CO (vol.%)	HC (ppm)	O_2 (vol.%)	NO_2 (ppm)	Lambda	AFR
4.95	6.02	178	0.8	678	0.82	12

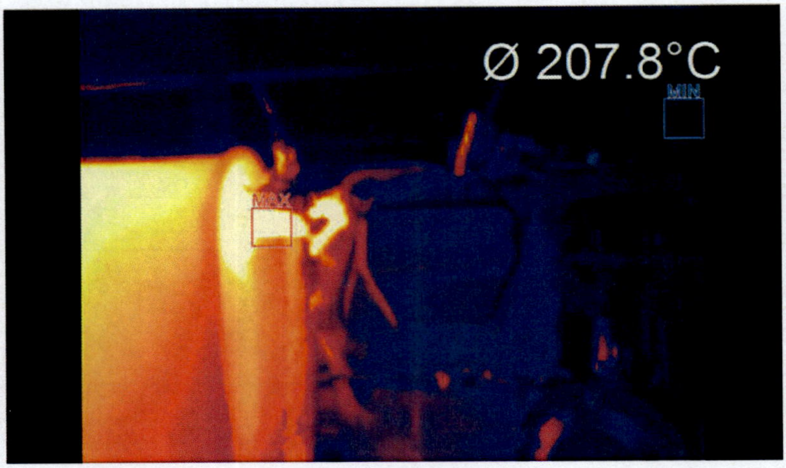

Fig. 4.52 The temperature distribution of the generator engine for the case with 40% propane–60% ammonia blend and 180 °C air inlet temperature

significant difference between temperature values measured, the electricity current values have increased to 17.9 and 18.7 A with the increasing air inlet temperature to 210 and 240 °C, respectively. Higher air inlet temperature up to 240 °C has increased PR value from 0.57 to 0.62. As seen from these observations, the air-preheating has enhanced the generator performance by almost 5%.

The experimental data for the case with 20% propane–80% ammonia fuel mixture and 180 °C air preheating is given in Fig. 4.53 and Table 4.38, respectively. The engine has operated for 32 min. Despite the temperature at the cylinder exit has reached up to 750 °C, it has dropped dramatically after 26th minute. The temperature measured from gas analyzer probe has reached up to 60 °C. As seen from Fig. 4.53b, the electricity current has measured to be 15.9 A which corresponds 1939 kW of power. The PR of the present case is 0.53. As seen from Table 4.38, the CO_2 and CO emissions are measured to be 3.36 and 3.98 vol.%, respectively. HC and NO_2 are measured to be 101 and 879 ppm, respectively. AFR is 12 and Lambda is 0.81.

Figure 4.54 demonstrates the temperature distribution of the generator engine for the case with 20% propane–80% ammonia blend and 180 °C air inlet temperature. The maximum temperature has seen in the cylinder exit like in previous cases. Additionally, the temperature trends have not changed significantly with the increasing air inlet temperature. On the other hand, the electricity current has measured to be 16.7 and 17.1 A for the cases with 210 and 240 °C air inlet temperatures. These current values correspond the 2037 and 2086 W, respectively. Thus, PR values are calculated to be 0.56 and 0.59. Consequently, as seen from these values, air preheating has improved the generator performance by 6%.

Figure 4.55 depicts the comparisons of the studied cases in the terms of power output and PR. As seen from Fig. 4.38, the 100% propane has reached 81.1% of the nominal power output. With the increasing ammonia content in the fuel blend, it has reduced. On the other hand, the generator performance has increased with the increasing air inlet temperature.

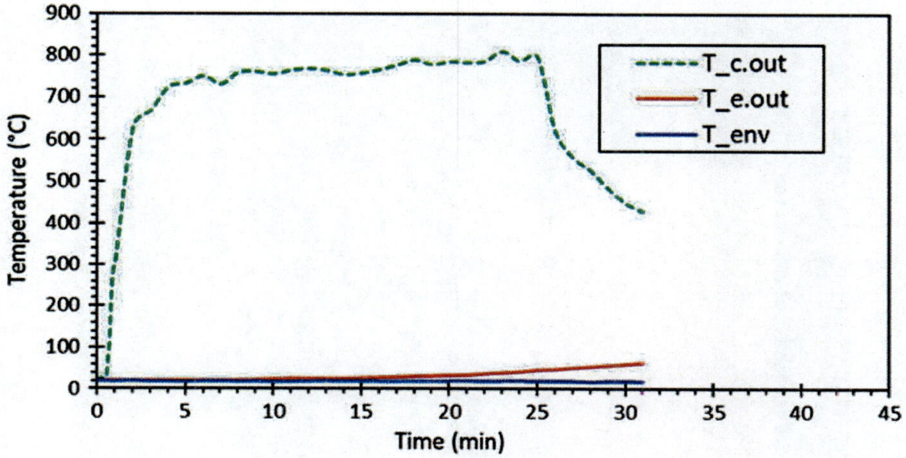

(a)

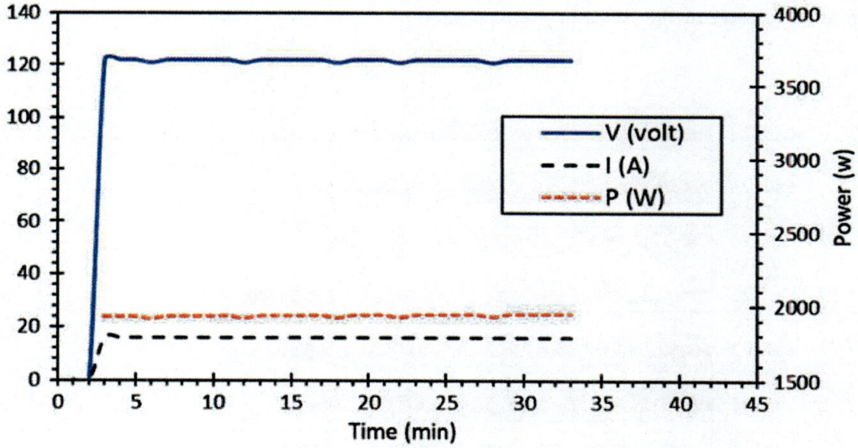

(b)

Fig. 4.53 Experimental data for 20% propane and 80% ammonia blend-fueled generator with 180 °C air-preheating. **a** Temperature changes and **b** power values

Table 4.38 The average emission values for 20% propane and 80% ammonia fueled generator

CO_2 (vol.%)	CO (vol.%)	HC (ppm)	O_2 (vol.%)	NO_2 (ppm)	Lambda	AFR
3.36	3.98	101	0.8	879	0.81	12

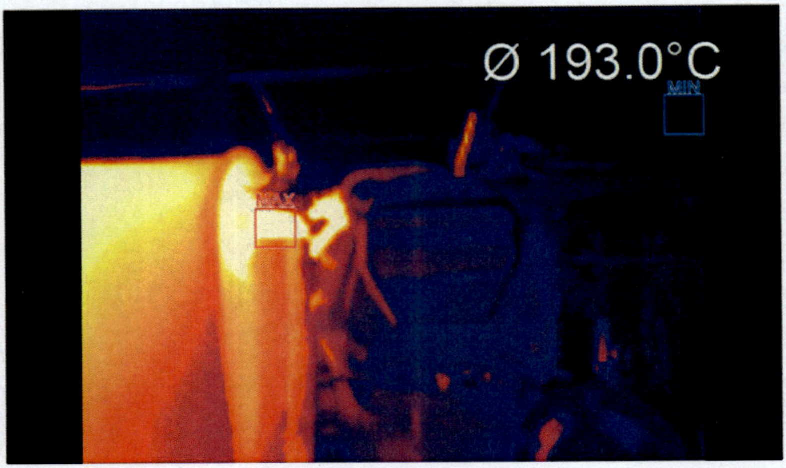

Fig. 4.54 The temperature distribution of the generator engine for the case with 20% propane–80% ammonia blend and 180 °C air inlet temperature

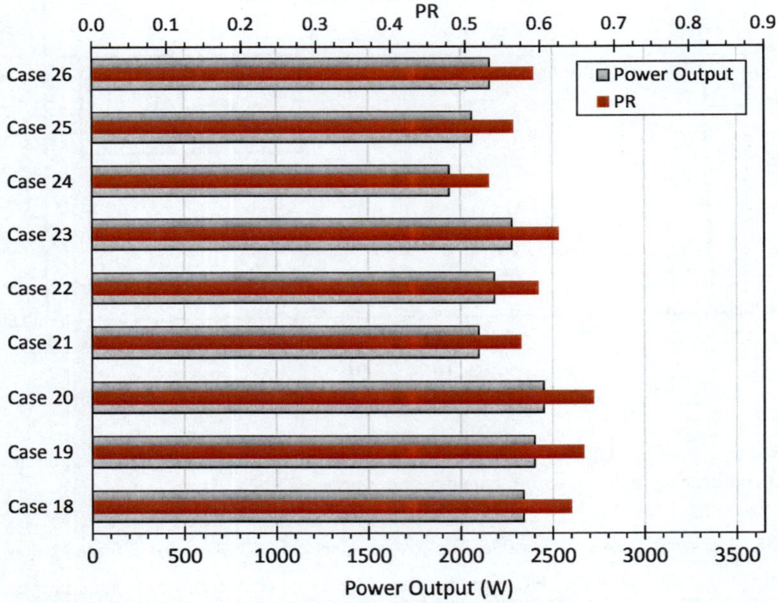

Case 18: 50% propane-50% ammonia and 180°C air inlet
Case 19: 50% propane-50% ammonia and 210°C air inlet
Case 20: 50% propane-50% ammonia and 240°C air inlet
Case 21: 40% propane-60% ammonia and 180°C air inlet
Case 22: 40% propane-60% ammonia and 210°C air inlet
Case 23: 40% propane-60% ammonia and 240°C air inlet
Case 24: 20% propane-80% ammonia and 180°C air inlet
Case 25: 20% propane-80% ammonia and 210°C air inlet
Case 26: 20% propane-80% ammonia and 240°C air inlet

Fig. 4.55 The comparison of the studied cases in terms of the power output and PR

4.9 Experimental Works with Ethanol and Ammonia Blends and Air Preheating

The effect of ethanol as a promoter for ammonia combustion has been researched experimentally. For this purpose, the gasoline has emptied from the generator trunk and a 10 L ethanol has been added to the trunk. The ethanol and ammonia flow has been provided as the same as gasoline and ammonia mixture tests. Figure 4.56 demonstrates the schematic of the experimental setup. The temperature controller on the hot air blower has been set to be 200 °C. Different ammonia and ethanol ratios have been studied experimentally. In the present term, we have included a thermal camera in the experiments in order to investigate the temperature distribution in generator surfaces.

In this section, we have worked on the ethanol and ammonia blends with a 200 °C air inlet temperature. In this regard, we have capability to perform the experiments for the cases with 25% ammonia and 75% ethanol (A25-E75), 50% ammonia and 50% ethanol (A50-E50), 75% ammonia and 25% ethanol (A75-E25).

In these experiments, preheating with gasoline has not been performed since propane can combust easily comparing to ammonia. Engine preheating has been performed with propane. The following procedures have been followed to perform the recent experiments accordingly:

1. The engine has been warmed up for 10 min with gasoline.
2. Gasoline has been emptied from the trunk.
3. Ethanol has been filled up.
4. The gas analyzer probe has been attached.

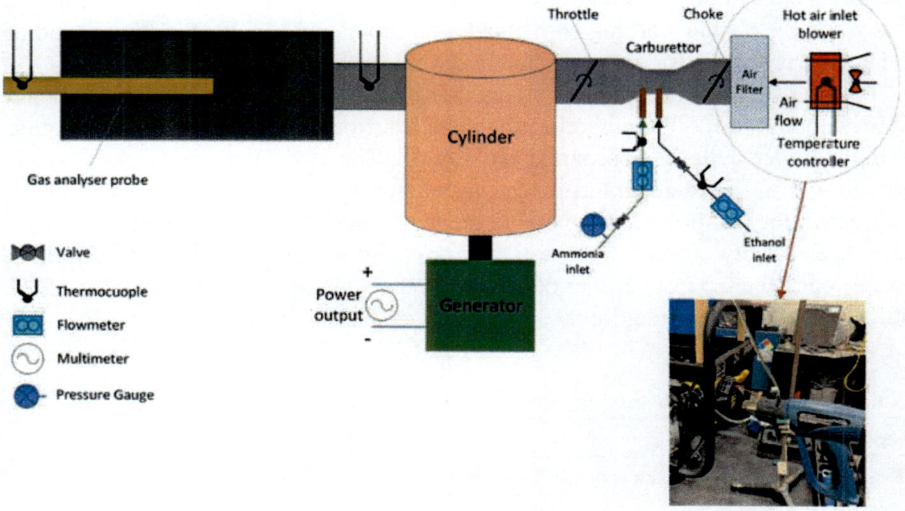

Fig. 4.56 The schematic view of the experimental setup

5. The flowrate and pressure of the ammonia have been set.
6. The flowrate of the ethanol has been set.
7. The temperatures of blower have been set.
8. Blower has been operated.
9. A quick check has been done for system stability and safety.
10. The experimental run has commenced, and the data have been collected for analysis and assessment.
11. Each experimental run has been conducted for 45 min.
12. Each experiment has been repeated at least two times for checking data reliability.

Results of Experiments with Ethanol and Ammonia Blends

All experimental data have been illustrated in three divisions, which are temperature and electricity versus time, and the average emission values like in the previous sections. In the present section, we have worked on the ammonia and ethanol mixture. Table 4.39 shows the operating periods of the generator which are studied the current and previous terms. At the first glance, it is observed that ethanol mixture has caused instability in the generator run. In all previous cases, the operating period of the engine is almost the same even if there is a little fluctuation. However, as seen from the ethanol mixture results, the generator has not completed the 45 min test period. For the case with 75% ethanol and 25% ammonia, the generator has run the periods which are from 28 to 41 min. There is 13-min difference. Similar results have been seen in the other ethanol-ammonia mixtures, too. On the other hand, the time difference on the operating period has reduced with the decreasing the ethanol content in the fuel mixture. While it is 10 min for the cases which are A50-E50 and A75-E25. When the operating periods are taken into consideration, we can say that ethanol is not a good promoter for ammonia combustion. Also, here, we should note that while the engine working with ethanol-ammonia mixture the working sound of the engine is very noisy. It is completely different from previous configurations.

Table 4.40 show the experimental data for the cases powered by 75% ethanol and 25% ammonia and 200 °C air preheating. It is clear from Fig. 4.57 that the temperature at the cylinder outlet is approximately 490 °C. The environmental temperature is around − 8 °C. The temperature at the gas analyzer probe has reached up to 70 °C. The generator has provided the stable power output. The generator has produced 15.8 A electricity current. So, the power provided by generator is 1927 W which corresponds the 0.52 of PR. On other words, the generator has reached about 52% of its ideal capacity with gasoline.

Table 4.39 Operating periods of the cases studied

Case	Test 1	Test 2	Test 3	Test 4
75% Ethanol–25% Ammonia and 200 °C air inlet	37	28	41	37
50% Ethanol–50% Ammonia and 200 °C air inlet	42	38	36	32
25% Ethanol–75% Ammonia and 200 °C air inlet	30	41	36	38

Table 4.40 The average emission values for 75% ethanol and 25% ammonia fueled generator

CO_2 (vol.%)	CO (vol.%)	HC (ppm)	O_2 (vol.%)	NO_2 (ppm)	Lambda	AFR
8.46	9.37	276	1.1	245	0.84	12

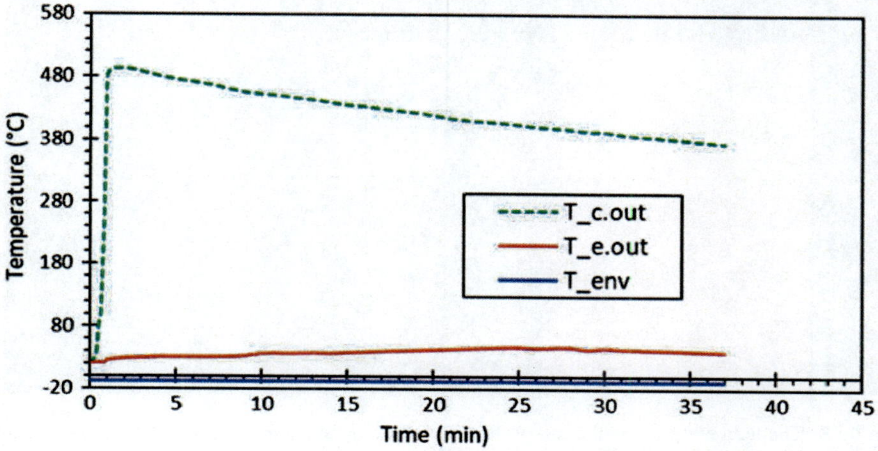

(a)

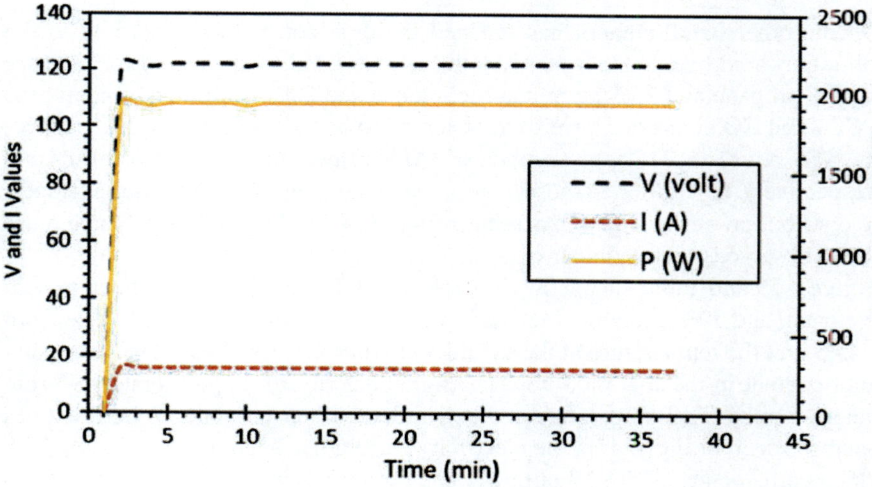

(b)

Fig. 4.57 Experimental data for 75% ethanol and 25% ammonia mixture-fueled generator with 200 °C air-preheating. **a** Temperature changes and **b** power values

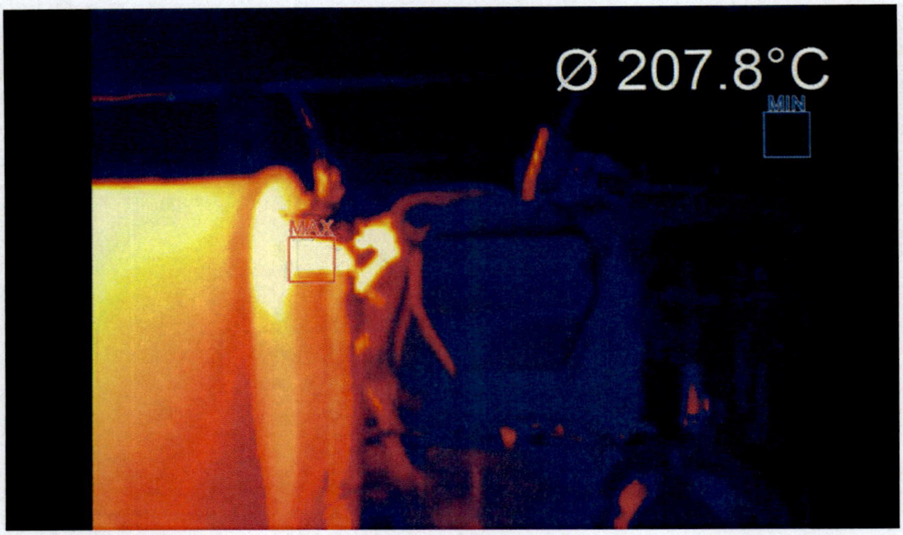

Fig. 4.58 The temperature distribution of the generator engine for the case with 75% ethanol and 25% ammonia blend and 180 °C air inlet temperature

On the other hand, ethanol has reduced the emission values according to the gasoline-powered one. Table 4.40 gives the emission values measured for the case with 75% propane and 25% ammonia fuel blends and 200 °C air inlet temperature. The CO_2 and CO emissions have been observed to be 8.46 and 9.37 vol.%, respectively. NO_2 values are 245 ppm. Lambda and AFR values are measured to be 0.84 and 12, respectively. In addition to those experimental data, Fig. 4.58 depicts the temperature distribution of the engine. As seen from Fig. 4.58, the maximum temperature has been observed at the cylinder exit.

Figure 4.59 and Table 4.41 show the experimental observations for the case with 50% ethanol and 50% ammonia fuel blend and 200 °C air preheating. It is clear from Fig. 4.45 that the temperature at the cylinder exit has increased with the decreasing ethanol content in the fuel mixture. The engine has run for approximately 35 min and then stopped. The temperature at the cylinder exit has not reduced before it has stopped. As seen in Fig. 4.45b, the electricity current has been measured to be 14.5 A which corresponds a 1769 W of power and 0.48 PR value.

Table 4.41 gives the emission values for the case with the case with 50% ethanol and 50% ammonia fuel blend and 200 °C air preheating. CO_2 and CO values have observed to be 6.95 and 7.02 vol.%, respectively. NO_2 emission is 478 ppm. While AFR is 12, HC has measured to be 169 ppm. In addition to these experimental data, Fig. 4.60 shows the temperature distribution of the generator engine for the case with 50% propane–50% ammonia blend and 180 °C air inlet temperature. The maximum temperature is seen in the cylinder outlet, as expected.

The experimental data for the case with 25% ethanol–75% ammonia fuel mixture and 200 °C air preheating is given in Fig. 4.61 and Table 4.42, respectively. The

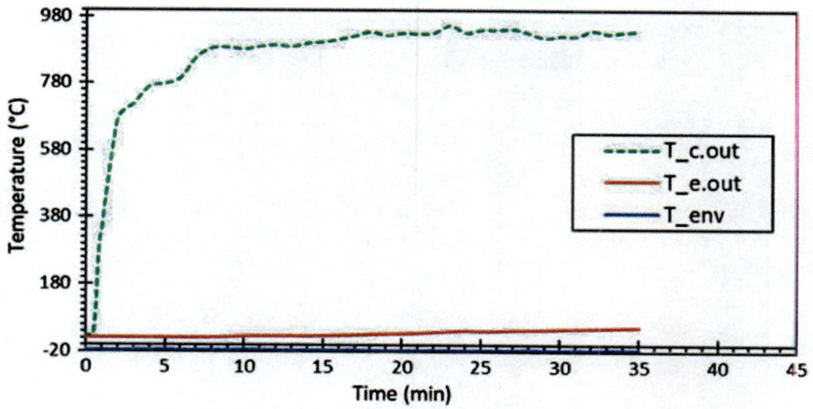

(a)

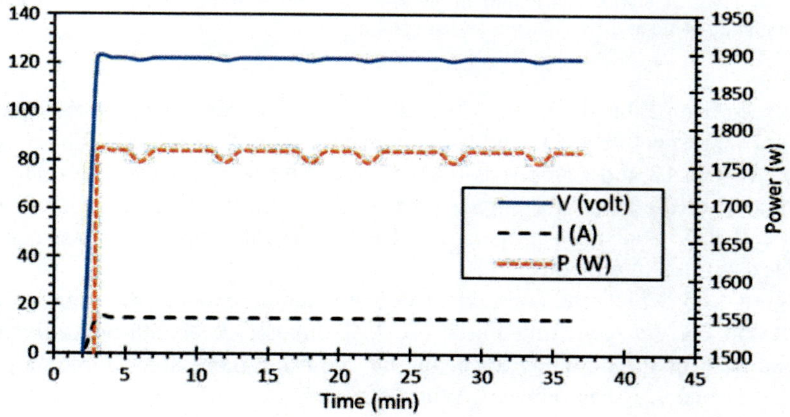

(b)

Fig. 4.59 Experimental data for 50% ethanol and 50% ammonia blend-fueled generator with 200 °C air-preheating. **a** Temperature changes and **b** power values

Table 4.41 The average emission values for 50% ethanol and 50% ammonia fueled generator

CO_2 (vol.%)	CO (vol.%)	HC (ppm)	O_2 (vol.%)	NO_2 (ppm)	Lambda	AFR
6.95	7.02	169	1.5	478	0.82	12

engine has operated for 33 min. The temperature at the cylinder exit has reached up to almost 500 °C, it has started to drop just after working. The temperature measured from gas analyzer probe has reached up to 60 °C. As seen from Fig. 4.61b, the electricity current has measured to be 12.6 A which corresponds 1537 W of power. The PR of the present case is 0.42.

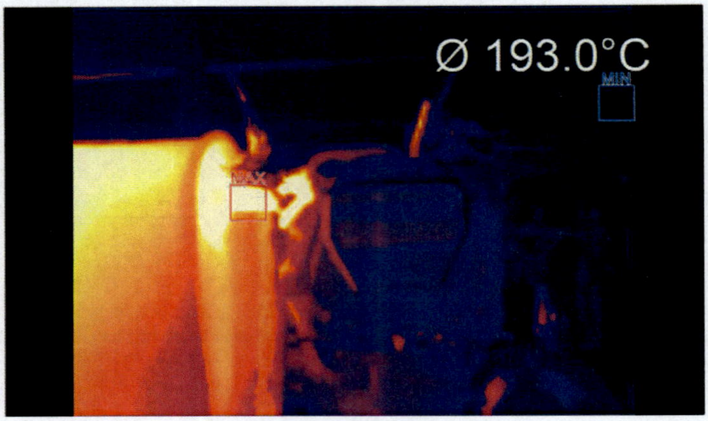

Fig. 4.60 The temperature distribution of the generator engine for the case with 50% ethanol and 50% ammonia blend and 180 °C air inlet temperature

As seen from Table 4.42, the CO_2 and CO emissions are measured to be 3.56 and 4.04 vol.%, respectively. HC and NO_2 are measured to be 95 and 729 ppm, respectively. AFR is 12 and Lambda is 0.81. Figure 4.62 demonstrates the temperature distribution of the generator engine for the case with 25% ethanol–75% ammonia blend and 200 °C air inlet temperature. The maximum temperature has seen in the cylinder exit like in previous cases.

Figure 4.63 depicts the comparisons of the studied cases in the terms of power output and PR. As seen from Fig. 4.63, 75% ethanol–25% ammonia mixture has reached 52% of the nominal power output. With the increasing ammonia content in the fuel blend, it has reduced. While 50% ethanol–50% ammonia mixture has provided 48% of the nominal power, 25% ethanol–75% ammonia mixture has provided 42% of the nominal power of the generator.

Consequently, when operating period and power output of the generator are compared with the previous cases, ethanol is not a convenient promoter for ammonia combustion. Ethanol/ammonia mixture has caused a loudly working for generator, too.

4.10 Experimental Works with Methanol and Ammonia Blends and Air Preheating

The effect of the methanol mixture with ammonia has been investigated experimentally. The gasoline and ethanol have emptied from the generator trunk and a 10 L methanol has been added to the trunk. The methanol and ammonia flow has been provided as the same as gasoline/ammonia blend tests. Figure 4.64 demonstrates the schematic of the experimental setup. The temperature controller on the hot air

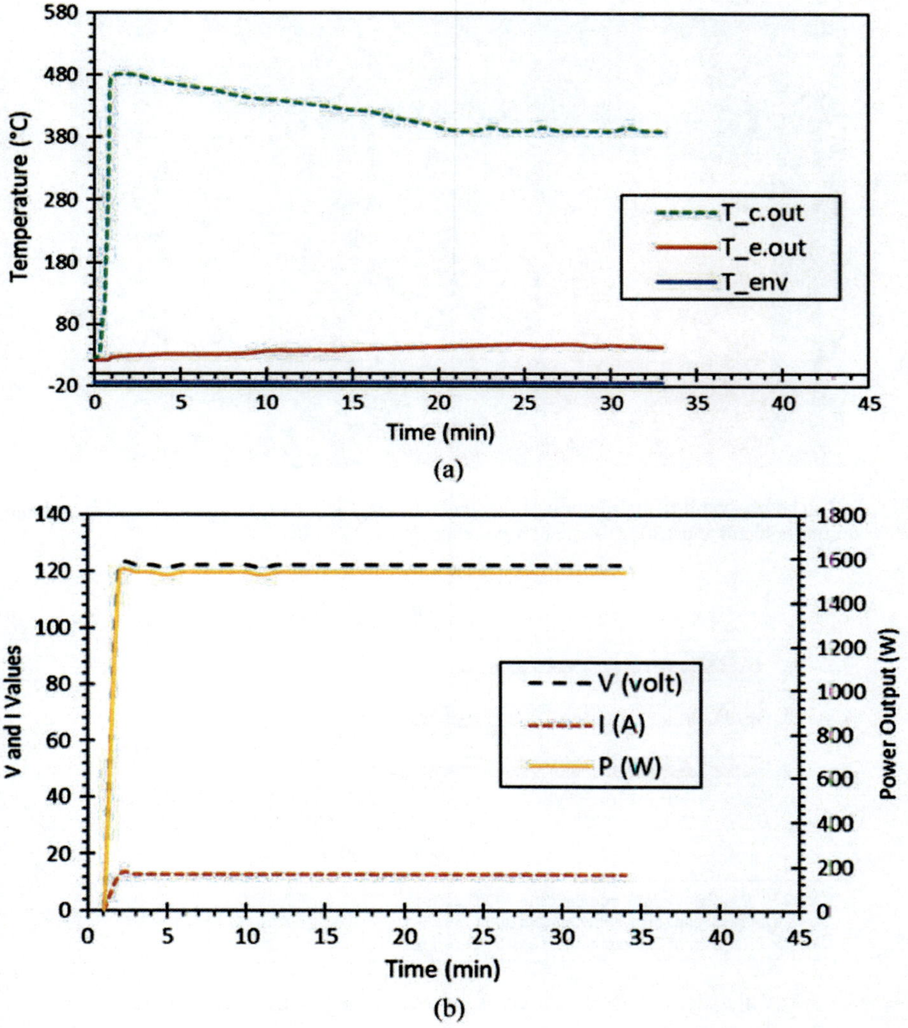

Fig. 4.61 Experimental data for 25% ethanol and 75% ammonia blend-fueled generator with 200 °C air-preheating. **a** Temperature changes and **b** power values

Table 4.42 The average emission values for 25% ethanol and 75% ammonia fueled generator

CO_2 (vol.%)	CO (vol.%)	HC (ppm)	O_2 (vol.%)	NO_2 (ppm)	Lambda	AFR
3.56	4.04	95	0.9	729	0.81	12

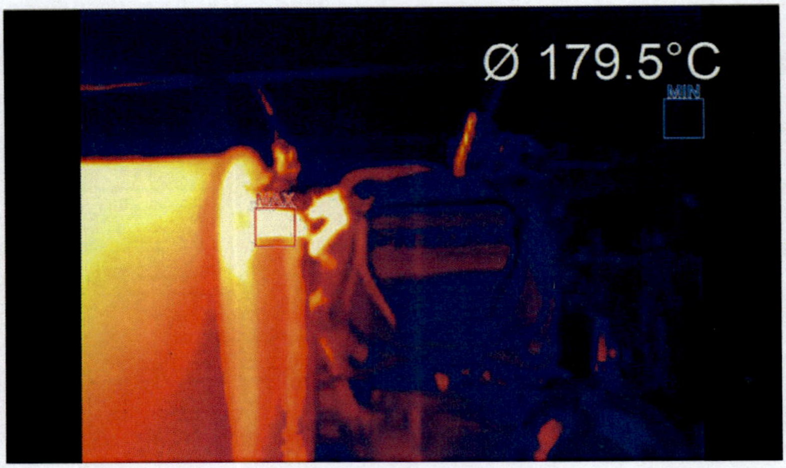

Fig. 4.62 The temperature distribution of the generator engine for the case with 25% ethanol and 75% ammonia blend and 180 °C air inlet temperature

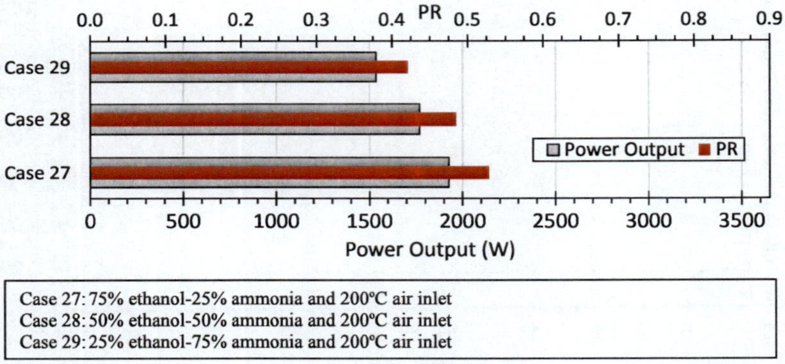

Fig. 4.63 The comparison of the studied cases in terms of the power output and PR

blower has been set to be 200 °C. Different ammonia and methanol ratios have been studied experimentally. In the present term, we have not used the thermal camera to investigate the temperature distribution of the generator engine.

In this term, we have worked on the ammonia/methanol blends with a 200 °C air inlet temperature. In this regard, we have capability to perform the experiments for the cases with 25% ammonia and 75% methanol (A25-M75) and 50% ammonia and 50% methanol (A50-E50).

The following procedures have been followed to perform the recent experiments accordingly:

1. The engine has been warmed up for 10 min with gasoline.
2. Gasoline has been emptied from the trunk.

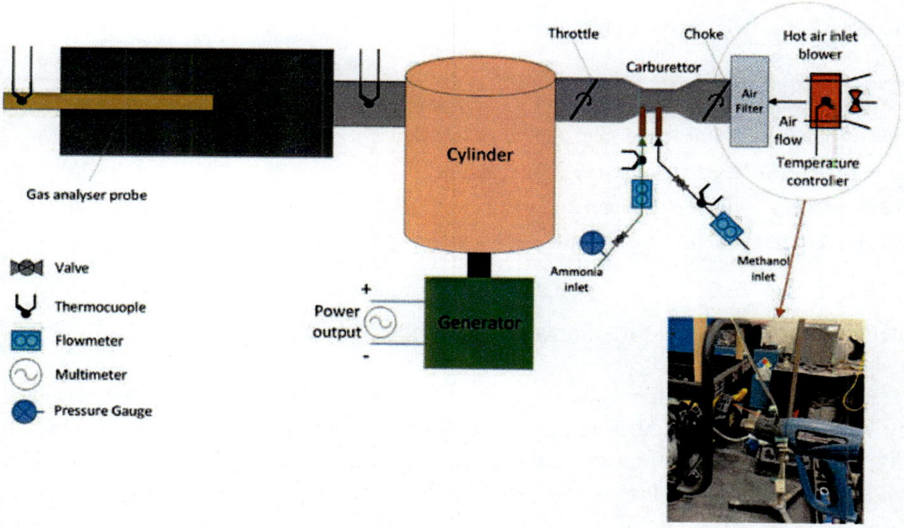

Fig. 4.64 The schematic view of the experimental setup

3. Methanol has been filled up.
4. The gas analyzer probe has been attached and gas analyzer has been reset.
5. The flowrate and pressure of the ammonia have been set.
6. The flowrate of the ethanol has been set.
7. The temperatures of blower have been set.
8. Blower has been operated.
9. A quick check has been done for system stability and safety.
10. The experimental run has commenced, and the data have been collected for analysis and assessment.
11. Each experimental run has been conducted for 45 min.
12. Each experiment has been repeated at least two times for checking data reliability.

Results of Experiments with Methanol and Ammonia Blends

Table 4.43 shows the operating periods of the generator which are studied the current and previous terms. As seen from Table 4.43, methanol/ammonia mixture has caused instability in the engine run like in the cases with ethanol/ammonia mixture. While the operating period of the engine is almost the same even if there is a little fluctuation, methanol mixture has caused unstable engine working conditions. Generator has not completed the 45 min test period. For the case with 75% methanol and 25% ammonia, the generator has run the periods which are from 25 to 32 min. It has varied from 26 to 33 min for the case with 50% methanol and 50% ammonia. Like in the cases with ethanol/ammonia mixture, we should note that while the engine working with

Table 4.43 Operating periods of the cases studied

Case	Test 1	Test 2	Test 3	Test 4
75% Methanol–25% Ammonia and 200 °C air inlet	25	32	23	30
50% Methanol–50% Ammonia and 200 °C air inlet	27	33	26	29
25% Methanol–75% Ammonia and 200 °C air inlet	30	21	23	26
100% Methanol with the modified carburetor	37	45	40	45
100% Methanol with the original carburetor	–	–	–	–

methanol-ammonia mixture the working sound of the engine is very noisy. There is a significant vibration.

Figure 4.65 and Table 4.44 show the experimental data for the cases powered by 75% methanol and 50% ammonia with 200 °C air preheating. It is clear from Fig. 4.65 that the temperature at the cylinder outlet is approximately 750 °C. The environmental temperature is around − 18 °C. The temperature at the gas analyzer probe has reached up to 50 °C. The generator has provided the stable power output. The generator has produced 15.8 A electricity current. So, the power provided by generator is 1927 W which corresponds the 0.53 of PR. On other words, the generator has reached about 53% of its ideal capacity with gasoline. Here, we should note that engine has run with noise and high vibration. High vibration has caused instability control of the throttle. In order to solve this issue, we have placed the spring in the throttle with a high dense one. However, there was no change in the working condition and operating duration of the engine.

On the other hand, methanol has reduced the emission values according to the gasoline-powered one. Table 4.44 gives the emission values measured for the case with 75% ethanol and 25% ammonia fuel blends with 200 °C air inlet temperature. The CO_2 and CO emissions have been observed to be 7.23 and 8.02 vol.%, respectively. NO_2 values are 291 ppm. Lambda and AFR values are measured to be 0.8 and 10, respectively.

Figure 4.66 and Table 4.45 show the experimental observations for the case with 50% methanol and 50% ammonia mixture with 200 °C air preheating. It is clear from Fig. 4.66 that the temperature at the cylinder exit has decreased with the increasing methanol content in the fuel mixture. The temperature at the cylinder exit has reached up to 585 °C. The temperature measured on the gas analyzer probe is around 70 °C. The engine has operated for about 29 min and then stopped. As seen in Fig. 4.66b, the electricity current generated by the generator has been measured to be 14.1 A which corresponds a 1720 W of power and 0.47 of the PR. Table 4.45 gives the emission values for the case with the case with 50% methanol and 50% ammonia fuel mixture with 200 °C air preheating. CO_2 and CO values have observed to be 6.71 and 7.14 vol.%, respectively. NO_2 emission is 481 ppm. While AFR is 10, HC has measured to be 202 ppm. Like in the previous case with 75% methanol and 25% ammonia, the engine has run with a noise and high vibration for both ordinary and dense spring in the throttle.

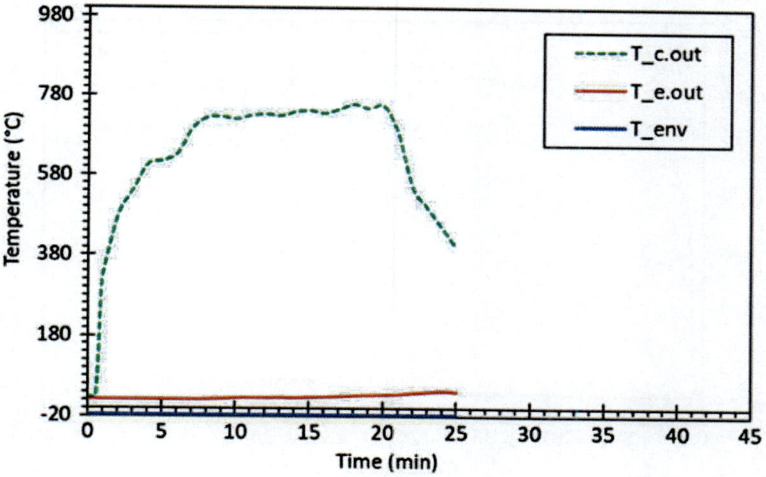

(a)

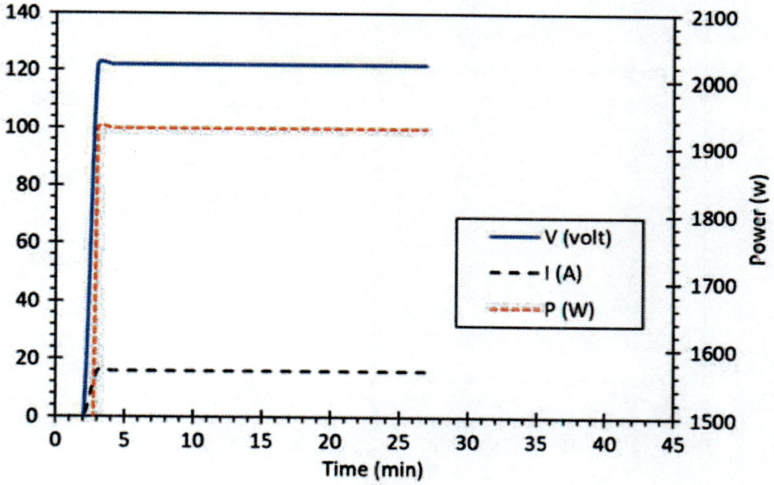

(b)

Fig. 4.65 Experimental data for 75% methanol and 25% ammonia mixture-fueled generator with 200 °C air-preheating. **a** Temperature changes and **b** power values

Table 4.44 The average emission values for 75% ethanol and 25% ammonia fueled generator

CO_2 (vol.%)	CO (vol.%)	HC (ppm)	O_2 (vol.%)	NO_2 (ppm)	Lambda	AFR
7.23	8.02	291	0.8	284	0.52	10

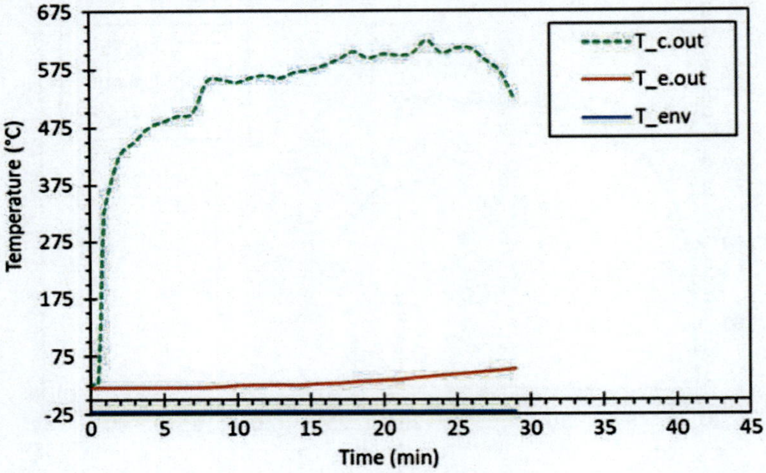

(a)

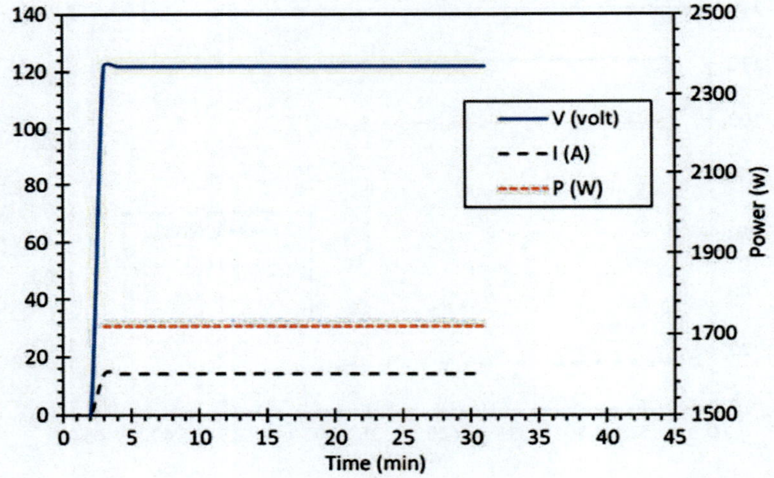

(b)

Fig. 4.66 Experimental data for 50% methanol and 50% ammonia blend-fueled generator with 200 °C air-preheating. **a** Temperature changes and **b** power values

Table 4.45 The average emission values for 50% methanol and 50% ammonia fueled generator

CO_2 (vol.%)	CO (vol.%)	HC (ppm)	O_2 (vol.%)	NO_2 (ppm)	Lambda	AFR
6.71	7.14	202	0.8	481	0.80	10

Figure 4.67 and Table 4.46 show the experimental data for the cases powered by 25% methanol and 75% ammonia with 200 °C air preheating. It is clear from Fig. 4.67 that the temperature at the cylinder outlet is approximately 650 °C. The environmental temperature is around − 13 °C. The temperature at the gas analyzer probe has reached up to 40 °C. The generator has provided the stable power output with a 14.1 A electricity current. The power output of the generator is 1720 W which corresponds the 0.47 of PR value. On other words, the generator has reached about 47% of its ideal capacity with gasoline. Like in the previous term, here, we should note that engine has run with noise and high vibration. High vibration has caused instability control of the throttle. In order to solve this issue, we have placed the spring in the throttle with a high dense one. However, there was no change in the working condition and operating duration of the engine. When the power values of the generator are taken into consideration, the power output of the generator has reduced with the ammonia content in the fuel mixture.

In terms of emission values, methanol has reduced the emission values according to the gasoline-powered one. Table 4.46 gives the emission values measured for the case with 25% methanol and 75% ammonia mixture and 200 °C air inlet temperature. The CO_2 and CO emissions have been observed to be 3.35 and 3.02 vol.%, respectively. NO_2 values are 157 ppm. Lambda and AFR values are measured to be 0.84 and 14, respectively.

Figure 4.68 and Table 4.47 show the experimental observations for the case with 100% methanol. As seen from Fig. 4.68 that the temperature at the cylinder exit has increased due to 100% methanol fuel content. In the steady conditions, it is measured to be approximately 750 °C which is almost 200 °C lower than the case with 100% gasoline. The temperature which is measured from the gas analyser probe has reached 120 °C. Engine has worked with higher vibration comparing to the cases with 100% gasoline. However, the vibration is considerably lower than the cases with ammonia/methanol and ammonia/ethanol mixtures. In a few cases, the engine has stopped in 40 and 42 min. It has happened due to instability in the throttle control. As seen from Fig. 4.68b, a 28.1 A of electric current has been generated which corresponds 3428 W of power output. The generator has reached 0.93 PR value with 100% methanol.

Table 4.47 gives the emission values for the case with the case with 100% methanol. CO_2 and CO values have observed to be 7.73 and 7.97 vol.%, respectively. NO_2 emission is 98 ppm and HC is 301 ppm. While AFR is 13, lambda is 0.88.

Additionally, we have tested 100% methanol with the original carburetor comes with the generator. In this case, engine has not operated. It has run 2–3 min and then stopped. Therefore, there is no experimental data for this case.

Figure 4.69 depicts the comparisons of the studied cases in the terms of power output and PR. As seen from Fig. 4.69, 25% methanol–75% ammonia mixture has reached 47% of the nominal power output. 100% methanol has reached 93% of the nominal power. 75% methanol–25% ammonia mixture has reached 53% of the nominal power output. With the increasing ammonia content in the fuel blend, it has reduced. While 50% methanol–50% ammonia mixture has provided 47% of the

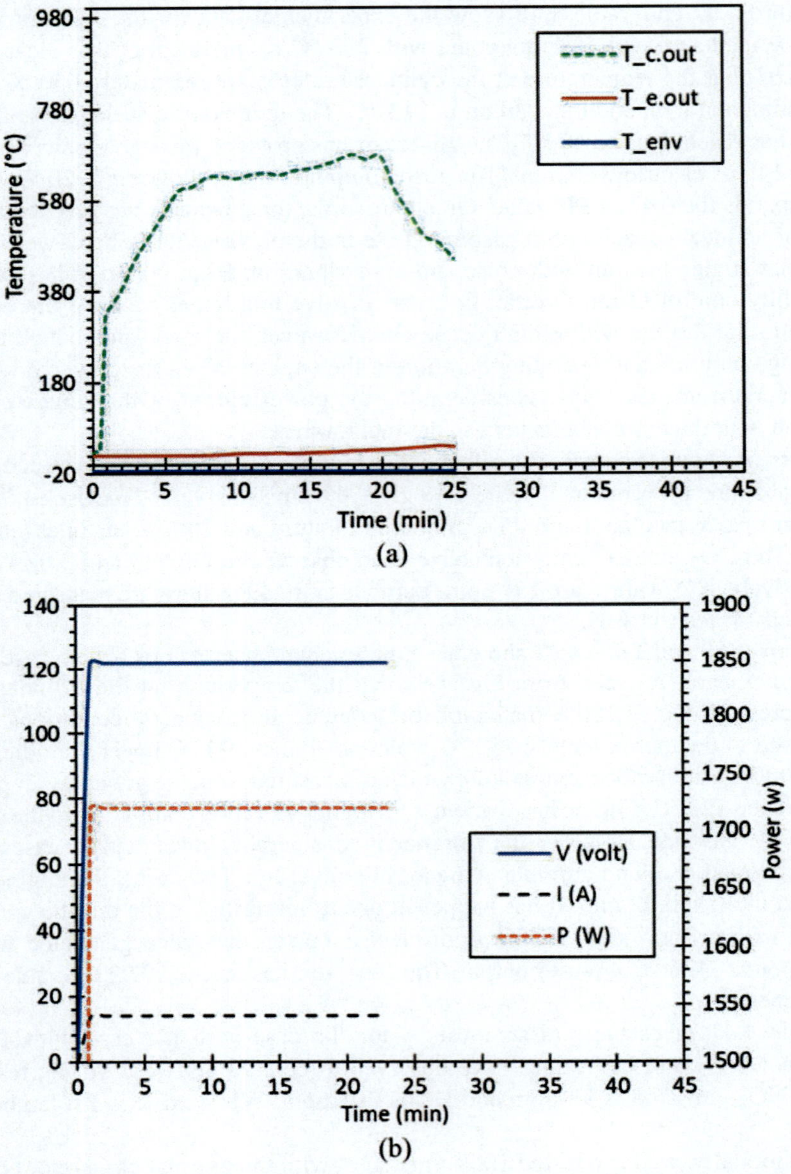

Fig. 4.67 Experimental data for 25% methanol and 75% ammonia mixture-fueled generator with 200 °C air-preheating. **a** Temperature changes and **b** power values

Table 4.46 The average emission values for 25% methanol and 75% ammonia-fueled generator

CO_2 (vol.%)	CO (vol.%)	HC (ppm)	O_2 (vol.%)	NO_2 (ppm)	Lambda	AFR
3.35	3.02	157	0.6	284	0.84	14

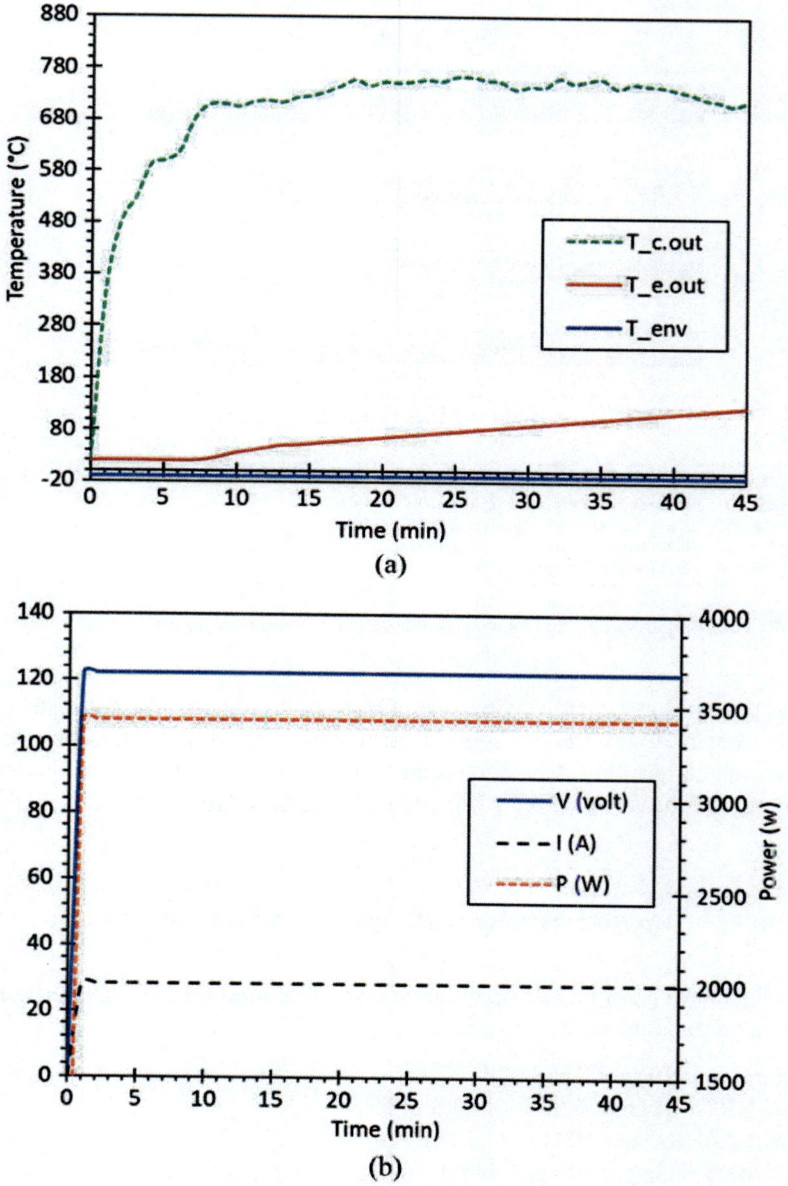

Fig. 4.68 Experimental data for 100% methanol-fueled generator. **a** Temperature changes and **b** power values

Table 4.47 The average emission values for 100% methanol-fueled generator

CO_2 (vol.%)	CO (vol.%)	HC (ppm)	O_2 (vol.%)	NO_2 (ppm)	Lambda	AFR
7.73	7.97	301	0.9	98	0.88	13

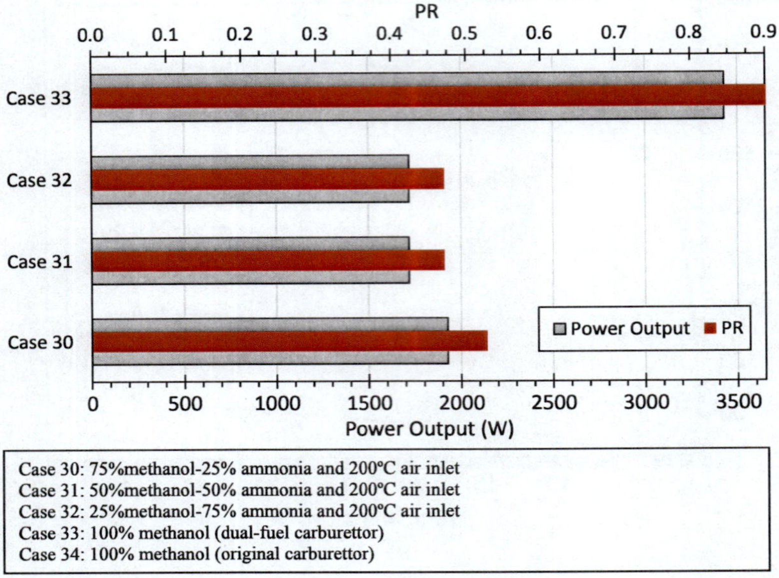

Fig. 4.69 The comparison of the studied cases in terms of the power output and PR

nominal power. Consequently, when operating period and power output of the generator are compared with the previous cases, methanol is not a convenient promoter for ammonia combustion. Methanol/ammonia mixture has caused a loudly working for generator. It has also caused a high vibration in the engine.

4.11 Experimental Works with Different Sparkplugs

Spark plugs have great impact on the ignition of fuel mixture. For this purpose, the following spark plugs have been tested.

- Champion RN9YC Gap: 0.028″ (0.7 mm)
- NGK BR2-LM Gap: 0.025″ (0.63 mm)
- NGK BR6HS Gap: 0.023″ (0.58 mm)
- NGK BR5-672 spark plug. It has a 0.032″ (0.8128 mm) gap
- NGK BR6-672 spark plug. It has a 0.030″ (0.7620 mm) gap.

In order to test different spark plugs, the experimental setup shown in Fig. 4.70 has been used. This experimental configuration is the same as the cases with the gasoline/ammonia blends and air preheater. Pure ammonia has been used as the fuel. The temperature controller on the hot air blower has been set to be 200 °C.

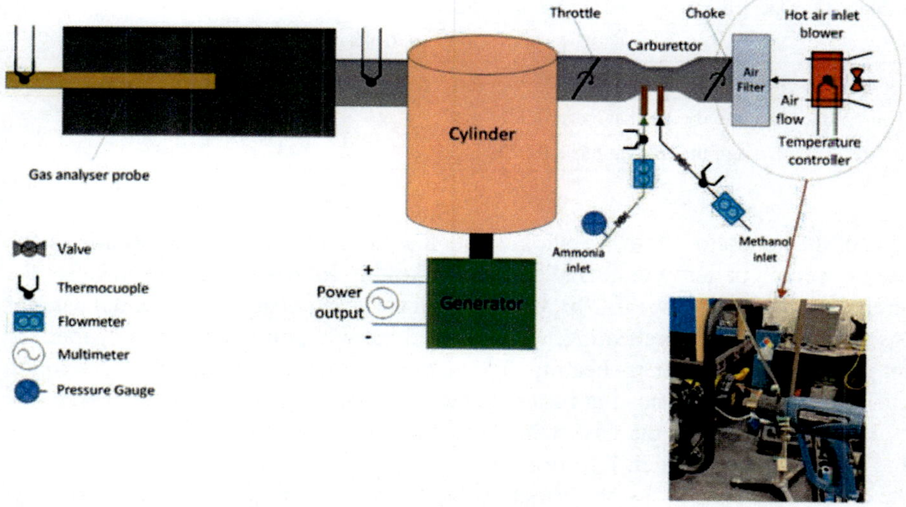

Fig. 4.70 A schematic view of the present experimental setup

The following test procedure has been implemented to perform the recent experiments accordingly:

1. The engine has been warmed up for 10 min with gasoline.
2. The gasoline flow to engine has been cut.
3. The gas analyzer probe has been attached and gas analyzer has been reset.
4. The flowrate and pressure of the ammonia have been set.
5. The temperatures of blower have been set.
6. Blower has been operated.
7. A quick check has been done for system stability and safety.
8. The experimental run has commenced, and the data have been collected for analysis and assessment.
9. Each experimental run has been conducted for 45 min.
10. Each experiment has been repeated at least two times for checking data reliability.

Results of Experiments with Different Sparkplugs

As carried out in the previous sections, all experimental data have been demonstrated in three divisions, which are temperature and electricity versus time, and the average emission values in the table form. In the present term, we have worked on pure ammonia combustion with different spark plug configurations. Table 4.48 shows the operating periods of the generator, which are studied in the current conditions. In this term, first, we have tested pure ammonia with the spark plug coming with the generator. Then, we changed the spark plug to the one that has smaller gaps. As seen in Table 4.48 that, the using smaller gap in the spark plug has extended the operating

Table 4.48 Operating periods of the cases studied

Case	Test 1	Test 2	Test 3	Test 4
100% Ammonia and the SP that has a 0.028″ gap	40	38	38	39
100% Ammonia and the SP that has a 0.025″ gap	39	42	40	41
100% Ammonia and the SP that has a 0.023″ gap	45	43	45	44

time of the generator. Also, we should note that the engine noise and vibration have been observed better with a smaller-gap spark plug. In these cases, we have set the temperature of the blower to 200 °C. Although we have not got experimental data for cases without the air preheating, we measured the operating time for the generator for the case without air preheating. The operating time of the generator is around 20–25 min. So, the preheating has provided better performance for the generator.

First, we have run the case with pure ammonia, 200 °C air inlet temperature and the spark plug which has a 0.028″ gap, which is the standard spark plug for the engine. This case was performed in the previous section however, the air inlet temperature was 180 °C. Figure 4.71 and Table 4.49 show the experimental data for the case with pure ammonia, 200 °C air inlet temperature and the spark plug which has a 0.028″ gap. The engine has run about 40 min. The temperature at the cylinder outlet has reach up to 500 °C and reduced slightly until the engine has stopped. The environmental temperature is around – 8 °C. The temperature at the gas analyzer probe has reached up to 55 °C. The generator has provided the stable power output with a 16.4 A electricity current. The power output of the generator is 2000 W which corresponds the 0.54 of PR value. Table 4.49 gives the emission values measured for the case with 25% methanol and 75% ammonia mixture and 200 °C air inlet temperature. The CO_2 and CO emissions have been observed to be 3.35 and 3.02 vol.%, respectively. NO_2 values are 157 ppm. The lambda and AFR values are measured to be 0.84 and 14, respectively.

Figure 4.72 and Table 4.50 demonstrate the experimental findings for the case with pure ammonia, 200 °C air-preheating and spark plug that has a 0.025″ gap. The temperature at the cylinder exit has increased with lowering gap in the spark plug. It has reached up to almost 520 °C and stayed at this temperature range for a while. Then, it has started to slightly decrease until the engine has stopped. The temperature taken from gas analyzer probe has reached about 95 °C. The environmental temperature is around – 5 °C. The generator has generated a 16.9 of electric current. So, the power output was 2061 W which corresponds 0.56 PR value. As seen from Table 4.50 that CO_2 and CO values have measured to be 1.01 and 0.62 vol.%, respectively. O_2 is 0.7 vol.% HC and NO_2 have observed to be 2.03 and 1224 ppm, respectively. While Lambda is 0.87, AFR are 14.

Figure 4.73 and Table 4.51 depict the experimental observations for the case with pure ammonia, 200 °C air-preheating and spark plug that has a 0.023″ gap. As seen from Fig. 4.73 that, the temperature at the cylinder exit has increased up to 550 °C which almost 50 °C higher than the regular spark plug. The temperature measured from the gas analyzer has increased to almost 100 °C. The ambient temperature

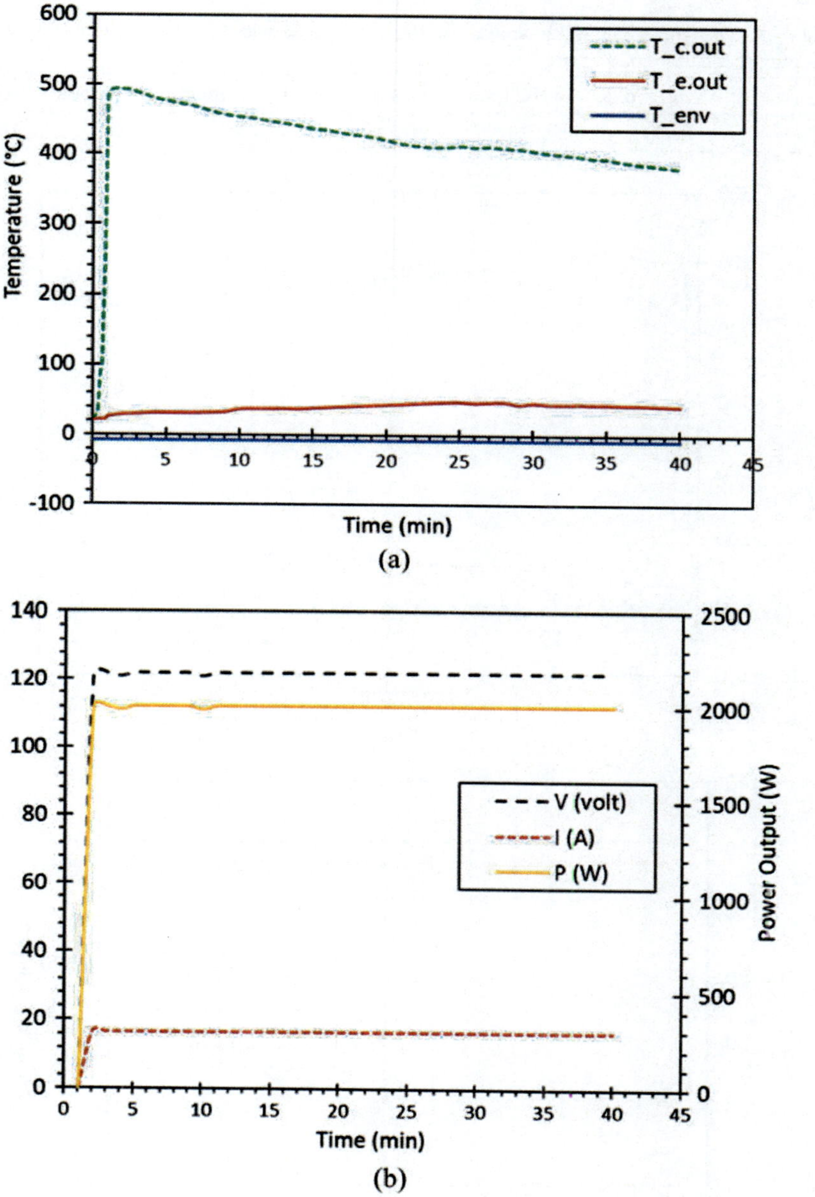

Fig. 4.71 Experimental data for 100% ammonia with 200 °C air-preheating and the spark plug that has a 0.028″ gap. **a** Temperature changes and **b** power values

Table 4.49 The average emission values for 100% ammonia with 200 °C air-preheating and the spark plug that has a 0.028″ gap

CO_2 (vol.%)	CO (vol.%)	HC (ppm)	O_2 (vol.%)	NO_2 (ppm)	Lambda	AFR
1.35	0.97	2.35	0.7	1359	0.88	14

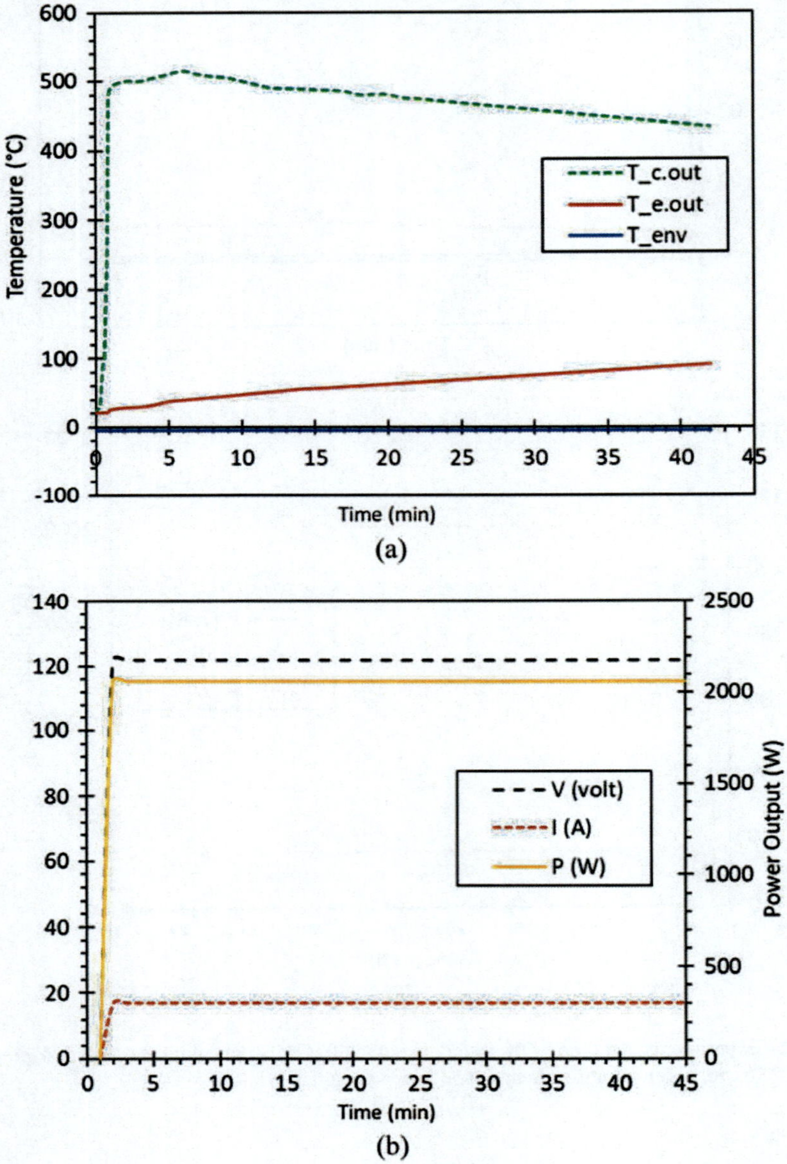

Fig. 4.72 Experimental data for 100% ammonia with 200 °C air-preheating and the spark plug that has a 0.025″ gap. **a** Temperature changes and **b** power values

Table 4.50 The average emission values for 100% ammonia with 200 °C air-preheating and the spark plug that has a 0.025″ gap

CO_2 (vol.%)	CO (vol.%)	HC (ppm)	O_2 (vol.%)	NO_2 (ppm)	Lambda	AFR
1.01	0.62	2.03	0.7	1224	0.87	14

is around − 6 °C. The generator has generated 17.7 A of electrical current. The power output is 2159 W which corresponds a 0.59 of PR. The engine has worked 45 min. Despite the engine is stopped working in two trials, it has completed 45 min test period remaining trials. Table 4.51 depits that CO_2 and CO values are 0.95 and 0.63 vol.%, respectively. HC and NO_2 values are observed to be 1.95 and 1174 ppm, respectively. While AFR is 13, lambda is 0.87.

As seen from the experimental observations, using a spark plug with a smaller gap has improved the combustion efficiency. The temperature at the cylinder exit has increased with the decreasing sparkplug gap. The power output has also improved from 2000 to 2160 W. Also, better combustion has reduced the NO_2 emission, too. The NO_2 emission has decreased from 1359 to 1174 ppm.

In addition to the spark plugs tried, the following spark plugs have also been studied. These spark plug has a wider gap than the ones tested previously. The main objective of using this spark plug is to observe the effect of a hotter ignition on ammonia combustion.

- NGK BR5-672 spark plug. It has a 0.032″ (0.8128 mm) gap.
- NGK BR6-672 spark plug. It has a 0.030″ (0.7620 mm) gap.

First, we have run the case with pure ammonia, 200 °C air inlet temperature and the spark plug which has a 0.030″ gap, which has a 0.002″ wider gap than the standard spark plug for the engine. Figure 4.74 and Table 4.52 show the experimental data for the case with pure ammonia, 200 °C air inlet temperature and the spark plug which has a 0.030″ gap. The engine has run 43–45 min. The temperature at the cylinder outlet has reach up to 540 °C which has 40 °C higher than the cases with a lower spark plug gap. The environmental temperature is around − 12 °C. The temperature at the gas analyzer probe has reached up to 62 °C. The generator has provided the stable power output with a 16.9 A electricity current. The power output of the generator is 2062 W which corresponds the 0.56 of PR value. Table 2 gives the emission values measured for the case with 25% methanol and 75% ammonia mixture and 200 °C air inlet temperature. The CO_2 and CO emissions have been observed to be 1.16 and 0.82 vol.%, respectively. NO_2 and HC are measured to be 1350 and 1.95 ppm. Lambda and AFR values are observed to be 0.70 and 14, respectively.

Figure 4.75 and Table 4.53 demonstrate the experimental findings for the case with pure ammonia, 200 °C air-preheating and spark plug that has a 0.032″ gap. The temperature at the cylinder exit has increased with a larger gap in the spark plug. It has reached up to almost 600 °C and stayed at this temperature range for 15 min. Then, it has started to slightly decrease. The temperature taken from gas analyzer probe has reached almost 100 °C. The environmental temperature is around − 14 °C.

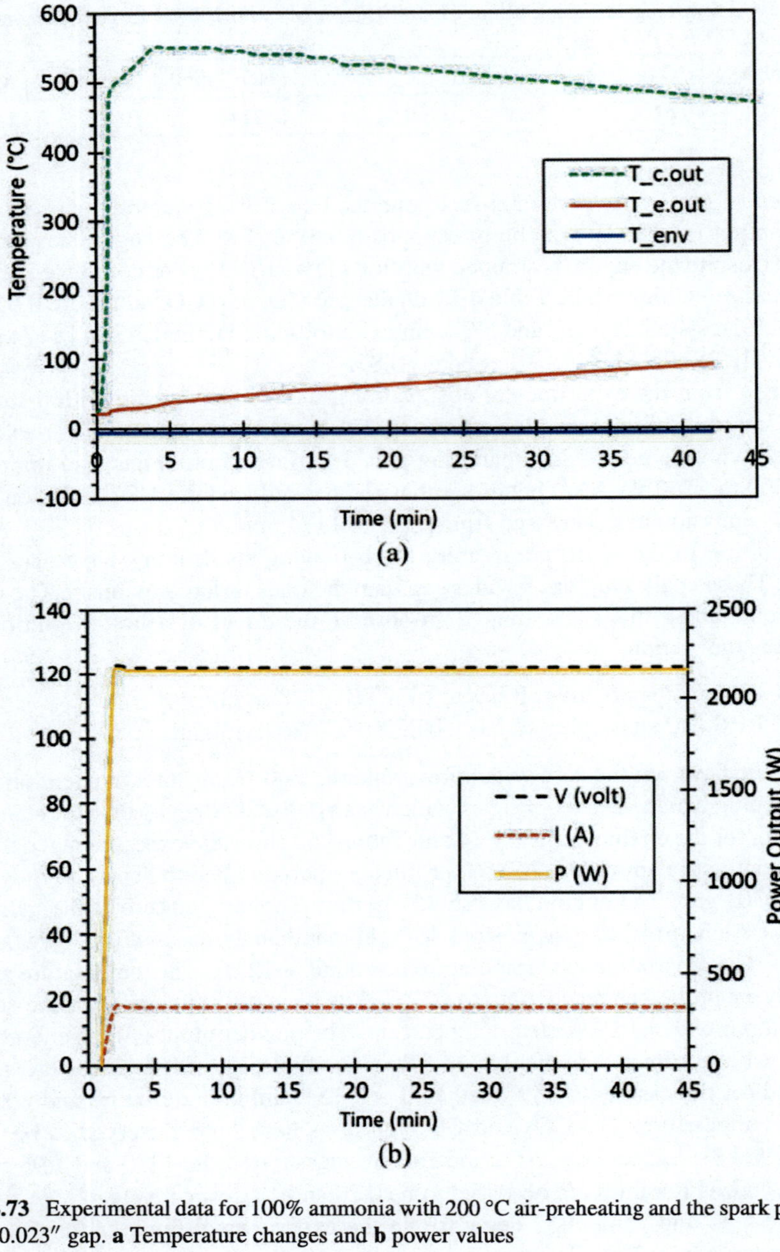

Fig. 4.73 Experimental data for 100% ammonia with 200 °C air-preheating and the spark plug that has a 0.023″ gap. **a** Temperature changes and **b** power values

Table 4.51 The average emission values for 100% ammonia with 200 °C air-preheating and the spark plug that has a 0.023″ gap

CO_2 (vol.%)	CO (vol.%)	HC (ppm)	O_2 (vol.%)	NO_2 (ppm)	Lambda	AFR
0.95	0.63	1.95	0.9	1174	0.87	13

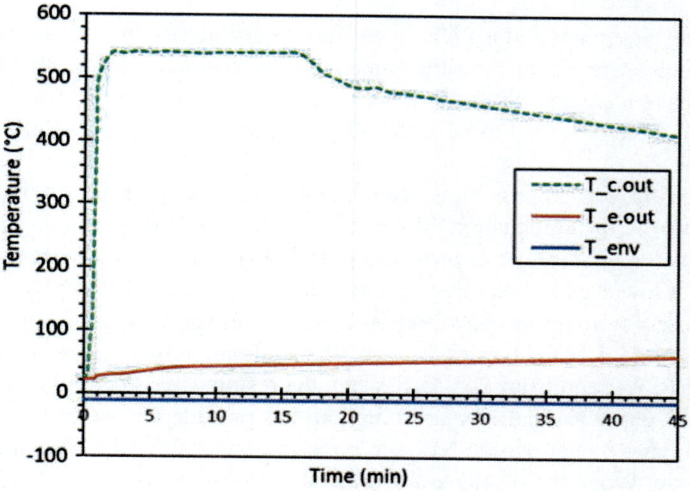

(a)

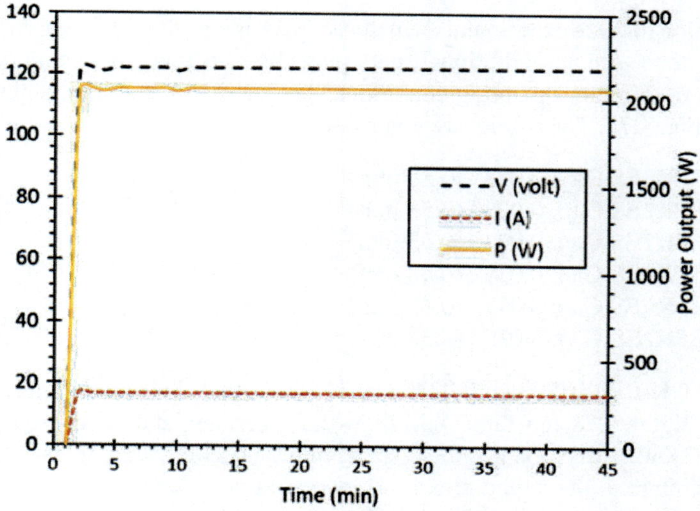

(b)

Fig. 4.74 Experimental data for 100% ammonia with 200 °C air-preheating and the spark plug that has a 0.030″ gap. **a** Temperature changes and **b** power values

Table 4.52 The average emission values for 100% ammonia with 200 °C air-preheating and the spark plug that has a 0.030″ gap

CO_2 (vol.%)	CO (vol.%)	HC (ppm)	O_2 (vol.%)	NO_2 (ppm)	Lambda	AFR
1.16	0.82	1.95	0.7	1350	0.88	14

The generator has generated a 17.1 A of electric current. So, the power output was 2086 W which corresponds 0.57 PR value. As seen from Table 4.53 that CO_2 and CO values have measured to be 1.01 and 0.72 vol.%, respectively. O_2 is 0.7 vol.% HC and NO_2 have observed to be 1.85 and 1238 ppm, respectively. While Lambda is 0.87, AFR are 14.

As seen from the experimental observations, using a spark plug with a wider gap has improved the combustion efficiency than the standard spark plug. However, although a wider spark plug has provided slightly better performance than the cases which have a lower gap, the enhancement is not significant level. The wider spark plug has caused more noise and vibration than a lower spark plug.

Figure 4.76 depicts the comparisons of the studied cases in the terms of power output and PR. As seen from Fig. 4.76, when the results are compared with the case with a 0.028″ gap, the smaller spark plug gap has provided an extended operating period and higher power output. NO_2 emission is also lower in the case of a smaller spark plug gap. When the results are compared with the case with a 0.028″ gap, the smaller spark plug gap has provided an extended operating period and higher power output.

In addition to these experiments with these spark plugs, the following experiments have been done with new additional spark plugs listed below. These park plugs have been used in different operating conditions. The view of the experimental setup is shown in Fig. 4.77. The results are given below.

- NGK-BPR6ES Gap: 0.023″ (0.58 mm)
- NGK-BPR2ES Gap: 0.023″ (0.58 mm)
- NGK-BPR2ES Gap: 0.021″ (0.53 mm)
- NGK-BPR2ES Gap: 0.019″ (0.48 mm)
- NGK-BPR2ES Gap: 0.017″ (0.43 mm)
- NGK-BPR2ES Gap: 0.015″ (0.38 mm)

First, NGK-BPR6ES Gap: 0.023″ is used with 200 °C air inlet temperature and $\lambda \cong 0.88$. Figure 4.78 and Table 4.54 show the experimental data for the case which is with NGK-BPR6ES Gap: 0.023″ spark plug and 200 °C air inlet temperature. The temperature at the cylinder exit has increased up to almost 600 °C and stayed at this temperature range for 17 min. Then, it has started to slightly decrease. The temperature taken from gas analyzer probe has reached almost 100 °C. The environmental temperature is around 8 °C. The generator has generated a 17.9 A of electric current. So, the power output was 2183 W which corresponds 0.59 PR value. As seen from Table 4.54 that CO_2 and CO values have measured to be 1.01 and 0.72 vol.%, respectively. O_2 is 0.7 vol.% HC and NO_2 have observed to be 1.81 and 1238 ppm, respectively. While Lambda is 0.87, AFR are 12.95.

Second, NGK-BPR2ES Gap: 0.023″ is used with 200 °C air inlet temperature and $\lambda \cong 0.88$. Figure 4.79 and Table 4.55 show the experimental data for this case. As seen from Fig. 4.79, the temperature at the cylinder exit has increased up to almost 620 °C and stayed at this temperature range for 21 min. Then, it has started to slightly

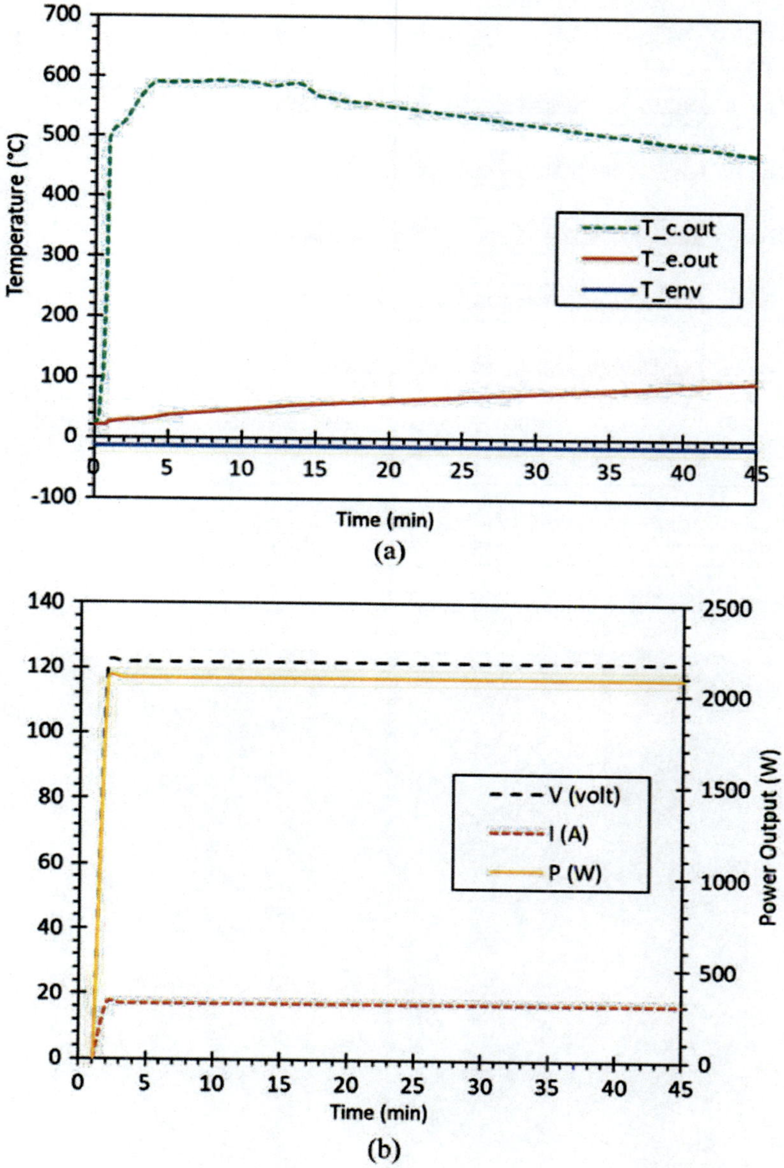

Fig. 4.75 Experimental data for 100% ammonia with 200 °C air-preheating and the spark plug that has a 0.032″ gap. **a** Temperature changes and **b** power values

Table 4.53 The average emission values for 100% ammonia with 200 °C air-preheating and the spark plug that has a 0.032″ gap

CO_2 (vol.%)	CO (vol.%)	HC (ppm)	O_2 (vol.%)	NO_2 (ppm)	Lambda	AFR
1.01	0.72	1.85	0.7	1238	0.87	14

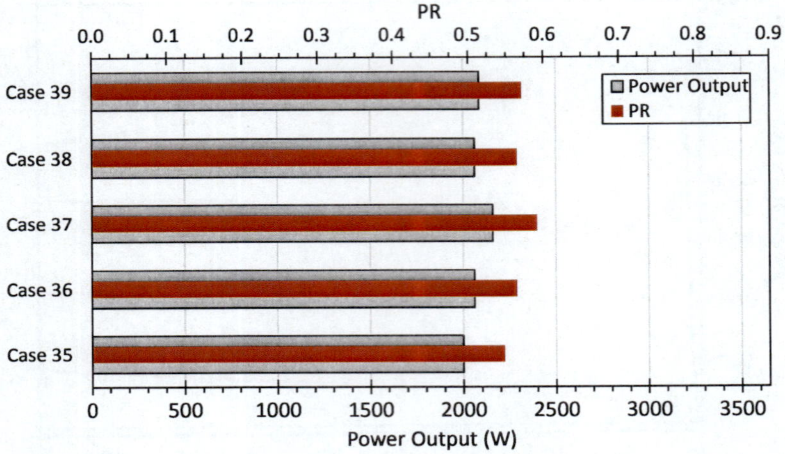

Case 34: 100% ammonia, 200°C air inlet and 0.028" spark plug gap
Case 35: 100% ammonia, 200°C air inlet and 0.028" spark plug gap
Case 36: 100% ammonia, 200°C air inlet and 0.028" spark plug gap

Fig. 4.76 The comparison of the studied cases in terms of the power output and PR

Fig. 4.77 A photo of the experimental setup during sparkplug tests

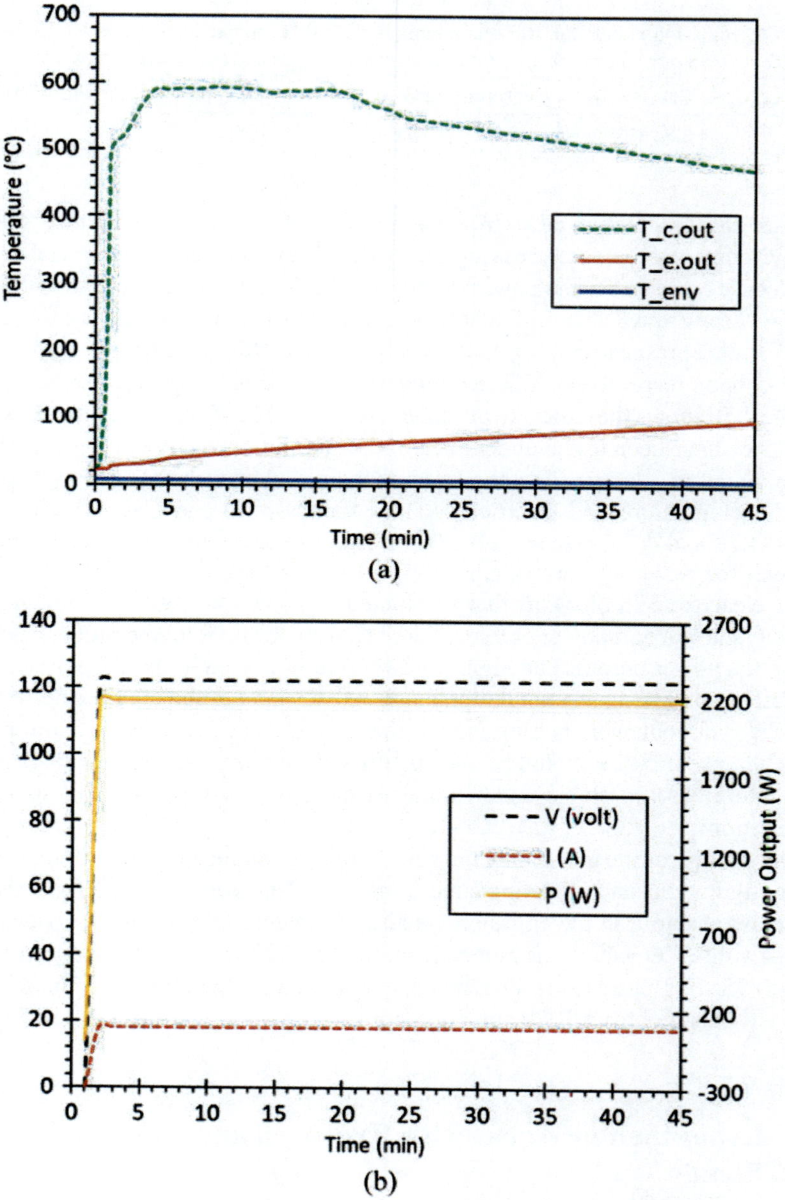

Fig. 4.78 Experimental data for 100% ammonia with 200 °C air-preheating and NGK-BPR6ES Gap: 0.023″ (0.58 mm) the spark plug. **a** Temperature changes and **b** power values

Table 4.54 Emission values for 100% ammonia with 200 °C air-preheating and NGK-BPR6ES Gap: 0.023″ (0.58 mm) the spark plug (a) temperature changes and (b) power values

CO_2 (vol.%)	CO (vol.%)	HC (ppm)	O_2 (vol.%)	NO_2 (ppm)	Lambda	AFR
1.07	0.52	1.90	0.8	1295	0.88	12.94

decrease. The temperature taken from gas analyzer probe has reached almost 100 °C. The environmental temperature is around 10 °C. The generator has generated an 18.5 A of electric current. So, the power output was 2257 W which corresponds 0.61 PR value. As seen from Table 4.55 that CO_2 and CO values have measured to be 1.07 and 0.52 vol.%, respectively. O_2 is 0.8 vol.% HC and NO_2 have observed to be 1.90 and 1295 ppm, respectively. While Lambda is 0.88, AFR is measured to be 12.94.

Table 4.56 shows the experiments conducted with BPR-6ES and BPR-2ES spark-plugs. They have been tested under different configuration and operating conditions. The temperature, electricity, and power trends are almost the same for each case studied. Temperature and electricity values for Case 40 and Case 41 are shown in Figs. 4.78 and 4.79, respectively. Therefore, it is presented that only significant parameter for generator performance and emission values.

It is clear from Table 4.56 that the increasing sparkplug gap has reduced the electricity current and power output. Also, lowering air inlet temperatures have also reduced the power output. The highest power output is seen in the Case 41 which is BPR-2ES Gap: 0.023, air inlet temperature 200 °C and Lambda 0.88. Additionally, while the fuel content is getting leaner, the power output reduced. Therefore, the fuel-rich intake mixture should be used in order to obtain a stable ammonia combustion. Engine has run stable and without extraordinary vibration in all sparkplug configurations.

Table 4.57 gives the emission values for the cases conducted with different spark-plug configurations and operating conditions. It is clear from Table 4.57 that there is no certain behaviour in the emission values. CO_2 values has changed between 1.01 and 1.14 vol.% CO values has ranged from 0.51 to 0.72 vol.% HC values has varied around 0.95–1.95 ppm. O_2 is almost constant and has ranged 0.8–0.7. Finally, NO_2 values has changed from 1270 and 1305 ppm.

4.12 Experimental Works with Hydrogen and Ammonia Blends

Hydrogen is a powerful promoter for ammonia combustion. In this study, in order to blend ammonia with hydrogen and other possible gaseous fuel, the apparatus, shown in Fig. 4.80 has been developed and manufactured. The apparatus has four inlets and one outlet. One of the inlet ports is used for the air which is provided via a portable air compressor. Hydrogen, ammonia, and additional gaseous fuel are provided via pressurized cylinders. Volumetric flowmeters are used for adjusting

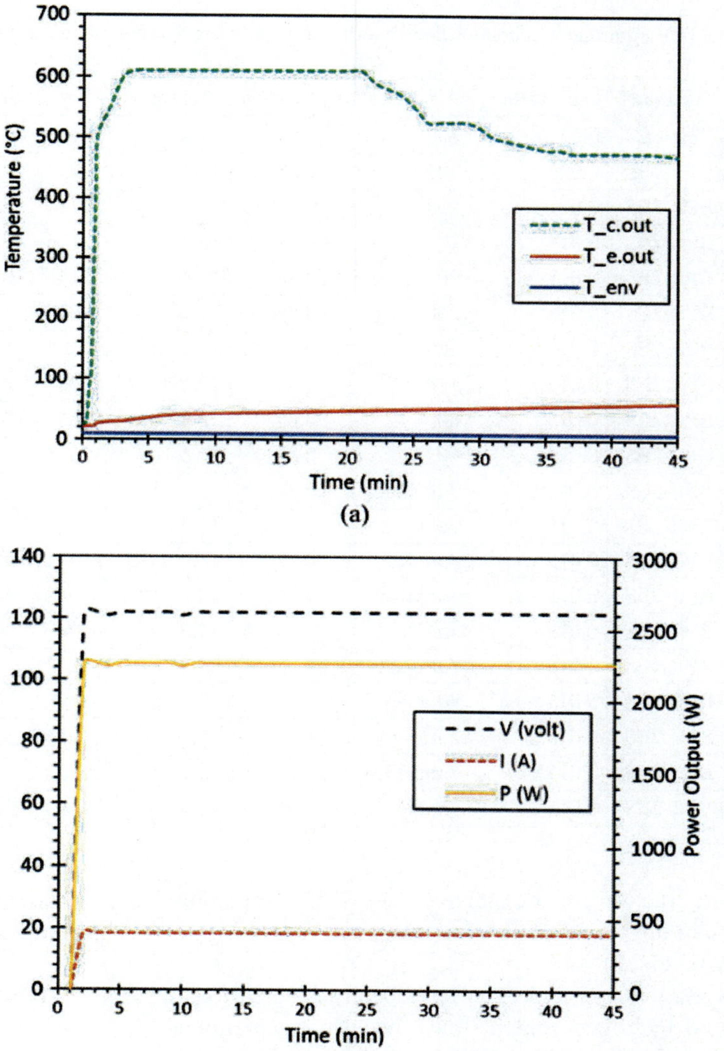

Fig. 4.79 Experimental data for 100% ammonia with 200 °C air-preheating and NGK-BPR2ES spark plug which has a 0.023″ (0.58 mm) gap. **a** Temperature changes and **b** power values

Table 4.55 Emission values for 100% ammonia with 200 °C air-preheating and NGK-BPR2ES Gap: 0.023″ (0.58 mm) the spark plug (a) temperature changes and (b) power values

CO_2 (vol.%)	CO (vol.%)	HC (ppm)	O_2 (vol.%)	NO_2 (ppm)	Lambda	AFR
1.01	0.72	1.81	0.7	1238	0.87	12.95

Table 4.56 The experiments conducted with different sparkplug configurations and operating conditions

Case	Sparkplug	Gap (inches)	Air inlet (°C)	Lambda	Current (A)	Power (W)	PR
Case 40	BPR-6ES	0.023	200	0.88	17.9	2183	0.598
Case 41	BPR-2ES	0.023	200	0.88	18.5	2257	0.618
Case 42	BPR-2ES	0.021	200	0.88	18.3	2233	0.612
Case 43	BPR-2ES	0.019	200	0.88	18	2196	0.602
Case 44	BPR-2ES	0.017	200	0.88	17.6	2147	0.588
Case 45	BPR-2ES	0.015	200	0.88	17.2	2098	0.575
Case 46	BPR-2ES	0.015	160	0.88	17.2	2098	0.575
Case 47	BPR-2ES	0.015	120	0.88	17.1	2086	0.572
Case 48	BPR-2ES	0.015	80	0.88	16.8	2050	0.562
Case 49	BPR-2ES	0.015	40	0.88	16.6	2025	0.555
Case 50	BPR-2ES	0.015	Ambient	0.88	16.4	2001	0.548
Case 51	BPR-2ES	0.015	Ambient	0.90	16.3	1989	0.545
Case 52	BPR-2ES	0.015	Ambient	0.92	16.1	1964	0.538
Case 53	BPR-2ES	0.015	Ambient	0.94	16	1952	0.535
Case 54	BPR-2ES	0.015	Ambient	0.96	15.9	1940	0.531
Case 55	BPR-2ES	0.015	Ambient	0.98	15.8	1928	0.528
Case 56	BPR-2ES	0.015	Ambient	1	15.8	1928	0.528
Case 57	BPR-2ES	0.015	Ambient	1.02	15.7	1915	0.525
Case 58	BPR-2ES	0.015	Ambient	1.04	15.6	1903	0.521
Case 59	BPR-2ES	0.015	Ambient	1.06	15.6	1903	0.521

the volume fractions of the intake contents. The apparatus has one outlet which is connected to the cylinder inlet via pipe, as shown in Fig. 4.81. The connection pipe stands by up to 250 °C.

Ammonia, air and hydrogen have been provided to the blender. A 2.5, 5, 7.5, and 10% of hydrogen contents have been test experimentally. These experiments have been performed with NGK-BPR2ES Gap: 0.023″ (0.58 mm) the sparkplug and ambient temperature.

Figure 4.82 and Table 4.58 show the experimental data for in the case of 2.5% hydrogen is included in the fuel mixture. As seen from Fig. 4.82a, the combustion inside the cylinder has been enhanced with hydrogen. The temperature at the cylinder exit has reached up to 680 °C. Also, the electricity current and power output have been observed to be 18.9 A and 2305 W, respectively. When the effect of 2.5% hydrogen adding to the fuel mixture has reduced the NO_2 emission. There is remarkable changes in the other components of the emission values.

The main impact of the hydrogen in the fuel mixture is on the combustion effectiveness. Therefore, the effect of the hydrogen content in the fuel mixture on the temperature at the cylinder outlet is shown in Fig. 4.83. It is clear from Fig. 4.83 that

Table 4.57 Emission values for the experiments case 40–59

Case	CO_2 (vol.%)	CO (vol.%)	HC (ppm)	O_2 (vol.%)	NO_2 (ppm)	Lambda	AFR
Case 40	1.07	0.52	1.90	0.8	1295	0.88	12.94
Case 41	1.01	0.72	1.81	0.7	1238	0.87	12.95
Case 42	1.04	0.63	0.95	0.8	1245	0.87	12.81
Case 43	1.10	0.67	1.21	0.7	1255	0.88	12.95
Case 44	1.05	0.64	1.85	0.7	1305	0.88	12.94
Case 45	1.09	0.62	1.75	0.7	1248	0.89	12.99
Case 46	1.12	0.60	1.94	0.8	1234	0.88	12.92
Case 47	1.05	0.58	1.82	0.8	1239	0.88	12.93
Case 48	1.04	0.63	1.42	0.8	1287	0.88	12.94
Case 49	1.05	0.51	1.63	0.8	1291	0.88	12.94
Case 50	1.05	0.62	1.82	0.7	1269	0.87	12.77
Case 51	1.11	0.59	1.86	0.7	1270	0.90	13.24
Case 52	1.03	0.51	1.84	0.8	1271	0.92	13.53
Case 53	1.14	0.49	1.82	0.8	1272	0.94	13.8
Case 54	1.13	0.54	1.78	0.8	1273	0.96	14.1
Case 55	1.09	0.61	1.79	0.7	1274	0.98	14.45
Case 56	1.05	0.51	1.73	0.7	1275	1	14.68
Case 57	1.04	0.53	1.81	0.7	1276	1.02	15.11
Case 58	1.02	0.54	1.86	0.8	1277	1.04	15.26
Case 59	1.08	0.61	1.84	0.8	1277	1.06	15.56

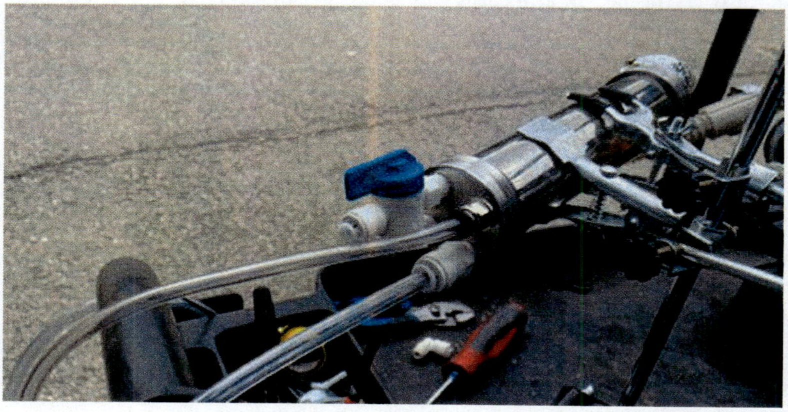

Fig. 4.80 A photo of the fuel mixture apparatus

Fig. 4.81 A photo of the fuel blender and connection to the engine

increasing hydrogen content has increased the combustion effectiveness. Because the temperature at the cylinder outlet has increased with the increasing hydrogen volume in the fuel mixture. The electricity current values are observed to be 18.9, 19.2, 19.6, and 19.7 A. The power outputs are calculated to be 2305, 2342, 2391, and 2403 W, respectively. As seen from these results, hydrogen has improved the generator performance significantly. However, it should be noted that almost 1.5 L engine oil is burned during the hydrogen tests. A proper material for piston and cylinder mechanism should be used for hydrogen combustions. Finally, as seen from Table 4.59, the only difference in emission values is seen in the NO_2. It has reduced with the increase hydrogen content. Also, there is slightly increase in the CO_2, CO and HC values due to the oil content in the cylinder.

4.13 Experimental Works with Oxygen-Rich Air

The use of O_2-rich air is one of the effective methods to enhance the combustion. In the present study, we have tested the effect of 40, 60, 80 and 100% oxygen content in the air. In these tests, it is assumed that ambient air has 20% oxygen content. For this purpose, the ammonia has been mixed with the O_2-rich air in the blender. In these test, NGK-BPR2ES Gap: 0.023″ (0.58 mm) sparkplug has been used. There is no preheating in these tests.

Figure 4.84 and Table 4.60 demonstrate the experimental data for the case that using 40% oxygen in the fuel mixture. As seen from Fig. 4.84a, the temperature at the cylinder exit has increased up to 700 °C and stayed stable with the improved combustion. The enhancement in the combustion has generated 18.7 A of electricity current, which corresponds the 2281 W of the power outlet. As seen from Table 4.60, CO_2, CO and HC values have slightly reduced according to the case with ambient

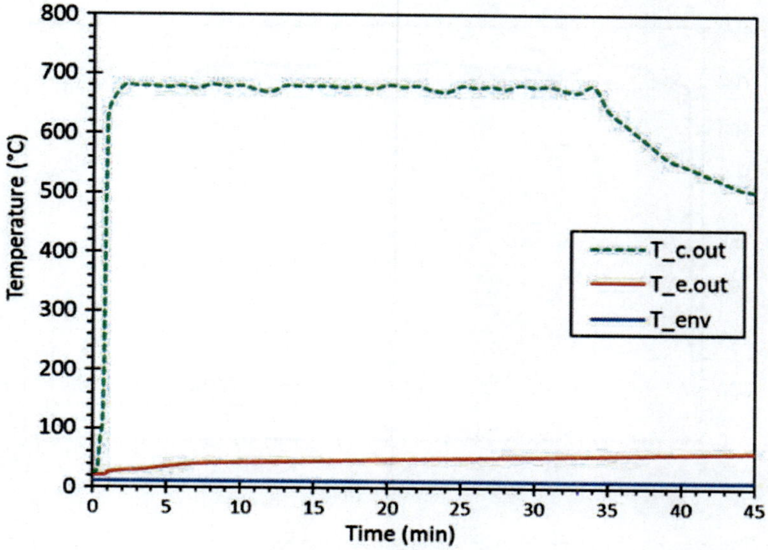

(a)

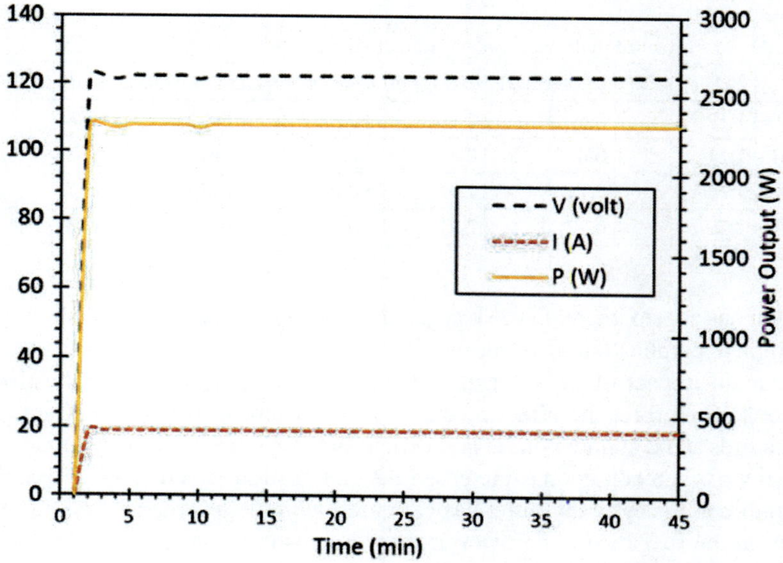

(b)

Fig. 4.82 Experimental data for 2.5% hydrogen adding into the fuel mixture. **a** Temperature changes and **b** power values

Table 4.58 Emission data for 2.5% hydrogen adding into the fuel mixture

CO_2 (vol.%)	CO (vol.%)	HC (ppm)	O_2 (vol.%)	NO_2 (ppm)	Lambda	AFR
1.01	0.92	1.21	0.8	1138	0.88	12.94

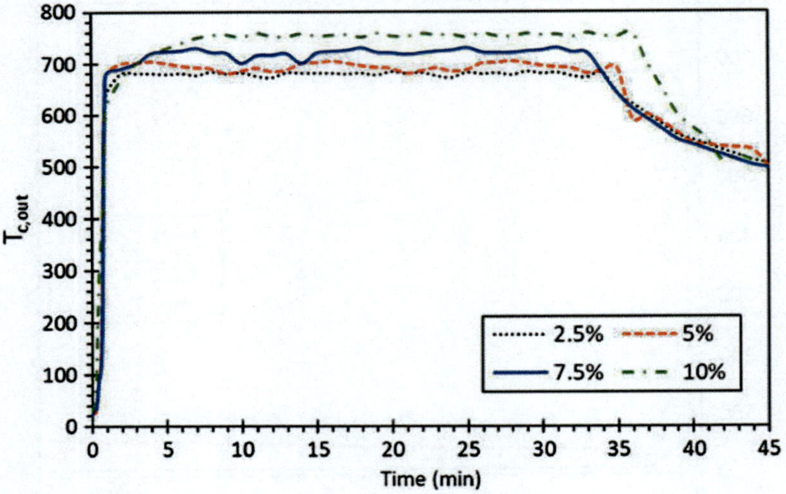

Fig. 4.83 The effect of hydrogen content in the fuel mixture on the temperature at the cylinder outlet

Table 4.59 The effect of hydrogen content in the fuel mixture on the emission values

Case	CO_2 (vol.%)	CO (vol.%)	HC (ppm)	O_2 (vol.%)	NO_2 (ppm)	Lambda	AFR
Case 60	1.17	0.62	2.90	0.8	1191	0.88	12.95
Case 61	1.11	0.82	2.81	0.7	1187	0.88	12.94
Case 62	1.14	0.73	1.95	0.8	1170	0.87	12.93
Case 63	1.20	0.77	2.21	0.7	1165	0.88	12.95

air which has approximately 20% oxygen. NO_2 emission has increased slightly with the complete combustion of ammonia.

The main impact of the oxygen in the fuel mixture is on the combustion and emissions. Therefore, the effect of the oxygen content in the fuel mixture on the temperatures at the cylinder outlet is shown in Fig. 4.85. It is clear from Fig. 4.85 that increasing oxygen content has increased the combustion effectiveness. Because the temperature at the cylinder outlet has increased with the increasing oxygen volume fraction in the fuel mixture. However, there is no significant increase in the outlet temperature with the increasing oxygen content. The electricity current values are observed to be 18.7, 19.1, 19.9, and 20 A. The power outputs are calculated to be 2281, 2330, 2427, and 2440 W, respectively. As seen from these results, oxygen has improved the generator performance significantly. As seen from Table 4.61, there is no remarkable difference in the emission values except from NO_2 values. It has increased slightly due to the complete combustion of ammonia with the increasing content of the oxygen in the fuel mixture.

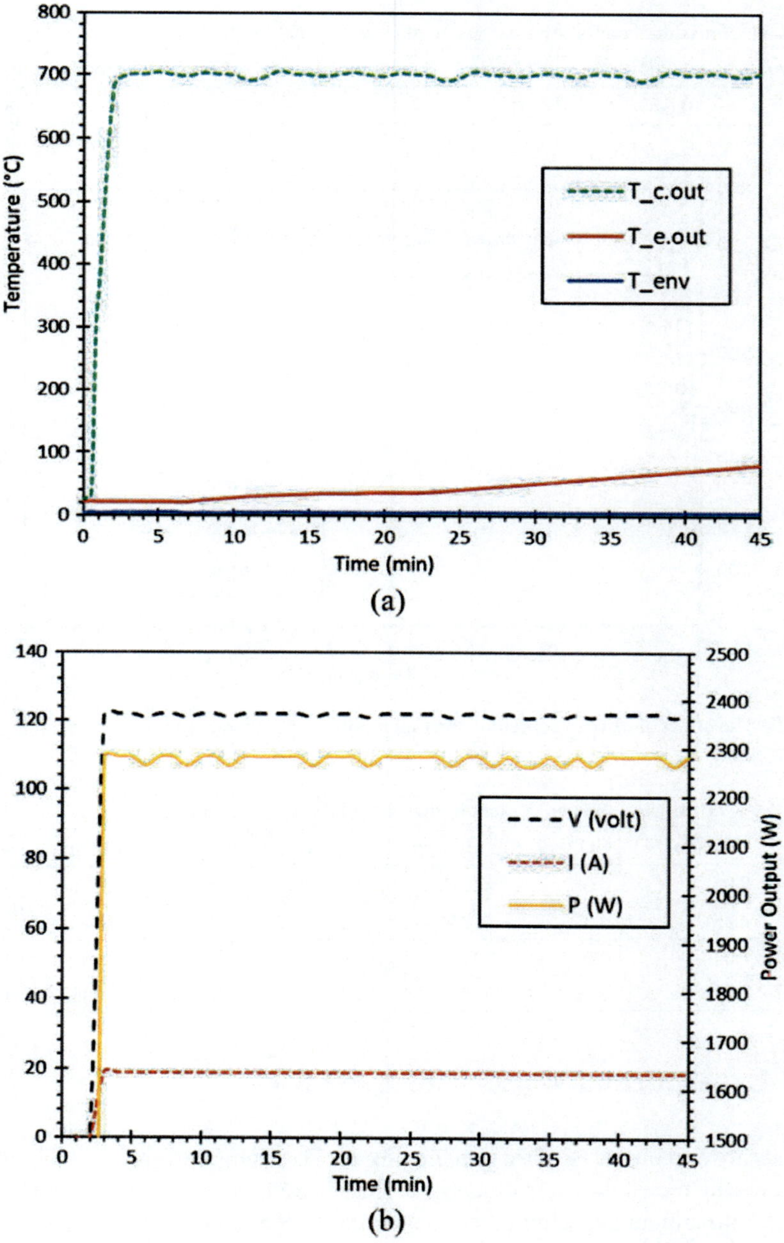

Fig. 4.84 Experimental data for 40% oxygen content in the fuel mixture. **a** Temperature changes and **b** power values

Table 4.60 Emission data for 40% oxygen content in the fuel mixture

CO_2 (vol.%)	CO (vol.%)	HC (ppm)	O_2 (vol.%)	NO_2 (ppm)	Lambda	AFR
0.95	0.62	1.01	1.1	1238	0.91	13.34

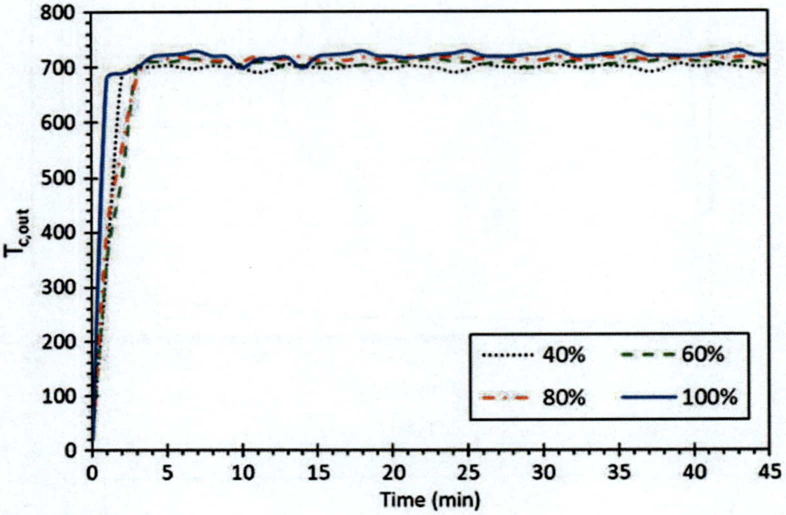

Fig. 4.85 The effect of oxygen content in the fuel mixture on the temperature at the cylinder outlet

Table 4.61 The effect of oxygen content in the fuel mixture on the emission values

Case	CO_2 (vol.%)	CO (vol.%)	HC (ppm)	O_2 (vol.%)	NO_2 (ppm)	Lambda	AFR
Case 64	0.95	0.62	1.01	1.1	1238	0.91	13.34
Case 65	0.89	0.57	0.98	1.1	1287	0.88	12.94
Case 66	1.01	0.61	1.21	1.4	1270	0.87	12.93
Case 67	0.97	0.59	1.07	1.6	1265	0.88	12.95

4.14 Experimental Works with HHO Kit

HHO kits are commonly used for reducing the fuel consumption and emission values and improving the engine performance. In this section, the effect of integration of a HHO kit into ammonia-fueled power generator. In the Sect. 4.12, the effect of use of hydrogen in fuel mixture is studied. HHO kits also generates hydrogen to be fed to engine. The main difference between both system is the way of hydrogen feeding to the engine. While hydrogen is fed with air from air intake, it is fed by the fuel blender. However, as the engine intake content is the same for both system, there is no significant changes in the generator performance and emissions values, when compared the results with the cases with hydrogen.

The experimental data for the case with HHO kit is shown in Fig. 4.86. It is clear from Fig. 4.86 that, the temperature changes are similar to cases with hydrogen. The temperature at the cylinder outlet has reached almost 700 °C. The power generator has produced 19.9 A of electricity current. The power output is 2910 W which corresponds to PR = 0.79. As given in Table 4.62, CO_2 and CO are measured to be 0.94 and 0.61, respectively. While HC is 1.22 ppm, O_2 content in the exhaust gases is 0.8 vol.% Lambda and AFR is found to be 12.95.

4.15 Comparative Analysis Studied Cases

PR and emission values for all cases studied in this chapter are presented in Table 4.63. It is clear from Table 4.63 that when ammonia content in the fuel mixture has increased, the CO_2 and CO emissions have decreased substantially. Results indicate that ammonia is a promising fuel in terms of emission values. NO_2 contents in the exhaust gases are more than 1000 ppm when 100% ammonia has been used. In order to reduce the NO_2 emissions, the solutions should be taken into consideration.

When PR values are considered, the generator has reached almost the PR of 0.70. In other words, 100% ammonia-fueled generator has reached about 70% of its full capacity (gasoline-fueled). Among the fuel promoters studied in this chapter, propane has presented promising results. Better sparkplug configurations have provided higher performance, despite they have not affected the emission values. Ethanol and methanol have not been observed as a suitable promoter for ammonia. While air preheating has improved the generator performance, the preheating of ammonia has not shown significant improvement in the power output. Results presented in Table 4.63 indicates that ammonia can be used in the power generators as the fuel.

4.16 A Comparative LCA Analysis of Ammonia-Fuel Blends in Power Generators

In this section, a comparative LCA analysis of ammonia-gasoline blends is presented. LCA methodology for assessing the environmental implications of a process or product from start to finish. The LCA's main ideas include taking a holistic view of energy and resource requirements, as well as their effects both inside and outside the process or production cycle. Setting the limits of the researched system effectively, following a consistent technique, and avoiding any bias caused by the practitioner are all required for a successful LCA study [11]. The LCA is carried out in accordance with ISO14040 and ISO14044 standards.

- The goal and scope of the LCA is defined.
- The inventory assessment is done based on the experimental and literature data.
- Research findings and emission factors are used to conduct the impact assessment.

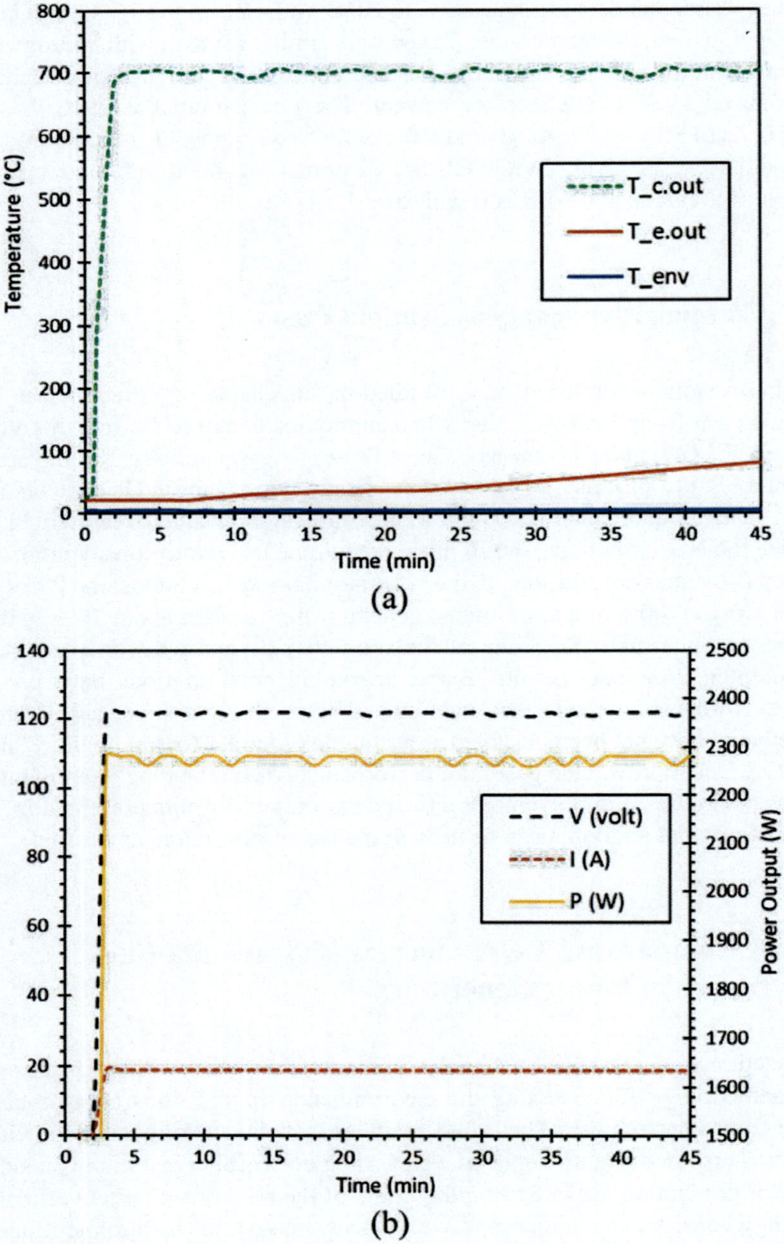

Fig. 4.86 Experimental data for the case with HHO kit. **a** Temperature changes and **b** power values

Table 4.62 Emission values for the case with HHO kit

Case	CO_2 (vol.%)	CO (vol.%)	HC (ppm)	O_2 (vol.%)	NO_2 (ppm)	Lambda	AFR
Case 68	0.94	0.61	1.22	0.8	1253	0.88	12.95

- Each phase is interpreted in order to create the final LCA.

The inventory of materials and energy flows is the foundation of LCA. There are several steps to follow in order to define the inventory, including data collection, data estimation based on literature, and data validation. As a result of the inventory analysis, a flow diagram will be created to depict the processes and their interrelationship. SimaPro software is used for LCIA calculations. In this particular study, CML-IA-baseline impact assessment methodology is used for the LCA calculations.

4.16.1 Goal and Scope Definition

The main purpose of this LCA is to find out how ammonia-fuel blends affect the environment depending on different combustion products. A functional unit is a measurement of a product system's performance that can be used as a reference unit. This makes determining reference flows for the systems under investigation much easier. The functional unit for final ammonia-fuel blends is 1 MJ of energy content based on the HHV delivered to the end user. The following are the environmental effect categories used in this LCA study:

- Acidification potential (AP)
- Eutrophication potential (EP)
- Global warming potential (GWP)
- Human toxicity potential (HTP)
- Ozone layer depletion potential (ODP)
- Photochemical oxidation potential (POP).

In LCA, system boundaries are defined to incorporate the major life cycle phases and unit processes, as well as the related environmental fluxes. The primary life cycle stages considered in this study is shown in Fig. 4.87; to provide the broad system limits for this LCA study.

4.16.2 Experimental Setup and Procedure

The data below are taken from Sect. 4.4 in order to carry out the LCA study. Table 4.5 shows list of experiments for 11 different configurations which is gasoline and ammonia blends. The photos of the experimental setup are shown in Fig. 4.88. Ammonia and gasoline flow rates were controlled with needle valves and, they

Table 4.63 PR and emission values for all cases studied

Case	Case label	PR	CO_2 (vol.%)	CO (vol.%)	HC (ppm)	O_2 (vol.%)	NO_2 (ppm)	Lambda	AFR
Base 1	G100	1	8.593	9.974	241.891	0.391	37.239	0.711	10.461
Base 2	G90-A10	0.933	8.336	9.774	232.216	0.587	74.478	0.711	10.461
Base 3	G80-A20	0.901	8.164	9.575	220.121	0.732	104.27	0.711	10.461
Base 4	G70-A30	0.887	7.906	8.905	215.283	0.747	115.441	0.711	10.461
Base 5	G60-A40	0.843	7.82	8.905	210.814	0.746	203.143	0.711	10.461
Base 6	G50-A50	0.785	7.648	8.713	209.236	0.775	260.061	0.709	10.485
Base 7	G40-A60	0.726	6.875	7.756	198.109	0.861	442.435	0.718	10.503
Base 8	G30-A70	0.641	6.273	6.702	174.162	1.213	635.152	0.711	10.461
Base 9	G20-A80	0.525	2.75	3.351	159.648	1.526	795.826	0.711	10.461
Base 10	G10-A90	0.513	2.047	1.725	141.618	1.976	1065.75	0.704	10.349
Base 11	A100	0.498	0.908	0.83	79.029	3.413	1562	0.681	10.011
Base 12	G100 at lower ambient temperature	1	9.28	9.29	147.27	0.4	61.07	0.74	10.69
Case 1	A100 at 20 °C ambient temperature	0.441	0.908	0.83	79.029	3.413	1562	0.681	10.011
Case 2	A100 at 12 °C ambient temperature	0.436	1.341	0.978	91.76	0.876	1591	0.681	10.011
Case 3	A100—80 °C air preheating	0.445	1.58	2.42	83.85	3.64	1481.17	0.73	10.75
Case 4	A100—100 °C air preheating	0.463	1.59	2.27	91.67	2.34	1367.91	0.74	10.512
Case 5	A100—120 °C air preheating	0.516	1.58	2.41	84.2	3.64	1304.89	0.73	10.73
Case 6	A100—140 °C air preheating	0.529	1.28	1.29	2.27	0.6	1290.07	0.74	15.69
Case 7	A100—160 °C air preheating	0.546	1.35	1.49	1.27	0.9	1245.91	0.79	16.1
Case 8	A100—180 °C air preheating	0.55	1.72	1.68	1.91	0.8	1210.07	0.79	17.21
Case 9	A100, 180 °C air heating and 50 cm heating tape	0.55	1.22	1.29	2.03	0.6	1375	0.79	17

(continued)

Table 4.63 (continued)

Case	Case label	PR	CO_2 (vol.%)	CO (vol.%)	HC (ppm)	O_2 (vol.%)	NO_2 (ppm)	Lambda	AFR
Case 10	A100, 180 °C air heating and 75 cm heating tape	0.555	1.28	1.32	2.65	0.9	1386	0.79	17
Case 11	A100, 180 °C air heating and 100 cm heating tape	0.56	1.31	1.39	2.45	0.7	1368	0.81	17
Case 12	P100—180 °C air heating	0.811	8.6	10.43	283	0.4	41	0.72	11.58
Case 13	P80-A20—180 °C air heating	0.755	7.98	9.8	267	0.8	215	0.79	12
Case 14	P60-A40—180 °C air heating	0.703	7.17	9.21	251	0.85	324	0.79	12
Case 15	P50-A50—180 °C air heating	0.644	6.43	8.42	245	1	417	0.83	13
Case 16	P40-A60—180 °C air heating	0.572	5.03	6.42	187	1.2	647	0.83	14
Case 17	P20-A80—180 °C air heating	0.54	3.47	4.13	98	1.1	850	0.85	15
Case 18	P50-A50—180 °C air heating	0.658	6.43	8.42	245	1	417	0.83	13
Case 19	P50-A50—210 °C air heating	0.671	6.46	8.37	251	0.7	434	0.81	14
Case 20	P50-A50—240 °C air heating	0.574	6.45	8.27	267	0.9	479	0.81	13.2
Case 21	P40-A60—180 °C air heating	0.598	5.03	6.42	187	1.2	647	0.83	14
Case 22	P40-A60—210 °C air heating	0.625	4.95	6.02	178	0.8	678	0.82	12
Case 23	P40-A60—240 °C air heating	0.531	4.78	5.96	183	0.9	691	0.8	11.9
Case 24	P20-A80—180 °C air heating	0.564	3.47	4.13	98	1.1	850	0.85	15
Case 25	P20-A80—210 °C air heating	0.591	3.36	3.98	101	0.8	879	0.81	12
Case 26	P20-A80—240 °C air heating	0.658	3.21	4.04	118	0.9	902	0.84	11.9
Case 27	E75-A25—200 °C air heating	0.528	8.46	9.37	276	1.1	245	0.84	12
Case 28	E50-A50—200 °C air heating	0.485	6.95	7.02	169	1.5	478	0.82	12
Case 29	E25-A75—200 °C air heating	0.421	3.56	4.04	95	0.9	729	0.81	12
Case 30	M75-A25—200 °C air heating	0.528	7.23	8.02	291	0.8	284	0.52	10

(continued)

Table 4.63 (continued)

Case	Case label	PR	CO_2 (vol.%)	CO (vol.%)	HC (ppm)	O_2 (vol.%)	NO_2 (ppm)	Lambda	AFR
Case 31	M50-A50—200 °C air heating	0.471	6.71	7.14	202	0.8	481	0.8	10
Case 32	M25-A75—200 °C air heating	0.471	3.35	3.02	157	0.6	284	0.84	14
Case 33	M100—modified carburetor	0.939	7.73	7.97	301	0.9	98	0.88	13
Case 34	M100—standard carburetor	–	–	–	–	–	–	–	–
Case 35	Champion RN9YC Gap: 0.028″ (0.7 mm)	0.548	1.35	0.97	2.35	0.7	1359	0.88	14
Case 36	NGK BR2-LM Gap: 0.025″ (0.63 mm)	0.565	1.01	0.62	2.03	0.7	1224	0.87	14
Case 37	NGK BR6HS Gap: 0.023″ (0.58 mm)	0.592	0.95	0.63	1.95	0.9	1174	0.87	13
Case 38	NGK BR5-672 Gap: 0.032″ (0.81 mm)	0.565	1.16	0.82	1.95	0.7	1350	0.88	14
Case 39	NGK BR6-672 Gap: 0.030″ (0.76 mm)	0.572	1.01	0.72	1.85	0.7	1238	0.87	14
Case 40	BPR-6ES, 0.023″, 200 °C	0.598	1.21	1.42	2.27	0.7	1369	0.88	12.726
Case 41	BPR-2ES, 0.023″, 200 °C	0.618	1.51	1.11	1.25	0.9	1442	0.88	12.803
Case 42	BPR-2ES, 0.021″, 200 °C	0.612	1.97	1.51	1.83	0.7	1272	0.88	12.368
Case 43	BPR-2ES, 0.019″, 200 °C	0.602	1.31	1.06	2.23	0.7	1533	0.88	12.462
Case 44	BPR-2ES, 0.017″, 200 °C	0.588	1.88	1.13	2.07	0.7	1580	0.88	12.136
Case 45	BPR-2ES, 0.015″, 200 °C	0.575	1.75	1.32	2.21	0.9	1668	0.88	12.987
Case 46	BPR-2ES, 0.015″, 160 °C	0.575	1.18	1.33	2.81	0.7	1464	0.88	12.458
Case 47	BPR-2ES, 0.015″, 120 °C	0.572	1.02	1.16	2.65	0.7	1614	0.88	12.986
Case 48	BPR-2ES, 0.015″, 80 °C	0.562	1.44	1.15	2.28	0.7	1104	0.88	12.826
Case 49	BPR-2ES, 0.015″, 40 °C	0.555	1.77	1.69	2.19	0.9	1443	0.88	12.781
Case 50	BPR-2ES, 0.015″, Ambient	0.548	1.39	1.28	2.76	0.7	1385	0.88	12.841
Case 51	BPR-2ES, 0.015″, Ambient	0.545	1.54	1.72	2.63	0.7	1455	0.9	12.572

(continued)

Table 4.63 (continued)

Case	Case label	PR	CO_2 (vol.%)	CO (vol.%)	HC (ppm)	O_2 (vol.%)	NO_2 (ppm)	Lambda	AFR
Case 52	BPR-2ES, 0.015″, Ambient	0.538	1.17	1.91	1.33	0.7	1423	0.92	12.47
Case 53	BPR-2ES, 0.015″, Ambient	0.535	1.34	1.32	2.93	0.9	1581	0.94	12.415
Case 54	BPR-2ES, 0.015″, Ambient	0.531	1.20	1.17	2.31	0.7	1468	0.96	12.431
Case 55	BPR-2ES, 0.015″, Ambient	0.528	1.97	1.13	1.74	0.7	1447	0.98	12.334
Case 56	BPR-2ES, 0.015″, Ambient	0.528	1.98	1.37	2.65	0.7	1358	1	12.742
Case 57	BPR-2ES, 0.015″, Ambient	0.525	1.98	1.40	1.94	0.9	1309	1.02	12.321
Case 58	BPR-2ES, 0.015″, Ambient	0.521	1.97	1.22	2.98	0.7	1391	1.04	12.089
Case 59	BPR-2ES, 0.015″, Ambient	0.521	1.89	1.99	2.57	0.7	1155	1.06	12.936
Case 60	2.5% hydrogen	0.632	1.17	0.62	2.9	0.8	1191	0.88	12.95
Case 61	5% hydrogen	0.642	1.11	0.82	2.81	0.7	1187	0.88	12.94
Case 62	7.5% hydrogen	0.655	1.14	0.73	1.95	0.8	1170	0.87	12.93
Case 63	10% hydrogen	0.658	1.2	0.77	2.21	0.7	1165	0.88	12.95
Case 64	40% oxygen	0.625	0.95	0.62	1.01	1.1	1238	0.91	13.34
Case 65	60% oxygen	0.638	0.89	0.57	0.98	1.1	1287	0.88	12.94
Case 66	80% oxygen	0.665	1.01	0.61	1.21	1.4	1270	0.87	12.93
Case 67	100% oxygen	0.668	0.97	0.59	1.07	1.6	1265	0.88	12.95
Case 68	HHO Kit	0.646	1.14	0.76	0.98	0.9	1263	0.89	12.99

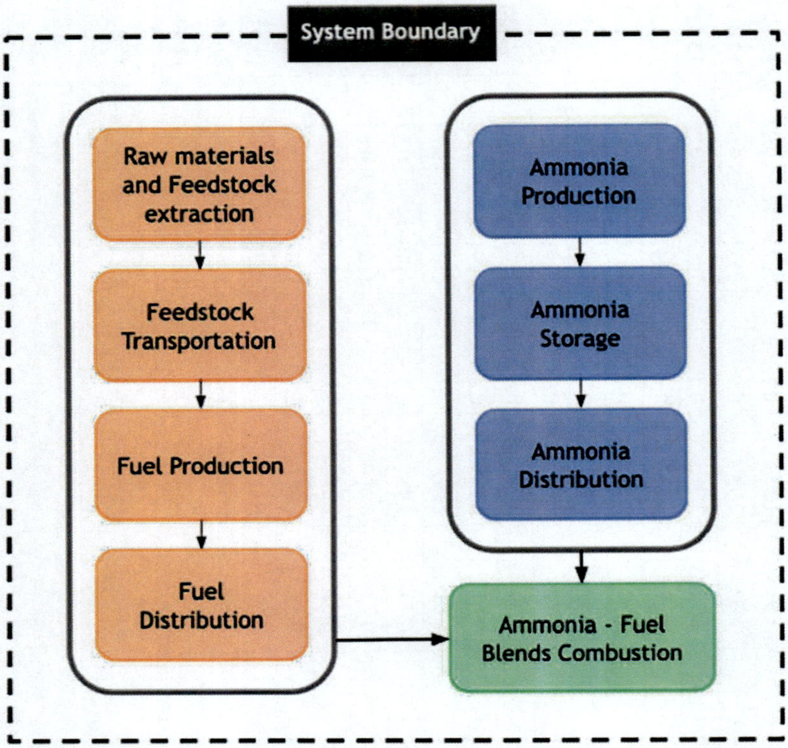

Fig. 4.87 The system boundary and life cycle phases examined in this study

were quantified with volumetric flow meters. All temperatures and emissions have been meticulously documented. All experimental data have been illustrated in three divisions, which are temperature, electricity, and emission values versus time.

The experiments showed that the power output has decreased as the volume of ammonia has increased due to poor combustion inside the cylinder. Temperature data at the cylinder exit have revealed this scenario. The temperature at the cylinder exit has dropped substantially as the ammonia content of the fuel blend has increased. While CO_2 and CO emissions have decreased as the ammonia rate has increased, NO_2 emissions have increased due to the higher N content of the blended fuels. In Table 5, experimentally measured average emission values are shown.

As can be observed from following findings, using gasoline as a promoter in ammonia combustion appears to be effective in terms of combustion performance. However, as the amount of gasoline in the blend has climbed, so the amount of CO_2 and CO emissions increased. The influence of ammonia volume in the fuel blend on emission values is clearly shown in Table 4.64.

Figure 4.89 shows a comparison of LCA results for global warming potential of gasoline and ammonia blends that were studied in Sect. 4.4. Using gasoline as a promoter in ammonia combustion appears to be effective in terms of combustion performance. However, in terms of environmental impacts we cannot say the same thing. As the amount of gasoline in the ammonia-fuel blend goes up, so the amount of carbon emissions goes up as well. When pure gasoline is combusted in the internal combustion engine the CO_2 emissions are at highest level in comparisons with all other cases. CO_2 emissions are at lowest level when pure ammonia is combusted. Typically, no CO_2 emissions are emitted during pure ammonia combustion since it is carbon free. However, since we are considering cradle-to-grave assessment in this study, a low amount of CO_2 emissions is emitted in the full life cycle. LCA is a cradle-to-grave assessment methodology that consists of a systematic set of processes for documenting and investigating the inputs and outputs of energy and materials, as well as the environmental consequences, that are directly attributable to a product or service over the course of its life cycle. The interconnected stages of a product and service system, from natural resource extraction through final disposal, are referred to as a life cycle. While pure gasoline emits 0.082 kg CO_2 eq per 1 MJ of energy content, pure ammonia emits only 0.002 kg CO_2 eq per 1 MJ over the course of its life cycle. In the case of 10% gasoline and 90% ammonia, the ammonia-fuel blend emits 0.0156 kg CO_2 eq per 1 MJ energy content. Ammonia-gasoline blends could cut down CO_2 emission up to 83% in comparison with diesel fuel. Therefore, to minimize environmental impact of gasoline combustion, ammonia-gasoline blends can be a potential solution to minimize global warming.

A comparison of LCA results for ozone layer depletion potential of ammonia-fuel blends using experimental results is presented in Fig. 4.90. Ozone layer depletion emission values are highest when amount of gasoline is higher in the fuel blend. The results significantly improve when amount of ammonia is increased, and amount of gasoline is lowered in the fuel blend. In the case of 90% gasoline and 10% ammonia ozone layer depletion emission are 1.42×10^{-8} kg CFC-11 eq per 1 MJ of energy content, and in the case of 10% gasoline and 90% ammonia ozone layer

Fig. 4.88 Photos of the experimental setup

Table 4.64 Average emission values for the case studies

	CO_2 (vol.%)	CO (vol.%)	HC (ppm)	O_2 (vol.%)	NO_2 (ppm)	Lambda	AFR
G100	8.593	9.974	241.891	0.391	37.239	0.711	10.461
G90-A10	8.336	9.774	232.216	0.587	74.478	0.711	10.461
G80-A20	8.164	9.575	220.121	0.732	104.270	0.711	10.461
G70-A30	7.906	8.905	215.283	0.747	115.441	0.711	10.461
G60-A40	7.820	8.905	215.283	0.747	115.441	0.711	10.461
G50-A50	7.648	8.713	209.236	0.775	134.061	0.711	10.461
G40-A60	6.875	7.756	198.109	0.861	223.435	0.711	10.461
G30-A70	6.273	6.702	174.162	1.213	335.152	0.711	10.461
G20-A80	2.750	3.351	159.648	1.526	595.826	0.711	10.461
G10-A90	2.047	1.725	141.618	1.976	1065.750	0.704	10.349
A100	0.908	0.830	79.029	3.413	1562.000	0.681	10.011

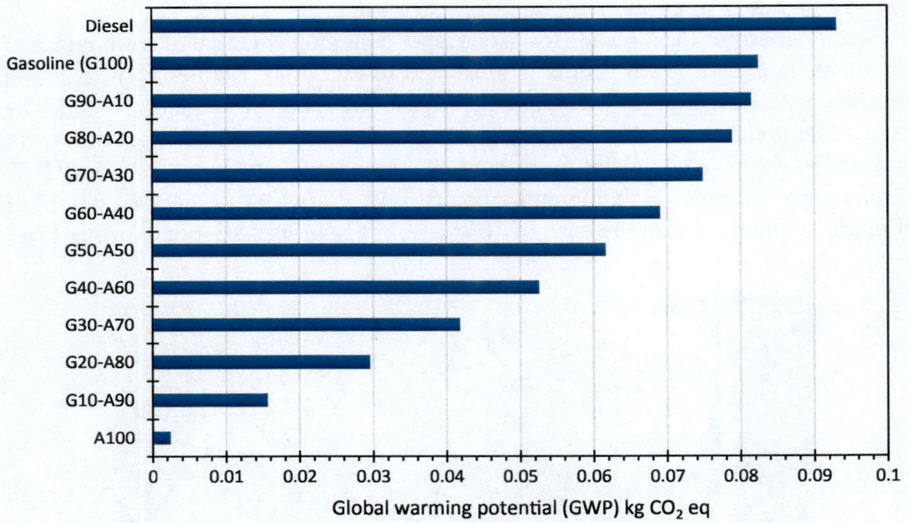

Fig. 4.89 Comparison of LCA results for global warming potential of ammonia-fuel blends using experimental results

depletion values drops to 2.71×10^{-9} kg CFC-11 eq per 1 MJ of energy content. Ammonia-gasoline fuel blends significantly lowers ozone layer depletion emissions in comparison with pure gasoline and diesel fuel.

Figure 4.91 compares LCA results for human toxicity potential of gasoline and ammonia blends using experimental results in Sect. 4.4. Ammonia exposure happens mainly through gas leaks from production, storage, or transportation facilities of

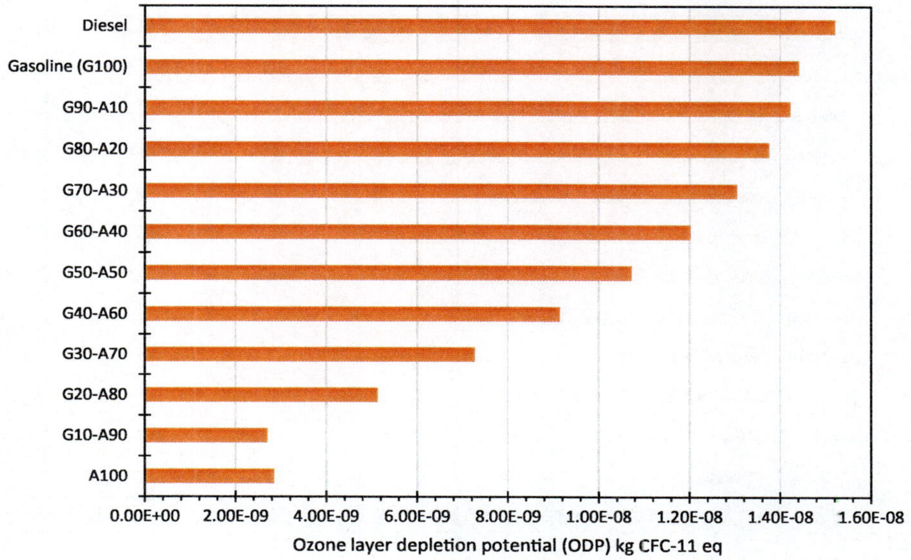

Fig. 4.90 Comparison of LCA results for ozone layer depletion potential of ammonia-fuel blends using experimental results

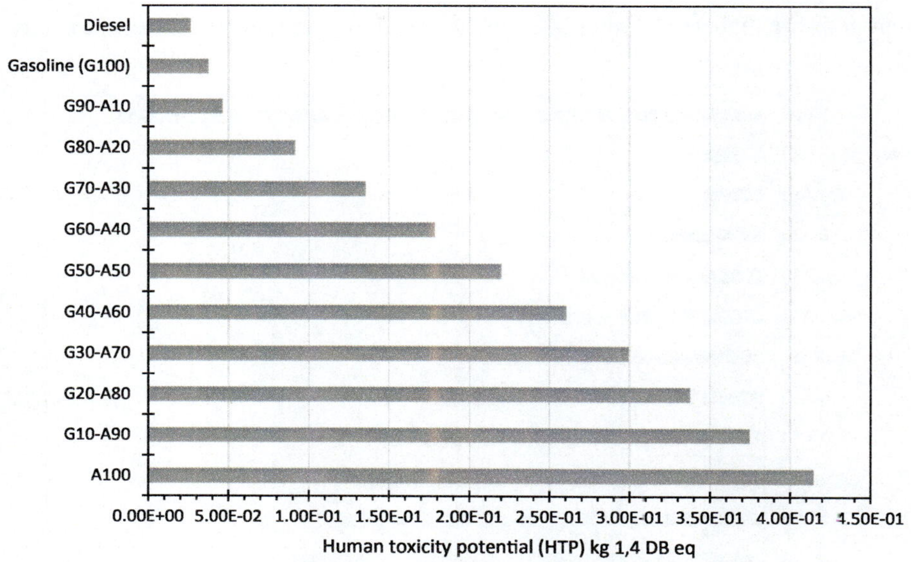

Fig. 4.91 Comparison of LCA results for human toxicity potential of ammonia-fuel blends using experimental results

ammonia. However, ammonia has a distinct smell; it is very easy to notice an ammonia leakage. Experiments reveal that when pure ammonia is combusted, human toxicity

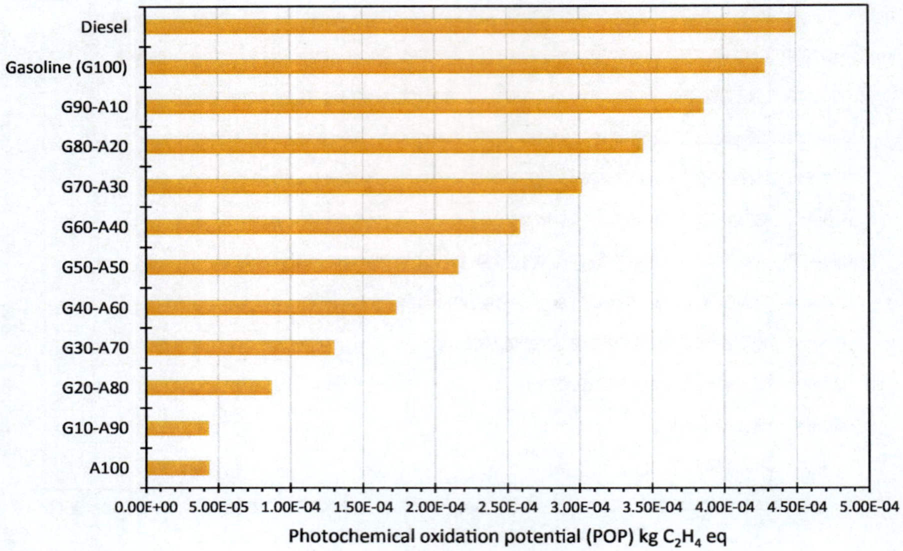

Fig. 4.92 Comparison of LCA results for photochemical oxidation potential of ammonia-fuel blends using experimental results

emissions are higher, and when gasoline is added to the fuel blend, human toxicity

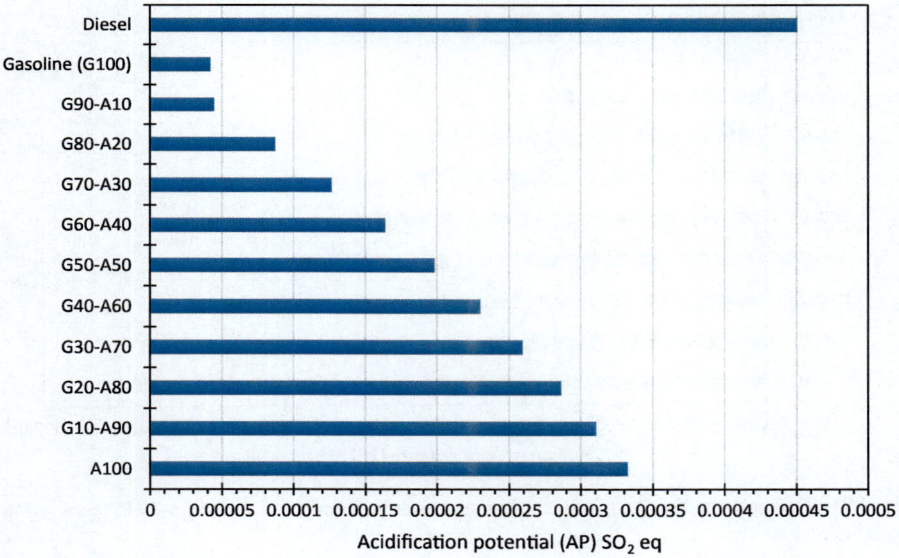

Fig. 4.93 Comparison of LCA results for acidification potential of ammonia-fuel blends using experimental results

emissions decrease. Ammonia-fuel blends can help reduce the impact of ammonia on human health. Figure 4.92 presents comparative LCA results of photochemical oxidation potential of ammonia-fuel blends using experimental results. Photochemical oxidation emissions are low when amount of ammonia in the fuel blend is high. The optimal result is obtained when pure ammonia is combusted without gasoline. In the case of 90% gasoline and 10% ammonia 3.86×10^{-4} kg C_2H_4 eq of photochemical oxidation is emitted per 1 MJ of energy content, and In the case of 10% gasoline and 90% ammonia 4.35×10^{-5} kg C_2H_4 eq of photochemical oxidation is emitted per 1 MJ of energy content. Pure ammonia and ammonia-fuel blends perform significantly better than diesel and gasoline in terms of photochemical oxidation potential.

Figure 4.93 shows comparison of LCA results for acidification potential of ammonia-fuel blends using experimental results. Acidification level is higher when pure ammonia is combusted and lower when pure gasoline is combusted. Experimental results demonstrate that ammonia-fuel blends tends to have higher acidification level when ammonia amount increase in the fuel blend. Livestock operations provide a substantial portion of ammonia to acid deposition. Ammonia makes up over half of the total acid deposition potential. In the case of 10% gasoline and 90% ammonia-fuel blend 0.00031 kg SO_2 eq is emitted per 1 MJ of energy content, and. In the case of 90% gasoline and 10% ammonia-fuel blend 4.48×10^{-5} kg SO_2 eq is emitted per 1 MJ of energy content. In the case of 50/50 ammonia-gasoline blend, 0.0002 kg SO_2 eq is emitted per 1 MJ of energy content. Pure ammonia and ammonia-gasoline fuel blends significantly improves acidification impact in comparison with diesel fuel. Sulfur dioxide, out of all the Sulphur oxides, is the most responsible for acid rain and particulate matter production. Figure 4.94 illustrates comparison

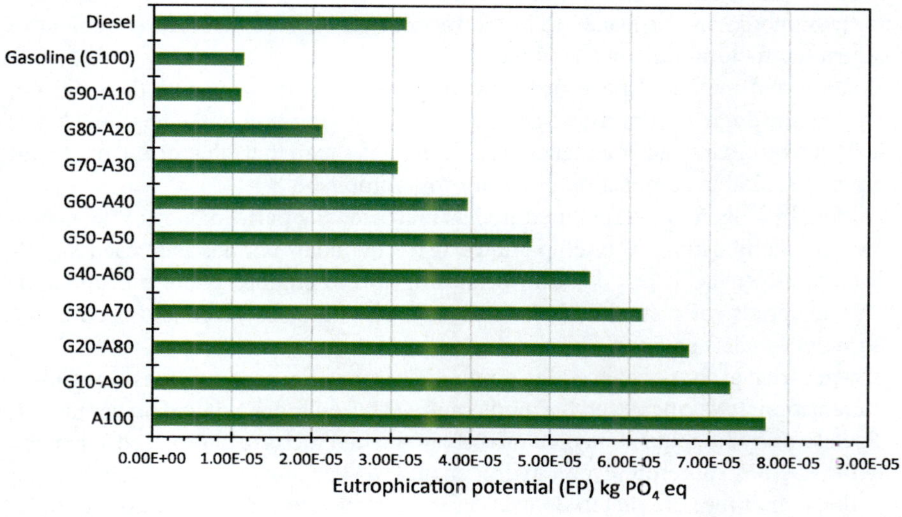

Fig. 4.94 Comparison of LCA results for eutrophication potential of ammonia-fuel blends using experimental results

of LCA results for eutrophication potential of ammonia-fuel blends using experimental results. The experimental LCA results of eutrophication potential is similar to results of acidification potential. As a nitrogen fertilizer, ammonia contributes to the eutrophication of forests and other natural vegetation. Eutrophication is a natural process that occurs when nutrients in lakes and other bodies of water build up.

Eutrophication emissions increases with amount of ammonia increase in the fuel blend. The lowest eutrophication value is calculated for G90-A10 fuel blend and highest eutrophication is calculated for G10-A90 fuel blend.

4.17 Closing Remarks

The main objective of this chapter was to investigate the attempts to convert a gasoline-powered engine into an ammonia-fueled one. Also, in order to enhance the ammonia combustion, propane, ethanol, methanol and hydrogen are used as the combustion promoter. Also, the effect of the oxygen-rich air on the combustion and generator performance have been researched experimentally. Performance of the generator is studied based on energy and exergy efficiencies. The effect of ammonia gasoline blends on the environmental impact categories is studied via LCA study. The following key findings have been reached in the present chapter.

- The generator has not operated with ammonia with its standard version. Therefore, a dual fuel carburetor has been included in the generator. The generator has not been completed 45 min test period with the only carburetor modification. The generator has run 4–5 min with 100% ammonia.
- Propane provides promising results than gasoline as a combustion promoter. Since both ammonia and propane are in the form of gaseous, it is more convenient to adjust the fuel mixture and send the engine.
- Ethanol and methanol have not provided promising results since they did not operate the engine properly. There is a significant vibration in the engine when it is fueled by ethanol and methanol. Therefore, it is found that ethanol and methanol are not a suitable combustion promoter for ammonia.
- Adding hydrogen to the fuel mixture has increased the performance of the generator and combustion. The temperature at the cylinder exit has increased significantly. However, it has caused oil burning in the engine. Therefore, the suitable materials for piston, cylinder and rings should be selected and used in the ammonia-fueled engines.
- Oxygen-rich air has enhanced the ammonia combustion as it provides a complete combustion. It has increased the temperature at the cylinder exit and power output.
- The use of hydrogen and oxygen directly in the fuel mixture has caused a damage in the engine. Therefore, when hydrogen and oxygen are used as the fuel content at the same time, the materials used in the engine should be selected carefully. Otherwise, the pistons rings will be damaged. Hydrogen caused an oil combustion in the engine.

- As the amount of gasoline in the ammonia-fuel blend goes up, so the amount of carbon emissions goes up as well. When pure gasoline is combusted in the internal combustion engine the CO_2 emissions are at highest level, CO_2 emissions are at lowest level when pure ammonia is combusted. Ammonia-gasoline blends could cut down CO_2 emission up to 83% in comparison with gasoline fuel.

References

1. Greenhouse Gas Emissions (2022) The Canadian environmental sustainability indicators. https://www.canada.ca/en/environment-climate-change/services/environmental-indicators/greenhouse-gas-emissions.html. Accessed 8 May 2022
2. Acar C, Dincer I (2019) Review and evaluation of hydrogen production options for better environment. J Clean Prod 218:835–849. https://doi.org/10.1016/j.jclepro.2019.02.046
3. Japan Bets on Ammonia as the Fuel of the Future (2020) https://www.ft.com/content/254e44c4-afec-11ea-94fc-9a676a727e5a. Accessed 20 Sept 2020
4. Ammonia-Fueled Gas Turbine Power Generation 2017 (n.d.) https://www.ammoniaenergy.org/articles/ammonia-fueled-gas-turbine-power-generation/#. Accessed 19 Sept 2020
5. California Opens 'Diesel Alternative' Discussion in Microgrid Proceeding (2020) https://microgridknowledge.com/california-microgrid-19-09-009/. Accessed 19 Sept 2020
6. Yazicioglu A (2018) Experimental investigation and evaluation of hybrid ammonia fuel options for power generators, MSc Thesis. University of Ontario Institute of Technology
7. Yapicioglu A, Dincer I (2019) A review on clean ammonia as a potential fuel for power generators. Renew Sustain Energy Rev 103:96–108. https://doi.org/10.1016/j.rser.2018.12.023
8. Erdemir D, Dincer I (2021) A perspective on the use of ammonia as a clean fuel: challenges and solutions. Int J Energy Res 45:4827–4834. https://doi.org/10.1002/er.6232
9. Zamfirescu C, Dincer I (2008) Using ammonia as a sustainable fuel. J Power Sources 185:459–465. https://doi.org/10.1016/j.jpowsour.2008.02.097
10. Dincer I, Acar C (2015) Review and evaluation of hydrogen production methods for better sustainability. Int J Hydrogen Energy 40:11094–11111. https://doi.org/10.1016/j.ijhydene.2014.12.035
11. Guinée JB, Gorrée M, Heijungs R, Huppes G, Kleijn R, Wegener Sleeswijk A et al (2001) Life cycle assessment: an operational guide to the ISO standards. LCA 2001:692. https://doi.org/10.1300/J082v38n04_05

Chapter 5
Conclusions

Not only the depletion of fossil fuels but also the environmental effects of using these fuels are one of the main reasons to turn to alternatives. In order to reduce environmental problems caused by energy production and to develop a truly sustainable approach, the systems should be evaluated in detail from source to end use. Ammonia and hydrogen are called as clean fuels as they have no carbon footprint. Due to environmental concerns, attention to them and their usage are ever-increase. The scientists and technologists have been intensely working on systems using ammonia and hydrogen in order to increase sustainability indexes of systems. Current studies have generally focussed on either developing novel systems used ammonia or hydrogen or converting present conventional systems into systems using ammonia or hydrogen as fuel. In the near future, it is expected that the use of ammonia and hydrogen will increase substantially with energy supply/demand issues and environmental concerns.

It is needless to mention that energy is one of the most significant needs for people. Energy consumption is ever-increasing globally due to increasing energy demand and consumerism. Every day, new energy-consuming items such as devices, machines, vehicles, etc. are entering our lives. These new devices can increase energy demands as well as change the type of energy consumed. For instance, electric vehicles have been changing the energy demand from traditional fossil fuels to electricity. Ever-increasing energy consumption has brought significant problems such as reaching energy sources, energy security, environmental concerns, etc. In order to reduce the environmental impacts regarding energy consumption, researchers and technologists are working on alternative fuels, energy sources, and more efficient devices.

It is now desperately necessary to set up the energy equation without hydrocarbon fuels, and it is therefore fully recognized by many researchers, scientists, organizations, companies, etc. that it is time to move to renewables and clean fuels (particularly with hydrogen and ammonia) which have been declared by Dincer [1] for many years. In a recent perspective article [1], he has declared that COVID-19 coronavirus: closing carbon age, but opening hydrogen age, and he considered the year of 2020 a turning point. This age is even more critical for human health and

© The Author(s), under exclusive license to Springer Nature Switzerland AG 2022

I. Dincer et al., *Ammonia Energy Technologies*, Lecture Notes in Energy 91,
https://doi.org/10.1007/978-3-031-13532-3_5

human welfare. In conjunction with this, there is a strong need to develop the technologies and economies to make a smooth transition as quickly as possible. That is why this perspective article takes ammonia into consideration for a fair evaluation and suggestions for better combustion practices.

Globally, about 235 million metric tonnes of ammonia was produced in 2019. Only a small portion of ammonia (approximately 4%) is used in direct applications, the remaining portion is used as a chemical in industrial applications or fertilizer in agriculture. Ammonia could be the key to finalize the increasing search for an alternative fuel. Ammonia is a chemical that can be used in many sectors and for many different purposes. It can serve as fuel in energy production, a key chemical in feedstock and chemicals production, main ingredient for cleaning materials, fuel for engines, refrigerant for cooling systems. Its versatile and wide range of application potential makes ammonia an important component.

Within the scope of this book, first, ammonia production systems and sub-systems required for ammonia production were examined. In order to determine the cost of ammonia production, the infrastructure needs of a facility that produces ammonia were determined and price estimates were made using costing methodology. In this context, alkaline electrolysis, and proton exchange membrane electrolysis units, in which the hydrogen required for ammonia production will be produced, were modeled and a price analysis was made based on the model data. Similarly, a pressure swing adsorption system was selected for nitrogen production and a price analysis was made.

- An economic analysis is made on the production of ammonia using surplus electricity generated in a microgrid designed for a remote community. According to the data, the annual hydrogen production was found to be 645, 2329, 1420 kg H_2 for case 1, case 2 and case 3, respectively. Accordingly, 3632, 13,116, 7997 kg of ammonia can be produced annually. Although the cost of hydrogen production can be reduced with surplus electricity generation, ammonia prices are negatively affected as the capacity of the Haber–Bosch unit is very low compared to industrial examples. The price of ammonia was found to be US$4.60/kg in the best case, and in the worst case, this price was up to US$14.28/kg.
- One of the biggest parameters determining the price of ammonia is the cost of hydrogen production. However, the costs can be significantly reduced with increasing the production capacity. The price of ammonia in the low-capacity systems analyzed within the scope of the book has been calculated between US$4.60 and 14/kg. These prices are very high compared to the prices of industrial production facilities, but the price of ammonia can be increased drastically especially in remote communities, due to transportation problems. For this reason, it may be feasible to direct the excess energy to ammonia production, especially in remote areas that meet the energy need with renewable energy sources that give electricity surplus.

- Ammonia is a chemical that is used in many sectors, especially in the agriculture sector, due to its high nitrogen content. Ammonia, which is used by the agriculture sector to meet the global food needs, is therefore the second most produced chemical in the world.
- Although 80% of the produced ammonia is used by the agricultural sector, it has a great potential due to its physical and chemical properties. Especially because of the 3 hydrogen atoms in its structure, it acts as a good energy carrier. Ammonia serves as a good alternative for the cost-effective and safe transportation of hydrogen. In addition, ammonia can be used directly in fuel cells or used as fuel in internal combustion engines alone or mixed with other fuels.

The importance of ammonia emerges with the scope of climate change measures. Ammonia stands out as a carbon-free chemical, fuel, and energy carrier from production to consumption with the utilization of clean hydrogen technologies. Reducing carbon emissions depends on minimizing fossil fuel consumption. Therefore, ammonia utilization is important for a sustainable future to find carbon-free alternatives to fossil fuels in both production technologies and human activities. As mentioned in Chaps. 1 and 2, in order to produce ammonia, gaseous hydrogen is needed, which is obtained by environmentally harmful methods. In the ammonia production industry, almost all hydrogen is produced by fossil-based fuels. Therefore, even if ammonia is a carbon-free fuel, its production methods will be critical in terms of its environmental impact when a cradle-to-grave aspect life cycle of ammonia is taken into consideration. Therefore, in Chap. 3, ammonia production and utilization techniques are studied over life cycle assessment. Ammonia will be a significant solution for a carbon-free and sustainable future once hydrogen and nitrogen are obtained with green sources and techniques. The following conclusions are reached in the life cycle assessment studies of ammonia.

The following key observations have been reached in this book:

- Ammonia from renewable resources has the lowest environmental impact. Ammonia from natural gas is most environmentally benign option among conventional methods. Ammonia from hydropower is best performing option in ammonia production methods with value of 0.0013 kg CO_2 eq/kg ammonia. Ammonia from wind energy emits 0.007 kg CO_2 eq/kg ammonia. Ammonia from solar tower power plant emits 0.014 kg CO_2 eq/kg ammonia, and ammonia from solar PV emits 0.02 kg CO_2 eq/kg ammonia. Ammonia from nuclear power emits only 0.0026 kg CO_2 eq/kg ammonia. Ammonia from geothermal power plant releases 0.0072 kg CO_2 eq/kg ammonia. Ammonia from biomass emits 0.013 kg CO_2 eq/kg ammonia. Ammonia from biofuels generates 0.0038 kg CO_2 eq/kg ammonia
- The highest global warming values are calculated for the diesel truck and lowest global warming values are calculated for ammonia-fueled truck followed by electric powered truck. Diesel truck corresponds to 0.447 kg CO_2 eq/t km propane fueled truck corresponds to 0.427 kg CO_2 eq/t km and CNG truck yield 0.381 kg CO_2 eq/t km, electric truck produces 0.158 kg CO_2 eq/t km, and ammonia fueled heavy duty truck emits only 0.125 kg CO_2 eq/t km.

- For 1 MJ energy storage, lithium-ion batteries produce 0.00115 kg SO_2 eq acidification impact while ammonia has nearly zero acidification impact. The highest eutrophication values are calculated for batteries and lowest for thermal energy storage. Battery energy storage corresponds to 0.31 kg CO_2 eq per 1 MJ of energy storage, the thermal energy storage corresponds to 0.11 kg CO_2 eq and ammonia energy storage emits only 0.011 kg CO_2 eq per 1 MJ.
- By eliminating the requirement for fossil fuels, biomass gasification will alleviate the environmental implications of ammonia manufacturing. The highest global warming values are calculated for straw production and transportation in biomass gasification, which corresponds to 0.431 kg CO_2 eq/kg of nitrogen fertilizer production.
- Global warming potential emission of wind electrolysis-based ammonia production comes mostly from transportation of ammonia and the parts that were used in construction of wind turbines. Copper production and steel production used in moving parts are the main reasons for high human toxicity emissions in wind-based ammonia production. Ammonia transportation emits roughly 2.4×10^{-6} kg C_2H_4 eq of photochemical oxidation potential per 1 kg of ammonia production.
- The results shows that the brown ammonia is opposite of green ammonia in the hydrogen spectrum, and it is the most environmentally harmful ammonia type. Brown ammonia emits 0.325 kg CO_2 eq per 1 kg of ammonia production while green ammonia produces only 0.0072 kg CO_2 eq per 1 kg of ammonia production. Blue ammonia is the most environmentally friendly option among the fossil-based ammonia production options.

One of the significant advantages of ammonia is that it can play an important role in the transition to the hydrogen economy. Now, it is clear that the transition to a hydrogen economy in all sectors is essential for a sustainable future, as hydrogen is a carbon-free and high-energy content fuel, energy carrier, energy storage medium and feedstock. However, this transition requires new systems or big revisions to the current systems which require high investment. At this point, ammonia, again, can play a critical role as a carbon-free fuel and energy storage medium. Therefore, we define ammonia as the "shortcut" to the hydrogen economy. One of the shortcuts that ammonia provides is using ammonia as the fuel in internal combustion engines since it is possible to use it in internal combustion engines with cost-friendly modifications. In this book, a commercial and gasoline-driven power generator is converted into an ammonia-fueled one. Various attempts are applied to convert the generator. The following key results are reached on the topic of power generator conversion:

- The generator has not operated with ammonia with its standard version. Therefore, a dual fuel carburetor has been included in the generator. The generator has not been completed 45 min test period with the only carburettor modification. The generator has run 4–5 min with 100% ammonia.
- Propane provides promising results than gasoline as a combustion promoter. Since both ammonia and propane are in a gaseous state, optimizing the fuel mixture is simpler and easier.

- Ethanol and methanol have not provided promising results since they did not operate the engine properly. There is a significant vibration in the engine when it is fueled by ethanol and methanol. Therefore, it is found that ethanol and methanol are not a suitable combustion promoter for ammonia.
- Adding hydrogen to the fuel mixture has increased the performance of the generator and combustion. The temperature at the cylinder exit has increased significantly. However, it has caused oil burning in the engine. Therefore, the suitable materials for piston, cylinder and rings should be selected and used in the ammonia-fueled engines.
- Oxygen-rich air has enhanced the ammonia combustion as it provides a complete combustion. It has increased the temperature at the cylinder exit and power output.
- The use of hydrogen and oxygen directly in the fuel mixture has caused a damage in the engine. Therefore, when hydrogen and oxygen are used as the fuel content at the same time, the materials used in the engine should be selected carefully. Otherwise, the pistons rings will be damaged. Hydrogen caused a oil combustion in the engine.
- While the combustion temperature of the fuel mixtures affects the power output of the generator, another factor that should also be considered is the effect of gaseous fuels on volumetric efficiency. As one might have noted, the output of an engine at a given engine speed is directly related to the amount of oxygen that it can ingest, which is why larger displacements and supercharging are effective. Gaseous fuels (such as propane, methane, and ammonia) displace air from the intake combustion charge with a corresponding decrease in power. Conversely, at part throttle, gaseous fuels' decrease in volumetric efficiency improves fuel efficiency because of reduced pumping losses.

Reference

1. Dincer I (2020) Covid-19 coronavirus: Closing carbon age, but opening hydrogen age. Int J Eng Res 44(8):6093–6097. https://doi.org/10.1002/er.5569